AF614751

Methods in Molecular Biology

For further volumes:
http://www.springer.com/series/7651

Plant Cold Acclimation

Methods and Protocols

Edited by

Dirk K. Hincha

Max-Planck-Institut für Molekulare Pflanzenphysiologie, Potsdam, Germany

Ellen Zuther

Max-Planck-Institut für Molekulare Pflanzenphysiologie, Potsdam, Germany

Editors
Dirk K. Hincha
Max-Planck-Institut für Molekulare
Pflanzenphysiologie
Potsdam, Germany

Ellen Zuther
Max-Planck-Institut für Molekulare
Pflanzenphysiologie
Potsdam, Germany

Videos to this book can be accessed at http://www.springerimages.com/videos/978-1-4939-0843-1

ISSN 1064-3745 ISSN 1940-6029 (electronic)
ISBN 978-1-4939-0843-1 ISBN 978-1-4939-0844-8 (eBook)
DOI 10.1007/978-1-4939-0844-8
Springer New York Heidelberg Dordrecht London

Library of Congress Control Number: 2014938067

Printed on acid-free paper

Humana Press is a brand of Springer
Springer is part of Springer Science+Business Media (www.springer.com)

Preface

It has frequently been remarked in the scientific literature that plants as sessile organisms are forced to deal with changes in environmental conditions by rapid acclimation, because they obviously cannot simply move to a more favorable location. Over evolutionary time scales this has provided plants with complex molecular, physiological, and morphological mechanisms to thrive and survive under stressful conditions. Prevalent stresses that plants are exposed to both in natural and agricultural environments are low temperatures and freezing, i.e., the crystallization of ice in living tissues. Not surprisingly, many plant species are able to survive tissue freezing. In addition, plants from temperate climate zones are able to adapt to low temperatures prior to an actual freezing event. This phenomenon of plant cold acclimation, i.e., the increase of freezing tolerance under conditions of low, but non-freezing temperatures, has been the subject of intensive scientific study for many decades.

Under natural conditions, cold acclimation helps plants to survive seasonal low winter temperatures. Cold acclimation and freezing tolerance are quantitative traits and cold acclimation is accompanied by complex changes in gene expression, enzyme activities, and the contents of a large number of proteins, primary and secondary metabolites, and lipids. In addition, even a few days under cold conditions can trigger irreversible morphological changes, particularly in growing plant tissues. Therefore, research on plant cold acclimation can (and quite often has to) be performed at different organizational levels (from populations to single genes and molecules). This requires experimental expertise in scientific disciplines such as ecology, plant breeding, genetics, physiology, or molecular biology and strongly favors interdisciplinary approaches. This volume of Methods in Molecular Biology combines a wide selection of experimental methods ranging from the whole plant level of ecology and breeding to molecular profiling and the detailed analysis of specific proteins, with many levels of investigated complexity in between. We hope that this collection of detailed experimental protocols will help researchers, both new and experienced, to enter this exciting field of research, or broaden the scope of their investigations.

Potsdam, Germany

Dirk K. Hincha
Ellen Zuther

Contents

Contributors

JESSICA ALPERS • *Applied Metabolome Analysis Research Group, Max-Planck-Institute of Molecular Plant Physiology, Potsdam, Germany*
MARGARETE BAIER • *Plant Physiology, Dahlem Center of Plant Sciences, Freie Universität Berlin, Berlin, Germany*
MADDALENA BAYER-GIRALDI • *Alfred Wegener Institute, Helmholtz Centre for Polar and Marine Research, Bremerhaven, Germany*
OLGA BEINE-GOLOVCHUK • *Applied Metabolome Analysis Research Group, Max-Planck-Institute of Molecular Plant Physiology, Potsdam, Germany*
ANNICK BERTRAND • *Agriculture and Agri-Food Canada, Soils and Crops Research and Development Centre, Québec, QC, Canada*
BRYNHILDUR BJARNADOTTIR • *Faculty of Humanities and Social Science, Department of Education, University of Akureyri, Akureyri, Iceland*
JOSÉE BOURASSA • *Agriculture and Agri-Food Canada, Soils and Crops Research and Development Centre, Québec, QC, Canada*
MELISSA BREDOW • *Department of Biology, Queen's University, Kingston, ON, Canada*
YVES CASTONGUAY • *Agriculture and Agri-Food Canada, Soils and Crops Research and Development Centre, Québec, QC, Canada*
LUIGI CATTIVELLI • *Consiglio per la Ricerca e la sperimentazione in Agricoltura, Genomics Research Centre, Fiorenzuola d'Arda, Italy*
PETER L. DAVIES • *Department of Biomedical and Molecular Sciences, Queen's University, Kingston, ON, Canada; Department of Biology, Queen's University, Kingston, ON, Canada*
FREDERIK DETHLOFF • *Applied Metabolome Analysis Research Group, Max-Planck-Institute of Molecular Plant Physiology, Potsdam, Germany*
STEFANIE DÖLL • *Department of Physiology and Cell Biology, Leibniz Institute of Plant Genetics and Crop Plant Research (IPK), Stadt Seeland, OT Gatersleben, Germany*
KEITA ENDOH • *Faculty of Agriculture, Hokkaido University, Sapporo, Japan*
ALEXANDER ERBAN • *Applied Metabolome Analysis Research Group, Max-Planck-Institute of Molecular Plant Physiology, Potsdam, Germany*
INES FEHRLE • *Applied Metabolome Analysis Research Group, Max-Planck-Institute of Molecular Plant Physiology, Potsdam, Germany*
SEIZO FUJIKAWA • *Faculty of Agriculture, Hokkaido University, Sapporo, Japan*
KIMBERLY GARLAND-CAMPBELL • *USDA-ARS, Wheat Genetics Unit, Washington State University, Pullman, WA, USA*
CARINE GÉRY • *Institut Jean-Pierre Bourgin, UMR1318 INRA-AgroParisTech, INRA Versailles, Versailles Cedex, France*
BJARNI E. GUDLEIFSSON • *Faculty of Environmental Sciences, Department of Nature and Environmental Science, The Agricultural University of Iceland, Akureyri, Iceland*
HUGH A.L. HENRY • *Department of Biology, University of Western Ontario, London, ON, Canada*
DIRK K. HINCHA • *Max-Planck-Institut für Molekulare Pflanzenphysiologie, Potsdam, Germany*

EONSEON JIN • *Hanyang University, Seoul, Republic of Korea*
ILONA JUSZCZAK • *Plant Physiology, Dahlem Center of Plant Sciences, Freie Universität Berlin, Berlin, Germany*
YUKIO KAWAMURA • *United Graduate School of Agricultural Sciences and Cryobiofrontier Research Center, Iwate University, Morioka, Iwate, Japan*
JOACHIM KOPKA • *Applied Metabolome Analysis Research Group, Max-Planck-Institute of Molecular Plant Physiology, Potsdam, Germany*
JUERGEN KREYLING • *Department of Biogeography, University of Bayreuth, Bayreuth, Germany*
EDITH KUPRIAN • *Unit Functional Plant Biology, Stressphysiology and Climate Resistance, Institute of Botany, University of Innsbruck, Innsbruck, Austria*
RICO LIPPMANN • *Department of Physiology and Cell Biology, Leibniz Institute of Plant Genetics and Crop Plant Research (IPK), Stadt Seeland, OT Gatersleben, Germany*
DAVID P. LIVINGSTON III • *USDA-ARS and North Carolina State University, Raleigh, NC, USA*
ANDREY V. MALYSHEV • *Department of Biogeography, University of Bayreuth, Bayreuth, Germany*
ADAM J. MIDDLETON • *Department of Biomedical and Molecular Sciences, Queen's University, Kingston, ON, Canada*
YUSHI MIKI • *Cryobiofrontier Research Center, Iwate University, Morioka, Iwate, Japan*
HANS-PETER MOCK • *Department of Physiology and Cell Biology, Leibniz Institute of Plant Genetics and Crop Plant Research (IPK), Stadt Seeland, OT Gatersleben, Germany*
TAKATO NAKAYAMA • *Faculty of Agriculture, Cryobiofrontier Research Center, Iwate University, Morioka, Iwate, Japan*
GILBERT NEUNER • *Unit Functional Plant Biology, Stressphysiology and Climate Resistance, Institute of Botany, University of Innsbruck, Innsbruck, Austria*
ISABEL ORF • *Applied Metabolome Analysis Research Group, Max-Planck-Institute of Molecular Plant Physiology, Potsdam, Germany*
CARLOS PEREA-RESA • *Departamento de Biología Medioambiental, Centro de Investigaciones Biológicas (CIB-CSIC), Madrid, Spain*
JULIO SALINAS • *Departamento de Biología Medioambiental, Centro de Investigaciones Biológicas (CIB-CSIC), Madrid, Spain*
STEFANIE SCHMIDT • *Applied Metabolome Analysis Research Group, Max-Planck-Institute of Molecular Plant Physiology, Potsdam, Germany*
JENS SCHWACHTJE • *Applied Metabolome Analysis Research Group, Max-Planck-Institute of Molecular Plant Physiology, Potsdam, Germany*
SUNITHA SHIVA • *Division of Biology, Kansas Lipidomics Research Center, Ackert Hall, Kansas State University, Manhattan, KS, USA*
DANIEL Z. SKINNER • *USDA-ARS, Wheat Genetics Unit, Washington State University, Pullman, WA, USA*
AARON SMALTER HALL • *K-INBRE Bioinformatics Core Facility, Structural Biology Center, University of Kansas, Lawrence, KS, USA*
DAISUKE TAKAHASHI • *United Graduate School of Agricultural Sciences, Iwate University, Morioka, Iwate, Japan*
EVELYNE TÉOULÉ • *Institut Jean-Pierre Bourgin, UMR1318 INRA-AgroParisTech, INRA Versailles, Versailles Cedex, France; UPMC, Paris Cedex 05, France*
ANJA THALHAMMER • *Max-Planck-Institut für Molekulare Pflanzenphysiologie, Potsdam, Germany*
HEATHER TOMALTY • *Department of Biology, Queen's University, Kingston, ON, Canada*

TAN D. TUONG • *USDA-ARS and North Carolina State University, Raleigh, NC, USA*
MATSUO UEMURA • *United Graduate School of Agricultural Sciences and Cryobiofrontier Research Center, Iwate University, Morioka, Iwate, Japan*
BARBARA VANDERBELD • *Department of Biology, Queen's University, Kingston, ON, Canada*
HIEU SY VU • *Division of Biology, Kansas Lipidomics Research Center, Ackert Hall, Kansas State University, Manhattan, KS, USA*
VIRGINIA K. WALKER • *Department of Biomedical and Molecular Sciences, Queen's University, Kingston, ON, Canada; Department of Biology, Queen's University, Kingston, ON, Canada*
RUTH WELTI • *Division of Biology, Kansas Lipidomics Research Center, Ackert Hall, Kansas State University, Manhattan, KS, USA*
PETER W. WILSON • *University of Tasmania, Hobart, Australia*
ELLEN ZUTHER • *Max-Planck-Institut für Molekulare Pflanzenphysiologie, Potsdam, Germany*

Chapter 1

Introduction: Plant Cold Acclimation and Freezing Tolerance

Dirk K. Hincha and Ellen Zuther

Abstract

This introductory chapter provides a brief overview of plant freezing tolerance and cold acclimation and describes the basic concepts and approaches that are currently followed to investigate these phenomena. We highlight the multidisciplinary nature of these investigations and the necessity to use methodologies from different branches of science, such as ecology, genetics, physiology, biochemistry, and biophysics, to come to a complete understanding of the complex adaptive mechanisms underlying plant cold acclimation.

Key words Cold acclimation, Experimental approaches, Freezing tolerance, Global climate change

The phenomenon of plant cold acclimation, i.e., the increase of freezing tolerance during exposure to low, but nonfreezing temperatures, has been described already in the nineteenth century (*see* ref. 1 for references and details) and has been the subject of intensive scientific study ever since (*see* refs. 1–4 for comprehensive reviews). The basic phenotypic readout for cold acclimation is the increased survival of plants, tissues or cells after a freeze–thaw cycle through a damaging temperature range. Unfortunately, not only in common usage but also in the scientific literature, the term freezing tolerance is often used synonymously with cold or low temperature tolerance. This is very unfortunate because freezing and cold/low temperature denote completely different concepts. While freezing is a clearly defined physical process (i.e., the crystallization of ice), cold is a completely subjective term, not only for humans but also for plants. As an example, many tropical and subtropical plants suffer severe damage at temperatures below approximately 15 °C, while Antarctic algae show a heat-shock response already at 5 °C [5]. In addition, mechanisms of injury are very different. Low temperature (or chilling) damage is a direct temperature effect. Freezing damage, on the other hand, is mainly the result of osmotic dehydration triggered by extracellular ice

Dirk K. Hincha and Ellen Zuther (eds.), *Plant Cold Acclimation: Methods and Protocols*, Methods in Molecular Biology, vol. 1166, DOI 10.1007/978-1-4939-0844-8_1, © Springer Science+Business Media New York 2014

crystallization that leads to the diffusion of water from the cells to the growing ice crystals [1–3].

Under natural conditions, cold acclimation is a response of plants to cope with seasonal low temperature. Since flowering plants were initially only adapted to the tropical climate that was prevalent in the Mesozoic era and only became exposed to temperate climates after severe climate cooling events in the Eocene and Oligocene, the ability to survive in temperate climates with seasonal cold had to evolve from tropical species. It has recently been estimated that less than half of all angiosperm families have members that are adapted for survival under seasonal cold conditions and that this trait has developed several times independently (*see* ref. 6 for a recent review).

In support of strong natural selection for freezing tolerance in plants, natural variation in this trait has been reported in both herbaceous and tree species. This variation is usually related to latitudinal and/or climatic gradients that are obvious candidates as potential driving forces of selection [7–11]. From a more practical point of view, such natural variation can be used to investigate the potential function of metabolic and physiological trait variation in determining the differential freezing tolerance of different genotypes within a species through correlation analysis [11, 12]. In addition, in crop plant species the corresponding variability may be used for breeding purposes, for the mapping of quantitative trait loci (QTL) and for the identification of molecular markers for marker-assisted breeding.

Obviously, global climate change has a major impact on the winter survival of plants and this impact is going to increase in the coming decades. In general, winters in the cold regions of the Earth are getting milder and naively one might assume that freezing tolerance will become less important for plant survival and geographical distribution, and also for crop yields, in the future. However, warmer winters are accompanied by reduced snow fall and a higher incidence of erratic early or late season frost. Since snow cover is a very effective insulator, plants under snow are exposed to significantly less-severe freezing temperatures than in the absence of snow. This can lead to the counterintuitive effect that plants are exposed to lower freezing temperatures as global warming progresses. In addition, premature warming in early spring can lead to loss of freezing tolerance (de-acclimation) making plants more prone to damage during later cold spells [13]. Therefore, research on all aspects of plant cold acclimation, including much neglected areas such as de-acclimation, will be more relevant than ever in times of global climate change [14].

For plants as sessile organisms, there are generally three strategies available to survive winter frost. The most obvious strategy that is used by many herbaceous annuals is seasonal avoidance by surviving winter as seeds, or as roots or rhizomes buried sufficiently deeply in the soil to evade subzero temperatures. Deep supercooling (i.e., preventing ice crystallization even at temperatures significantly below 0 °C) is a frequently observed winter survival strategy for example in cold acclimated trees [15]. The third strategy, that is embraced by all plants that cold acclimate, is to modify their cellular constituents in a way that allows survival at lower subzero temperatures in the presence of extracellular ice.

The term supercooling usually refers to a solution (cellular or otherwise) that shows no ice crystallization when cooled below its melting point. It should be emphasized here that the freezing point of pure water is, contrary to common perception, not identical to its melting point at 0 °C. Ice crystallization at or slightly below 0 °C requires impurities that can serve as seeds. Pure water only crystallizes at about −42 °C, the homogeneous nucleation temperature. In any biological system it is probably impossible to remove all molecules and structures that could serve to seed ice crystallization. In addition, colligative freezing point depression from common metabolites such as sugars or amino acids only has a very limited potential to achieve significant supercooling as a concentration of one molar of any (nondissociating) solute will depress the freezing/melting point only by 1.84 °C and the ability of living cells to accumulate solutes is obviously limited. Therefore, supercooling relies on biological antifreezes, such as specific proteins [16, 17] and more complex solutes such as flavonoids and tannins [18]. These substances allow some plants and algae to remain unfrozen at subzero temperatures, similar to Arctic and Antarctic fish and some species of insects that produce highly efficient anti-freeze or thermal hysteresis proteins [19, 20]. In addition to their thermal hysteresis activity, many such proteins also exhibit ice re-crystallization inhibition activity, i.e., the ability to suppress the growth of large ice crystals at the expense of smaller crystals. In particular in the plant proteins, thermal hysteresis activity is very low (0.1–0.5 °C) and they are mainly referred to as ice binding proteins (IBP). Their main function is probably the regulation of ice crystal size in the intercellular spaces of plant tissues [21], or in the case of sea ice algae in the narrow water channels inhabited by these cells.

Cold acclimation and freezing tolerance are genetically complex, quantitative traits. The increase in freezing tolerance during cold acclimation is accompanied by complex physiological changes that are, at least to a large extent, based on complex changes in gene expression (*see* ref. 4 for a recent review). While our understanding of temperature perception in plants is still quite fragmentary [22], work from many laboratories primarily in *Arabidopsis thaliana* has shed light on the signal transduction cascades regulating the expression of important target genes encoding potential protective proteins such as the COR/LEA proteins. In addition, genes encoding enzymes either directly involved in reactive oxygen species (ROS) detoxification or the biosynthesis of low molecular weight antioxidants, and enzymes responsible for the biosynthesis of compatible solutes are upregulated during cold acclimation [23]. For instance, in Arabidopsis and Thellungiella compatible solutes such as sugars and proline are strongly accumulated during plant cold acclimation and their content is highly correlated with freezing tolerance across diverse genotypes [11, 24, 25]. In fact, compatible solutes can have strong predictive value for freezing tolerance in Arabidopsis [12].

In addition to compatible solutes, many other primary and secondary metabolites are accumulated during cold acclimation, but for most of these solutes no specific role in cellular freezing tolerance has been proposed or experimentally shown. However, for flavonoids a correlation of the expression levels of biosynthetic genes and of some key compounds with freezing tolerance in different Arabidopsis genotypes was found [24, 26]. Also, cellular lipid composition is strongly modified during cold acclimation, both for membrane (diacyl) and storage (triacyl) lipids [27, 28]. Of course, all these metabolic changes are also evident on the gene expression (e.g., refs. 26, 29–32) and protein abundance [33–35] levels. Investigation of knockout mutants with altered freezing tolerance or cold signal transduction behavior is an invaluable tool to clarify the function of specific genes/proteins or key metabolic pathways or compounds in the process of cold acclimation (*see* refs. 36–38 for reviews).

The brief overview given above should give an idea about the many different organizational levels (from populations to single genes and molecules) and the corresponding scientific disciplines that are involved in research of plant cold acclimation and freezing tolerance, with their respective focus on ecology, breeding, genetics, physiology or molecular biology, or any combination of these specialties. Obviously, this area of research strongly favors interdisciplinary approaches. At the same time this means that researchers very often have to combine experimental methods and concepts from different areas of science. We therefore hope that this volume of Methods in Molecular Biology, that comprises a large range of experimental protocols covering all the mentioned organizational levels and disciplines, will help not only new researchers starting in this exciting field but also those already working in a particular area of cold acclimation and freezing tolerance research who are looking to expand their range of experimental approaches.

References

1. Steponkus PL (1984) Role of the plasma membrane in freezing injury and cold acclimation. Annu Rev Plant Physiol 35:543–584
2. Guy CL (1990) Cold acclimation and freezing stress tolerance: role of protein metabolism. Annu Rev Plant Physiol Plant Mol Biol 41: 187–223
3. Levitt J (1980) Responses of plants to environmental stresses, volume I: chilling, freezing, and high temperature stresses. Academic Press, Orlando
4. Hincha DK, Espinoza C, Zuther E (2012) Transcriptomic and metabolomic approaches to the analysis of plant freezing tolerance and

cold acclimation. In: Tuteja N, Gill SS, Toburcio AF, Tuteja R (eds) Improving crop resistance to abiotic stress. Wiley-Blackwell, Berlin, pp 255–287

5. Vayda ME, Yuan M-L (1994) The heat shock response of an antarctic alga is evident at 5 °C. Plant Mol Biol 24:229–233
6. Preston JC, Sanve SR (2013) Adaptation to seasonality and the winter freeze. Front Plant Sci 4:167
7. Holliday JA, Ritland K, Aitken SN (2010) Widespread, ecologically relevant genetic markers developed from association mapping of climate-related traits in Sitka spruce (*Picea sitchensis*). New Phytol 188:501–514
8. Kreyling J, Thiel D, Simmnacher K, Willner E, Jentsch A, Beierkuhnlein C (2012) Geographic origin and past climatic experience influence the response to late spring frost in four grass species in central Europe. Ecography 35:268–275
9. Kreyling J et al (2012) Cold hardiness of *Pinus nigra* Arnold as influenced by geographic origin, warming, and extreme summer drought. Environ Exp Bot 78:99–108
10. Zhen Y, Ungerer MC (2008) Clinal variation in freezing tolerance among natural accessions of *Arabidopsis thaliana*. New Phytol 177: 419–427
11. Zuther E, Schulz E, Childs LH, Hincha DK (2012) Natural variation in the non-acclimated and cold-acclimated freezing tolerance of *Arabidopsis thaliana* accessions. Plant Cell Environ 35:1860–1878
12. Korn M, Gärtner T, Erban A, Kopka J, Selbig J, Hincha DK (2010) Predicting Arabidopsis freezing tolerance and heterosis in freezing tolerance from metabolite composition. Mol Plant 3:224–235
13. Pagter M, Arora R (2013) Winter survival and deacclimation of perennials under warmer climate: physiological perspectives. Physiol Plant 147:75–87
14. Kalberer SR, Wisniewski M, Arora R (2006) Deacclimation and reacclimation of cold-hardy plants: current understanding and emerging concepts. Plant Sci 171:3–16
15. Kuroda K, Kasuga J, Arakawa K, Fujikawa S (2003) Xylem ray parenchyma cells in boreal hardwood species rspond to subfreezing temperatures by deep supercooling that is accompanied by incomplete desiccation. Plant Physiol 131:736–744
16. Duman JG, Wu DW, Olsen TM, Urrutia M, Tursman D (1993) Thermal-hysteresis proteins. In: Steponkus PL (ed) Advances in low temperature biology. JAI Press, London, pp 131–182
17. Urrutia ME, Duman JG, Knight CA (1992) Plant thermal hysteresis proteins. Biochim Biophys Acta 1121:199–206
18. Kuwabara C, Wang D, Endoh K, Fukushi Y, Arakawa K, Fujikawa S (2013) Analysis of supercooling activity of tannin-related polyphenols. Cryobiology 67:40–49
19. DeVries AL, Cheng CHC (1992) The role of antifreeze glycopeptides and peptides in the survival of cold-water fishes. In: Somero GN, Osmond CB, Bolis CL (eds) Water and life. Comparative analysis of water relationships at the organismic, cellular, and molecular levels. Springer, Berlin, pp 301–315
20. Duman JG (2001) Antifreeze and ice nucleator proteins in terrestrial arthropods. Annu Rev Physiol 63:327–357
21. Griffith M, Yaish MWF (2004) Antifreeze proteins in overwintering plants: a tale of two activities. Trends Plant Sci 9:399–405
22. Knight MR, Knight H (2012) Low-temperature perception leading to gene expression and cold tolerance in higher plants. New Phytol 195:737–751
23. Guy CL, Kaplan F, Kopka J, Selbig J, Hincha DK (2008) Metabolomics of temperature stress. Physiol Plant 132:220–235
24. Korn M, Peterek S, Mock H-P, Heyer AG, Hincha DK (2008) Heterosis in the freezing tolerance, and sugar and flavonoid contents of crosses between *Arabidopsis thaliana* accessions of widely varying freezing tolerance. Plant Cell Environ 31:813–827
25. Lee YP, Babakov A, de Boer B, Zuther E, Hincha DK (2012) Comparison of freezing tolerance, compatible solutes and polyamines in geographically diverse collections of *Thellungiella spec.* and *Arabidopsis thaliana* accessions. BMC Plant Biol 12:131
26. Hannah MA, Wiese D, Freund S, Fiehn O, Heyer AG, Hincha DK (2006) Natural genetic variation of freezing tolerance in Arabidopsis. Plant Physiol 142:98–112
27. Degenkolbe T, Giavalisco P, Zuther E, Seiwert B, Hincha DK, Willmitzer L (2012) Differential remodeling of the lipidome during cold acclimation in natural accessions of *Arabidopsis thaliana*. Plant J 72:972–982
28. Wang X, Li W, Li M, Welti R (2006) Profiling lipid changes in plant responses to low temperatures. Physiol Plant 126:90–96
29. Fowler S, Thomashow MF (2002) Arabidopsis transcriptome profiling indicates that multiple regulatory pathways are activated during cold acclimation in addition to the CBF cold response pathway. Plant Cell 14:1675–1690

30. Hannah MA, Heyer AG, Hincha DK (2005) A global survey of gene regulation during cold acclimation in *Arabidopsis thaliana*. PLoS Genet 1:e26
31. Kreps JA, Wu Y, Chang HS, Zhu T, Wang X, Harper JF (2002) Transcriptome changes for Arabidopsis in response to salt, osmotic, and cold stress. Plant Physiol 130:2129–2141
32. Oono Y et al (2006) Monitoring expression profiles of Arabidopsis genes during cold acclimation and deacclimation using DNA microarrays. Funct Integr Genomics 6:212–234
33. Amme S, Matros A, Schlesier B, Mock H-P (2006) Proteome analysis of cold stress response in *Arabidopsis thaliana* using DIGE-technology. J Exp Bot 57:1537–1546
34. Kawamura Y, Uemura M (2003) Mass spectrometric approach to identifying putative plasma membrane proteins of *Arabidopsis* leaves associated with cold acclimation. Plant J 36: 141–154
35. Kjellsen TD, Shiryaeva L, Schröder W, Strimbeck GR (2010) Proteomics of extreme freezing tolerance in Siberian spruce (*Picea obovata*). J Proteomics 73: 965–975
36. Chinnusamy V, Zhu J, Zhu J-K (2007) Cold stress regulation of gene expression in plants. Trends Plant Sci 12:444–451
37. Medina J, Catala R, Salinas J (2011) The CBFs: three Arabidopsis transcription factors to cold acclimate. Plant Sci 180:3–11
38. Thomashow MF (2010) Molecular basis of plant cold acclimation: insights gained from studying the CBF cold response pathway. Plant Physiol 154:571–577

Chapter 2

Measuring Freezing Tolerance: Survival and Regrowth Assays

Daniel Z. Skinner and Kimberly Garland-Campbell

Abstract

Screening plants for freezing tolerance under tightly controlled conditions is an invaluable technique for studying freezing tolerance and selecting for improved winterhardiness. Artificial freezing tests of cereal plants historically have used isolated crown and stem tissue prepared by "removing all plant parts 3 cm above and 0.5 cm below the crown tissue" (Fowler et al., Crop Sci 21:896–901, 1981). Here, we describe a method of conducting freezing tolerance tests using intact plants grown in small horticultural containers, including suggested methods for collecting and analyzing the data.

Key words Whole-plant freezing, Probit, Zero-inflated Poisson

1 Introduction

Crops that are planted in the autumn and harvested the following summer take advantage of fall and winter precipitation and yield significantly more than their spring-planted counterparts, but must survive the rigors of the winter months. Selecting for improved winter hardiness in the field is seldom effective because the unpredictability of winter weather often results in either death or survival of most plants within a trial [1], and tremendous variation in survival may occur within short distances within a field [2], often making comparisons among entries essentially meaningless.

Artificial freezing tests that enable the determination of the LT_{50}, the temperature predicted to be lethal to 50 % of the plants, lead to much more reproducible results, can be conducted throughout the year, and require a relatively small amount of space. The LT_{50} provides a measure that can be used to compare freezing tolerance of plant lines tested at different times and in different places, obviating the need for side-by-side comparisons, especially if standard check varieties are included in each trial. Here, we present a method for carrying out freezing tolerance trials, and suggestions for collecting and analyzing the resulting data. These methods are based largely on

Dirk K. Hincha and Ellen Zuther (eds.), *Plant Cold Acclimation: Methods and Protocols*, Methods in Molecular Biology, vol. 1166, DOI 10.1007/978-1-4939-0844-8_2, © Springer Science+Business Media New York 2014

our experience with winter wheat (*Triticum aestivum* L.), but also have been successfully applied to several other monocotyledonous and dicotyledonous plant species with minor modifications.

2 Materials

1. Potting mix such as Sunshine Mix LC1 planting medium (Sun Gro Horticulture, Bellevue, WA, USA).
2. Horticultural cell packs such as 6-cell packs (ca. 100 mL capacity per cell) for model 1020 trays (Blackmore Co., Belleville, MI, USA).
3. Snomax snow inducer (Snomax LLC, Centennial, CO, USA).
4. Liquid fertilizer such as Peters Professional all purpose plant food with trace elements (Scotts Co., Camarillo, CA, USA).
5. Programmable freezer, such as Model LU-113, Espec Corp., Hudsonville, MI, USA.
6. Temperature probes and monitor such as Model E-16, Sensatronics, Bow, NH, USA.
7. Growth chambers.
8. Greenhouse.

3 Methods

3.1 Artificial Freezing Protocol

In preparation for freezing survival tests, plants are grown in a soilless potting mix in horticultural "cell packs" with about 100 mL capacity per cell. Twenty seeds of each line to be tested are planted in each cell. Seeds are germinated and seedlings grown at 22 °C in a growth chamber under cool white fluorescent lights (about 300 μmol/m^2/s at the soil surface) with a 16 h photoperiod until the seedlings reach the three-leaf stage. Relative humidity is not controlled. The plants are then transferred to a growth chamber at 4 °C with a 16 h photoperiod (about 250 μmol/m^2/s at mid-plant height) for 35 days to induce cold acclimation prior to freezing survival tests (*see* **Note 1**). The length of time the plants are acclimated greatly impacts freezing tolerance [3, 4] and should be standardized for all tests. Plants are irrigated weekly with nutrient solution containing macro and micronutrients. Just prior to freezing, plants in each cell are counted, the leaves are removed at the top of the lowermost leaf sheath, the flats are drenched with a solution of 10 mg/L Snomax snow inducer maintained at 4 °C, and allowed to drain until drainage has essentially ceased, a layer of crushed ice is placed on the soil surface to nucleate ice formation, and freezing is carried out in a programmable freezer. The temperature of the plant growth medium in each container near the crowns of the plants is monitored using food-piercing temperature probes and an Internet-enabled temperature monitor. The temperature is recorded every 2 min using a data capture script running on a remote computer. The temperature in the freezer is reduced from 4 °C to −3 °C and maintained at −3 °C for 16 h to allow all liquid water to be converted to ice and all heat of ice formation to dissipate, and is then lowered to a target temperature at 2 °C/h. The temperature is held at the target temperature for 1–2 h then raised to 0 °C at a rate of 2 °C/h. Following freezing, the plants are held at 4 °C for 24 h, are then moved to a greenhouse or growth chamber with the temperature at about 20 °C (*see* **Note 2**). Survival is scored as the proportion of plants in each cell that have regrown after 3–5 weeks.

3.2 Target Test Temperatures

The minimum temperature the plants are exposed to in artificial freezing tests is by far the most significant determinant of survival [5, 6]. Therefore, selection of appropriate test temperatures is vital to successful testing. A range of temperatures in two-degree increments is recommended in most cases [7]. Ideally, the test temperatures should range from the lowest temperature that results in no mortality, to the highest temperature that results in 100 % mortality. These temperatures depend entirely on the freezing tolerance of the test material, may differ by as little as 4 °C [8], and must be empirically determined. When screening breeding lines for freezing tolerance, the temperature range can be estimated from the freezing tolerance, expressed as LT_{50}, of the parent lines. However, transgressive segregation does occur [9] and may dictate adjustment of the test temperatures to be used, and again, must be empirically determined.

3.3 Data Analysis

Objectives of the data analysis of this kind of freezing survival data may be to calculate an LT_{50} score for experimental lines, to compare survival proportions of those lines, or to assess the impact of portions of the freezing process on survival. Suggestions for each of these kinds of analyses are presented below.

Due to size constraints of programmable freezing chambers, we typically design each study as a randomized incomplete block design with checks included in all freezing trials. One block consists of 48 cells in eight, 6-cell packs. Each block contains 40 experimental lines and eight check varieties. One check variety is included in each 6-cell pack. Each block is tested under each set of test conditions independently three times. To correct for variation between blocks and replications, the percent survival is standardized as follows. First, the grand mean survival of all check varieties over all replications at each test condition is calculated. Then, the mean survival of all check varieties within a replication is calculated, and a proportional correction factor is generated to adjust the mean of the check varieties within that replication to be equal to the grand mean of the check varieties under those test conditions. This correction factor is applied to the experimental lines within that replication to create a standardized percent survival score for each experimental line within each experimental condition. These standardized percent survival scores are then used as response variables in the data analysis.

3.4 Comparing Survival Means

For data analysis purposes, the response is expressed as the arcsin of the square root of the standardized survival to homogenize the variance of proportions [10]. A combined inter- and intra-block analysis of variance for incomplete blocks is conducted according to Littell et al. [11] by, for example, using PROC MIXED of SAS (Statistical Analysis System, http://www.sas.com) with blocks treated as random effects and plant lines treated as fixed effects. Blocks are adjusted for plant lines and plant lines are adjusted for blocks [11]. Survival means comparisons are conducted with *t*-tests on the least squares means.

3.5 Assessing the Impact of Portions of the Freezing Process on Survival

In order to assess the impact of portions of the freezing process on survival, the temperature record for each freezing episode for each container is parsed with a computer script to describe the (a) cooling rate from −3 °C to the minimum temperature reached, (b) minimum temperature, (c) time the plants were held at the minimum temperature, and (d) warming rate as the freezing episode ends. From these measures, the total time frozen is calculated, as is degree minutes, a summation of the temperature multiplied by the time at that temperature, and essentially the area bounded by the curve of the plot of time vs. temperature. Comparisons of the impact of each of these factors on survival are effected through comparisons of partial regression slopes associated with each component in each population [6], or comparisons of odds ratios determined with logistic regression [5].

3.6 Calculating LT_{50}

3.6.1 Probit Analysis

Survival of the plants after exposure to a range of low temperatures represents dose-response, dichotomous (alive vs. dead) data and therefore is amenable to probit analysis [12, 13]. Probit analysis is easily carried out using PROC PROBIT of SAS. This software generates a table of predicted percent survival vs. temperature and the temperature associated with 50 % survival is taken as the LT_{50} for a given plant line. Fiducial limits for the 50 % estimate, which may be used to infer significant differences between plant lines [14], are also generated by PROC PROBIT. Graphical representations of the probit regression of temperature vs. predicted survival can illustrate the relative responsiveness of the plant lines to temperature (e.g., ref. 9). Similarly, probit analysis can be used to model survival of plants at a single low temperature over time (e.g., ref. 15).

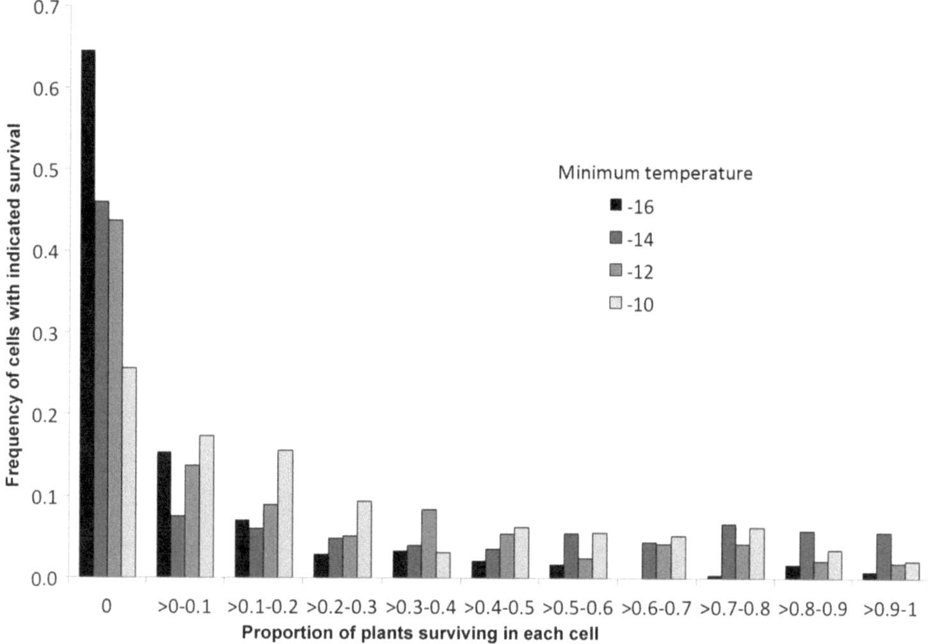

Fig. 1 Frequencies of horticultural container cells with indicated proportions of acclimated winter wheat plants surviving in freezing tolerance tests conducted to four subzero temperatures. Proportions are based on about 20 plants per cell

3.6.2 Zero-Inflated Poisson (ZIP) Analysis

Survival measurements are based on counts of live plants per cell in the cell packs and thus comprise "counts data." Survival is ultimately expressed as the percentage of plants surviving; however, we have noted that an excess of cells with no surviving plants is observed in almost all trials; some examples are shown in Fig. 1. These distributions are classic zero-inflated counts data [16, 17] and therefore may be correctly modeled using a zero-inflated Poisson (ZIP) model. Figure 2 shows predicted LT_{50} scores based on probit analysis compared to LT_{50} scores based on ZIP analysis of the same data. The two methods predict very similar LT_{50}s when overall mean survival ranges from about 30 to 65 % but diverge in cases where mean survival is outside of that range, especially where mean survival is very low (Fig. 2). In our experience, the LT_{50} scores predicted by the ZIP analysis are more likely to reflect real-world survival of the poorly cold-tolerant lines under the artificial testing as described above.

Therefore, we recommend using ZIP analysis for LT_{50} prediction if preliminary analysis shows an overabundance of cells with no survivors, such as shown in Fig. 1.

3.6.3 A Word About Prolonged Freezing Tests

A rather simple test emulating exposure to winter temperatures can be performed if the proper equipment is available. Plants are grown in cell packs as above to the 3-leaf stage at 20–25 °C, with a 16 h

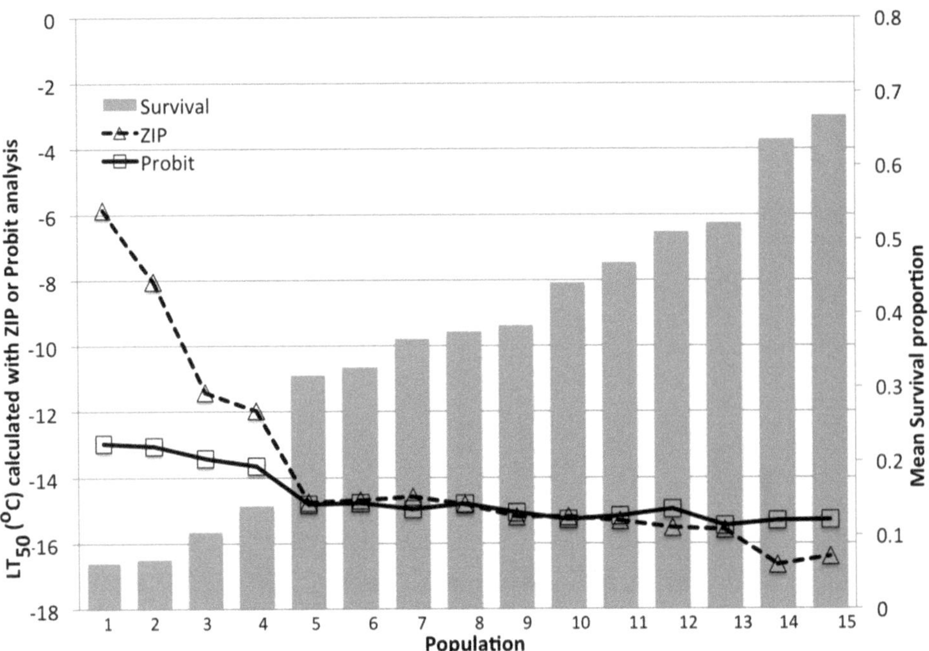

Fig. 2 LT_{50}s of 15 winter wheat plant populations tested to a range of subzero temperatures calculated with probit or ZIP analysis

photoperiod, then cold acclimated for 5 weeks at 4 °C with a 12 h photoperiod. The number of plants per pack is then recorded, the leaf blades are removed, leaving the stems, flats are covered with a layer of crushed ice to simulate snow cover and to nucleate ice formation (care is taken to maintain the plants in an upright position) and the flats are transferred to a walk-in freezer set to −5 °C. The flats are placed on wire racks or wooden planks to allow air circulation around the flats. No light is provided. The flats are maintained in this state for 15 weeks, approximately the length of time plants are exposed to subzero temperatures during the winter months. The ice may be lost through sublimation and is replenished as needed.

After 15 weeks, the flats are moved to a growth chamber at 4 °C, allowed to thaw for 7 days, and then moved to a greenhouse or growth chamber at 20–25 °C. Survival is scored as the number of plants that have begun to grow after 5 weeks.

This method was shown to result in survival that was significantly correlated with the LT_{50} scores determined as above. Additionally, this method showed that a 1 °C lower LT_{50} score resulted in 29.3 % greater survival of the plants after 15 weeks at −5 °C [18], remarkably close to the 30 % greater winter survival associated with 1 °C lower LT_{50} reported from field studies [19].

4 Notes

1. The intensity, quality, and photoperiod of the light the plants are exposed to also greatly impact freezing tolerance. Therefore, in our system, flats of plants are physically moved in a consistent path through the growth chamber such that they are each exposed to the same area of the chamber for the same length of time before being subjected to the freezing test.
2. The temperature during the recovery period following the freezing test cannot be greater than about 22 °C for at least about the first week following freezing. In some areas during the summer months, greenhouses may become too warm to be conducive to reproducible recovery. In this case, recovery should be carried out in a growth chamber.

References

1. Fowler DB, Gusta LV (1979) Selection for winterhardiness in wheat. I. Identification of genotypic variability. Crop Sci 19:769–772
2. Fowler DB (1979) Selection for winterhardiness in wheat. II. Variation within field trials. Crop Sci 19:773–775
3. Fowler DB, Limin AE, Wang SY, Ward RW (1996) Relationship between low-temperature tolerance and vernalization in wheat and rye. Can J Plant Sci 76:37–42
4. Fowler DB, Limin AE, Ritchie JT (1999) Low-temperature tolerance in cereals: model and genetic interpretation. Crop Sci 39:626–633
5. Skinner DZ, Bellinger BS (2011) Differential response of wheat cultivars to components of the freezing process in saturated soil. Crop Sci 51:69–74
6. Skinner DZ, Mackey BE (2009) Freezing tolerance of winter wheat plants frozen in saturated soil. Field Crop Res 113:335–341
7. Brule-Babel AL, Fowler DB (1989) Use of controlled environments for cereal cold hardiness evaluation: controlled freeze tests and tissue water content as prediction tests. Can J Plant Sci 69:355–366
8. Pomeroy MK, Fowler DB (1973) Use of lethal dose temperature estimates as indices of frost tolerance for wheat cold acclimated under natural and controlled environments. Can J Plant Sci 53:489–494
9. Skinner DZ, Garland-Campbell KA (2008) Evidence of a major genetic factor conditioning freezing sensitivity in winter wheat. Plant Breed 127:228–234
10. Snedecor GW, Cochran WG (1967) Statistical methods applied to experiments in agriculture and biology, 6th edn. Iowa State University Press, Ames, IA
11. Littell RC, Milliken GA, Stroup WW, Wolfinger RD, Schabenberger O (2006) SAS for mixed models, 2nd edn. SAS Institute Inc., Cary, NC
12. Finney DJ (1971) Probit analysis, 3rd edn. Cambridge University Press, New York, NY, 333 pp
13. Throne JE, Weaver DK, Baker JE (1995) Probit analysis—assessing goodness-of-fit based on back transformation and residuals. J Econ Entomol 88:1513–1516
14. Payton ME, Greenstone MH, Schenker N (2003) Overlapping confidence intervals or standard error intervals: what do they mean in terms of statistical significance? J Insect Sci 3:34–39
15. Thomas JB, Schaalje GB, Roberts DWA (1988) Prolonged freezing of dark-hardened seedlings for rating and selection of winter wheats for winter survival ability. Can J Plant Sci 68:47–55
16. Lambert D (1992) Zero-inflated Poisson regression, with an application to defects in manufacturing. Technometrics 34:1–14
17. Ridout MS, Demetrio CGB, Hinde JP (1998) Models for counts data with many zeros. Proceedings of the XIXth international biometric conference, Cape Town, South Africa. Invited Papers, pp. 179–192
18. Skinner DZ, Garland-Campbell KA (2008) The relationship of LT_{50} to prolonged freezing survival in winter wheat. Can J Plant Sci 88: 885–889
19. Fowler DB, Gusta LV, Tyler NJ (1981) Selection for winterhardiness in wheat. III. Screening methods. Crop Sci 21:896–901

Chapter 3

Measuring Freezing Tolerance: Electrolyte Leakage and Chlorophyll Fluorescence Assays

Anja Thalhammer, Dirk K. Hincha, and Ellen Zuther

Abstract

Quantitative assessment of freezing tolerance is essential to unravel plant adaptations to cold temperatures. Not only the survival of whole plants but also impairment of detached leaves after a freeze–thaw cycle can be used to accurately quantify plant freezing tolerance in terms of LT_{50} values. Here we describe two methods to determine the freezing tolerance of detached leaves using different physiological parameters. Firstly, we illustrate how to assess the integrity of (predominantly) the plasma membrane during freezing using an electrolyte leakage assay. Secondly, we provide a chlorophyll fluorescence imaging protocol to determine the freezing tolerance of the photosynthetic apparatus.

Key words Freezing tolerance, LT_{50}, Electrolyte leakage, Chlorophyll fluorescence, F_v/F_m

1 Introduction

Robust experimental approaches for the precise quantification of plant freezing tolerance are of fundamental importance to understand the genetic and molecular mechanisms underlying and determining this complex trait. Cellular membranes are widely accepted as primary sites of freezing damage (*see* ref. 1 for a comprehensive review). Therefore, next to plant survival, methods assessing cellular membrane integrity are frequently used to determine plant freezing tolerance. The two methods we describe here provide stable and highly reproducible LT_{50} values, defined as the temperatures at which 50 % of damage occurs. Moreover, combining the two protocols allows to discriminate the site of freezing damage between plasma membrane and chloroplasts. This constitutes a powerful tool to investigate the mechanistics of plants altered in their freezing tolerance by breeding or genetic engineering and to test the site of activity of known or novel cellular protectants. In addition, we have used both methods also to assess the natural variation in the freezing tolerance of different *Arabidopsis thaliana* accessions [2–5]. Moreover, both protocols can easily be adapted

Dirk K. Hincha and Ellen Zuther (eds.), *Plant Cold Acclimation: Methods and Protocols*, Methods in Molecular Biology, vol. 1166, DOI 10.1007/978-1-4939-0844-8_3, © Springer Science+Business Media New York 2014

for use with other plant species, such as electrolyte leakage assays in *Thellungiella salsuginea* [6].

The first part of the protocol we describe concerns the controlled-rate freezing of detached leaves. Subsequently, the thawed leaves can be used for conductivity measurements to assess electrolyte leakage. Various forms of this method have been used for many years, not only to determine freezing damage but also in the context of many other stresses that may have an impact on cellular membranes, such as drought or reactive oxygen species. The method we use here was originally described in [7]. It not only reports on the intactness of the plasma membrane as a semipermeable barrier for intracellular ions but also on the integrity of the vacuole as the major storage compartment for inorganic ions [2, 8].

In addition to the plasma membrane, also chloroplast membranes are susceptible to freezing damage. Linear electron transport is interrupted, which finally results in the inactivation of photosynthesis [9]. Therefore, chlorophyll a fluorescence measurements are a suitable tool to study freezing damage. Values of F_v/F_m determined with dark-adapted leaves reflect the potential quantum use efficiency of photosystem II (PSII) and have been widely used for assessing stress damage to the photosynthetic apparatus [10, 11]. The second protocol therefore describes the use of chlorophyll fluorescence imaging [12, 13] to quantify leaf freezing damage. Its use is advantageous over classical chlorophyll fluorescence measurements because it is not limited to single-point measurements but allows integration of F_v/F_m over the whole leaf area. This turned out to be extremely important, as the basal leaf parts are damaged at milder freezing temperatures than the upper parts [8].

Here we give a comprehensive description of experimental design, plant cultivation, and leaf freezing procedures, which are identical for both assays. From this point on, the freeze-thawed leaves can be either used for electrolyte leakage measurement or subjected to chlorophyll fluorescence imaging.

2 Materials

2.1 Equipment

1. Programmable cooling bath thermostats (CC130; Huber, Offenburg, Germany, or similar) with a large opening to allow handling of three metal racks for 48 glass tubes each. If possible, the thermostats should be placed in a 4 °C chamber to prevent overheating during cooling of the baths to low subzero temperatures.
2. Silicon oil (Thermal HY; Julabo, Seelbach, Germany, or similar) to fill the cooling baths.
3. Lab 960 Conductometer equipped with LF613T Conductivity electrode (SI Analytics, Mainz, Germany).

4. Water bath with the capacity for boiling.
5. Imaging Pam Chlorophyll Fluorometer IMAG-C with STANDARD measuring head (Walz, Effeltrich, Germany).
6. Bottletop dispenser, volume range 1–10 ml (optional).
7. Automatic pipette suitable for 300 μl volumes.

2.2 Software

1. Graph Pad Prism 3.0 (Graph Pad, La Jolla, USA).
2. Imagewin v2.32 (Walz, Effeltrich, Germany).

2.3 Consumables

1. Glass tubes in metal racks (10 cm height, 1.5 cm diameter, 48 tubes per rack, 3 or 6 racks per experiment)—can be washed and reused.
2. ddH_2O.
3. Metal lids for glass tubes—can be washed and reused.
4. Razor blades or scalpels.
5. Blunt-end forceps.
6. Round-bottom 15 ml falcon tubes without lid—can be washed and reused.
7. Standard microscopy glass slides.

3 Methods

3.1 Freezing Experiment [2, 14]

3.1.1 Design the Experiment

The use of two cooling baths in parallel will provide you with sufficient space to process 288 samples in one experiment. The freezing temperature course of each experiment should be carefully planned. For non-acclimated leaves of the moderately freezing tolerant *A. thaliana* accession Col-0 a range of −1 to −15 °C and for cold acclimated leaves a range of −1 to −21 °C are recommended. It is advisable to take samples in steps of 1–2 °C and to condense the temperature interval to 0.5–1 °C steps in the range of the expected LT_{50} value in order to reach an optimal resolution. To monitor the comparability between single experiments an internal control line (e.g., *A. thaliana* Col-0) should be used in each independent experiment. Each plant line in an experiment should be covered by four to five technical replicates, so using the setup given here will enable you to assess three to four independent plant lines in a single experiment.

3.1.2 Grow the Plants

1. Seeds are sown in soil and vernalized in a phytotron for 1 week under cold-night conditions (12/12-h day/night cycle, 20 °C/6 °C). After one additional week in short-day conditions (8/16-h day/night cycle, 20 °C/16 °C), plants are pricked (three plants per pot (Ø 10 cm)) and kept in short-day conditions for 2 more weeks. Four weeks after sowing, plants

are transferred to long-day conditions (16/8-h day/night cycle, 20 °C/18 °C) with light supplementation to reach at least 200 μE/m^2/s. Non-acclimated plants are thus used when 6 weeks old.

2. For cold acclimation, the same quantity of plants is transferred to a 4 °C growth cabinet (16/8-h day/night cycle, 4 °C) with 90 μE/m^2/s for an additional 14 days [2]. The additional cold treatment will minimize developmental processes, so relative leaf age will be hardly changed between non-acclimated and cold acclimated leaves. Ideally, plants should be in a developmental stage where the leaf rosette is fully expanded, but the plant is not yet bolting. Avoid the use of already flowering plants. Roughly, 40–45 plants will provide enough leaf material to cover all technical replicates of one plant line.

3.1.3 Conduct the Experiment

1. Prepare and label a sufficient amount of glass tubes in metal racks, and put them on ice. Additionally prepare a rack containing tubes for unfrozen control leaves which will be kept on ice during the entire experiment.
2. Fill 300 μl of ddH_2O into each glass tube for freezing treatment and 600 μl of ddH_2O into the control tubes.
3. Cut three to six rosettes, and arrange the single fully expanded leaves in a way to generate stacks of three leaves stemming from different rosettes each (*see* **Note 1** for additional information). The total amount of leaf material should be about the same in each stack. Exclude apparently damaged or senescent leaves (*see* **Note 2** for additional information). For chlorophyll fluorescence experiments carefully remove soil crumbs from the leaves as these will block the fluorescence signal. Prepare a sufficient number of stacks to cover all temperature steps of one technical replicate and the control. Work quickly to avoid wilting of the leaves.
4. Cut the petioles with a sharp razor blade to generate a common base of each stack (Fig. 1a).
5. Put each leaf stack to the bottom of the appropriate test tube with a forceps so that the petioles are enclosed by water (Fig. 1b). Take care not to press or fissure the leaves. Put the tubes on ice immediately.
6. Prepare all replicates of all plant lines correspondingly.
7. Put lids on the control samples, and store them on ice for the duration of the experiment.
8. Put the other tubes into the respective racks in the precooled cooling baths (−1 °C), and allow the samples to equilibrate to −1 °C for 30 min.
9. Grind some ice with mortar and pestle to obtain small ice crystals.

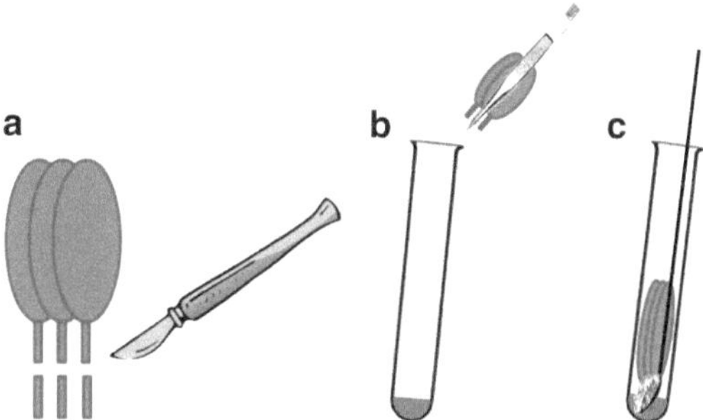

Fig. 1 Schematic overview of sample handling during the preparation for a freezing experiment. Stacking of three leaves from different leaf rosettes and cutting of petioles (**a**), insertion of leaf stacks into glass tubes containing water (**b**), addition of small ice crystals to initiate ice nucleation in the leaves (**c**)

10. Carefully add a small ice crystal to the bottom of each tube with a small spatula (Fig. 1c) to initiate freezing, close the tube lids, and incubate for another 30 min at −1 °C to allow ice nucleation in the leaves and temperature equilibration.
11. Start the program with a cooling rate of 4 °C/h, take out respective samples at the desired temperatures, and store them on ice. If needed, lower cooling rates, such as 2 °C/h, can be used as well, but higher cooling rates should be avoided, as sample temperature will then not be able to follow bath temperature.
12. Remove samples from the bath at the predetermined steps in the temperature protocol, and transfer them immediately to an ice bath. Leave all samples on ice in the 4 °C chamber overnight to thaw. Samples are then ready to process either for electrolyte leakage or chlorophyll fluorescence measurements.

3.2 Electrolyte Leakage [2, 7, 14]

3.2.1 Experimental Setup

1. Add 7 ml of ddH_2O to each tube, including the control tubes, so that the leaves are completely immersed. If you are using large leaves it might be necessary to use more water.
2. Incubate the samples on a shaker at 150 rpm at 4 °C for approximately 24 h (compare **Note 3**).
3. Prepare metal racks with 15 ml falcon tubes in the same setup as in your freezing experiment. Fill 4.5 ml of ddH_2O into each tube, and add 1 ml of the respective sample.
4. Insert the electrodes carefully into the sample tubes. Measure the electrical conductivity of each sample after mixing the solution thoroughly by moving the electrodes up and down for 10–12 times (compare **Note 4**). Wait until a stable value is displayed (several seconds). Measure each sample twice, and note the higher value. Before transferring the electrodes to the next sample, clean them by swaying in ddH_2O and tap them dry on a paper towel (*see* **Note 5** for additional information).

5. Empty the falcon tubes, and rinse them 2–3 times with deionized water.
6. Incubate the tubes containing the leaves in a boiling water bath for 30 min for determination of the total electrolyte content.
7. Allow the samples to cool down to room temperature, and repeat **steps 1–4**. Use the same tube setup as for the unboiled samples.

3.2.2 Data Analysis

1. Collect the data in a suitable data processing software (e.g., MS Excel). Calculate the percentage of electrolyte leakage (% EL) relative to the conductivity of the boiled samples:

$$\% \, EL = \frac{Conductivity\left[unboiled\ sample\right]}{Conductivity\left[boiled\ sample\right]} \times 100$$

2. As the leakage of electrolytes in the control samples is not caused by freezing, normalize the % EL of each sample (% EL_{sample}) to the average % EL of the control samples (% $EL_{control}$) within each replicate line. For a better comparability of the graphs, normalize your data to a maximum electrolyte leakage of 100 %. For this purpose, use the % EL of the lowest freezing temperature (% EL_{max}):

$$\% \, EL_{normalized} = \frac{\left(\% \, EL_{sample} - \% \, EL_{control}\right)}{\left(\% \, EL_{max} - \% \, EL_{control}\right)} \times 100$$

3. Import the % $EL_{normalized}$ values into the GraphPad Prism 3.0 software (or any other software able to perform the necessary calculations) indicating the appropriate number of replicates. Analyze the data using nonlinear regression (curve fit) with a sigmoidal dose–response. This will give you the LT_{50} value over all technical replicate lines as LOGEC50 (Fig. 2) [2] (*see* **Note 6** for additional information). For control purposes we strongly recommend to compare the leakage curves of all single replicates of one line before calculating the LT_{50} value.

3.3 Chlorophyll Fluorescence Imaging [8, 15]

3.3.1 Experimental Setup

1. Keep samples on ice during the whole experiment.
2. Dark adapt samples, e.g., by completely covering with a black sheet or cloth for at least 20 min in order to bring all PSII reaction centers into the open state.
3. Set up the measuring facilities in a darkroom to keep the samples dark adapted during the measurement, and start the Imagewin software for data acquisition.
4. Arrange one of the three leaves of one replicate sample carefully on a microscopy slide in the Live Video mode (NIR-measuring

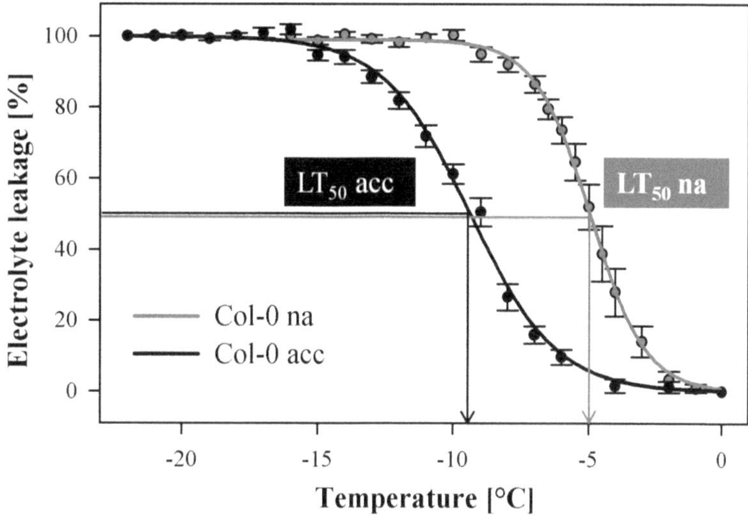

Fig. 2 Electrolyte leakage from non-acclimated (na) and cold-acclimated (acc) *A. thaliana* Col-0 leaves frozen to a range of different temperatures. Curves were fitted using a logistic regression model, and LT_{50} values were calculated as the temperatures at which 50 % electrolyte leakage occurred. Error bars represent ± SEM from at least four replicate measurements, with each replicate including three leaves from different plants

light pulses, frequency 1 Hz) (compare **Note 7**). Focus the leaf, and fix the distance of the camera to the sample. This distance must be kept the same during all measurements.

5. Quit the Live Video mode by pressing Exit, and press the F_o, F_m button to trigger a saturation pulse. This measurement will give you the variable (F_v) and maximal (F_m) chlorophyll a fluorescence of the leaf.
6. Select the F_v/F_m image, reflecting the maximal PSII quantum yield of a dark-adapted leaf.
7. Save the picture as PAM Image (PIM) file for later analysis, press the New Record button, and continue with the next sample.

3.3.2 Data Analysis

1. Use the Imagewin software for data processing. You can extract the F_v/F_m pictures displayed in a false color scale representation as shown in Fig. 3.
2. In addition, you can quantify F_v/F_m values by integration over the whole leaf area or desired leaf sections. For this purpose, open the respective PIM file with the Imagewin software and select F_v/F_m. By using the area of interest (AOI) routine, you can set user-defined limits within which the average F_v/F_m value will be calculated. This will be displayed in the Report menu (compare **Note 8**).

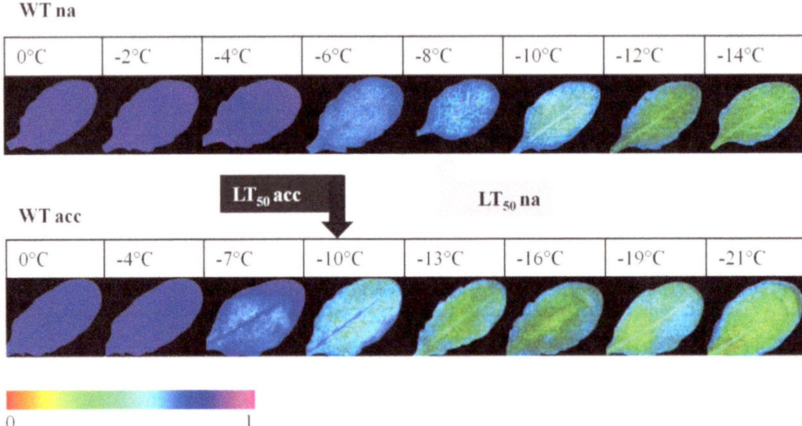

Fig. 3 Chlorophyll fluorescence imaging of *A. thaliana* Col-0 leaves. Detached leaves of non-acclimated (na—*upper panel*) and cold-acclimated (acc—*lower panel*) plants were frozen to different temperatures and thawed slowly. Maximum quantum yield of photosystem II (F_v/F_m) is shown in false color images (as specified by the scale bar) at the indicated freezing temperatures. The LT_{50} values calculated from the numeric F_v/F_m values derived from logistic regression models are indicated by *arrows*

3. Determination of the LT_{50} values is done analogous to the electrolyte leakage procedure (*see* **Note 9** for additional information). Determine relative F_v/F_m ($F_v/F_{m\,rel}$) values by normalizing the F_v/F_m values of each replicate line ($F_v/F_{m\,sample}$) to that of the unfrozen control sample ($F_v/F_{m\,control}$) as well as to the F_v/F_m value of the sample exposed to the lowest freezing temperature ($F_v/F_{m\,max}$):

$$Fv\,/\,Fm_{rel} = \frac{\left(Fv\,/\,Fm_{sample} - Fv\,/\,Fm_{control}\right)}{\left(Fv\,/\,Fm_{max} - Fv\,/\,Fm_{control}\right)} \times 100$$

4. Import the $F_v/F_{m\,rel}$ values into the GraphPad Prism 3.0 software, and continue as described for % $EL_{normalized}$ values.

4 Notes

1. We found no significant difference in freezing tolerance between old and young fully expanded leaves from plants of the same age [2]. This may, however, be different for small, very young leaves.
2. Working with leaves that are too long and stick out above the surface of the cooling fluid (silicon oil) should be avoided, as these parts will not reach the same temperatures as the lower, fully immersed parts. This will lead to a gross distortion of the

LT_{50} values. If you have to work with such leaves, you need to cut them to a suitable length.

3. If you work with other plants than Arabidopsis, this may need to be modified, as thicker leaves may need longer incubation times to reach equilibrium with the surrounding water (*see* for example ref. 16).
4. For a higher sample throughput it is recommended to use two programmable baths and two conductivity meters simultaneously. Measuring the conductivity can then be done in two racks in parallel. Make sure to use the same electrode on a respective sample before and after boiling to exclude systematic errors caused by differences in the electrodes.
5. From time to time the conductivity electrodes should be cleaned by washing in 1 % SDS in a glass beaker under gentle stirring for 1.5 h. Be careful to fix the electrodes in a way that they will not slip into the beaker and get broken by the stirring bar.
6. For orientation purposes, it is helpful to know that in *A. thaliana* Col-0, the LT_{50} of cold acclimated leaves derived from the electrolyte leakage assay is about −9.5 °C and for non-acclimated leaves about −5.5 °C [2, 5]. These numbers reflect the ability of the plant to adapt to freezing temperatures by exposure to low but nonfreezing temperatures.
7. Measuring the chlorophyll fluorescence of all three leaves in a tube is not feasible with the number of samples generated in these experiments. However, the measurement of one leaf is sufficient to provide stable results, when you stick to the suggested number of four to five technical replicates per sample. However, if one leaf is providing an obviously unreliable picture, you should assess another leaf from the same tube. Do not measure the same leaf twice, since the saturating pulse will close all of the reaction centers of PSII. A second measurement will therefore not provide the F_v/F_m anymore.
8. In healthy, non-stressed leaves F_v/F_m usually has a value of around 0.83 [17] and declines with increasing stress levels.
9. Although there is a tight linear correlation between LT_{50} values from both assays, those derived from chlorophyll fluorescence imaging are generally lower than those from electrolyte leakage measurements. This is mainly due to secondary damage resulting from the preincubation of electrolyte leakage samples in distilled water. However, biological reasons in terms of a higher freezing tolerance of photosynthetic membranes in comparison to the plasma membrane cannot be completely excluded [8].

References

1. Steponkus PL (1984) Role of the plasma membrane in freezing injury and cold acclimation. Annu Rev Plant Physiol 35:543–584
2. Rohde P, Hincha DK, Heyer AG (2004) Heterosis in the freezing tolerance of crosses between two *Arabidopsis thaliana* accessions (Columbia-0 and C24) that show differences in non-acclimated and acclimated freezing tolerance. Plant J 38:790–799
3. Hannah MA, Wiese D, Freund S, Fiehn O, Heyer AG, Hincha DK (2006) Natural genetic variation of freezing tolerance in Arabidopsis. Plant Physiol 142:98–112
4. Korn M, Peterek S, Mock H-P, Heyer AG, Hincha DK (2008) Heterosis in the freezing tolerance, and sugar and flavonoid contents of crosses between *Arabidopsis thaliana* accessions of widely varying freezing tolerance. Plant Cell Environ 31:813–827
5. Zuther E, Schulz E, Childs LH, Hincha DK (2012) Clinal variation in the non-acclimated and cold-acclimated freezing tolerance of *Arabidopsis thaliana* accessions. Plant Cell Environ 35:1860–1878
6. Lee YP, Babakov A, de Boer B, Zuther E, Hincha DK (2012) Comparison of freezing tolerance, compatible solutes and polyamines in geographically diverse collections of *Thellungiella* sp. and *Arabidopsis thaliana* accessions. BMC Plant Biol 12:131
7. Ristic Z, Ashworth EN (1993) Changes in leaf ultrastructure and carbohydrates in *Arabidopsis thaliana* L. (Heyn) cv. Columbia during rapid cold acclimation. Protoplasma 172: 111–123
8. Ehlert B, Hincha DK (2008) Chlorophyll fluorescence imaging accurately quantifies freezing damage and cold acclimation responses in Arabidopsis leaves. Plant Methods 4:12
9. Krause GH, Grafflage S, Rumich-Bayer S, Somersalo S (1988) Effects of freezing on plant mesophyll cells. Symp Soc Exp Biol 42: 311–327
10. Woo N, Badger M, Pogson B (2008) A rapid, non-invasive procedure for quantitative assessment of drought survival using chlorophyll fluorescence. Plant Methods 4:27
11. Maxwell K, Johnson GN (2000) Chlorophyll fluorescence—a practical guide. J Exp Bot 51: 659–668
12. Oxborough K (2004) Imaging of chlorophyll a fluorescence: theoretical and practical aspects of an emerging technique for the monitoring of photosynthetic performance. J Exp Bot 55: 1195–1205
13. Lichtenthaler HK, Miehe JA (1997) Fluorescence imaging as a diagnostic tool for plant stress. Trends Plant Sci 2:316–320
14. McKhann HI, Gery C, Berard A, Leveque S, Zuther E, Hincha DK, de Mita S, Brunel D, Teoule E (2008) Natural variation in CBF gene sequence, gene expression and freezing tolerance in the Versailles core collection of *Arabidopsis thaliana*. BMC Plant Biol 8:105
15. Schreiber U, Bilger W (1987) Rapid assessment of stress effects on plant leaves by chlorophyll fluorescence measurement. In: Tenhunen JD, Catarino FM, Lange OL, Oechel WC (eds) Plant response to stress. Functional analysis in Mediterranean ecosystems. Springer, Berlin, pp 27–53
16. Hincha DK, Pfüller U, Schmitt JM (1997) The concentration of cryoprotective lectins in mistletoe (*Viscum album* L.) leaves is correlated with leaf frost hardiness. Planta 203:140–144
17. Hunt S (2003) Measurements of photosynthesis and respiration in plants. Physiol Plant 117:314–325

Chapter 4

Conducting Field Trials for Frost Tolerance Breeding in Cereals

Luigi Cattivelli

Abstract

Cereal species can be damaged by frost either during winter or at flowering stage. Frost tolerance per se is only a part of the mechanisms that allow the plants to survive during winter; winterhardiness also considers other biotic or physical stresses that challenge the plants during the winter season limiting their survival rate. While frost tolerance can also be tested in controlled environments, winterhardiness can be determined only with field evaluations. Post-heading frost damage occurs from radiation frost events in spring during the reproductive stages. A reliable evaluation of winterhardiness or of post-heading frost damage should be carried out with field trials replicated across years and locations to overcome the irregular occurrence of natural conditions which satisfactorily differentiate genotypes. The evaluation of post-heading frost damage requires a specific attention to plant phenology. The extent of frost damage is usually determined with a visual score at the end of the winter.

Key words Frost tolerance, Winterhardiness, Vernalization, Post-heading frost damage, Barley, Wheat

1 Introduction

Cereal species can be damaged by frost either during winter or at flowering stage. The tolerance of winter cereals to low temperatures depends on the physiological process known as hardening or cold acclimation that occurs when plants are exposed to temperatures ranging from 0 to 5 °C prior to winter freezing. There is large genetic variation for the ability to survive freezing temperatures among the cereal species, with winter-habit rye cultivars having the best freezing tolerance followed by hexaploid winter wheat and winter barley and oat [1]. Nevertheless, it should be noticed that rye, wheat, barley, and oat genotypes are all capable to cold acclimate, to some extent, in response to low temperatures.

Frost tolerance is intimately connected with vernalization. The cereal genotypes have traditionally been classified into three main groups: spring types, which pass to the reproductive phase quickly, without vernalization and even in short days; winter types, which

Dirk K. Hincha and Ellen Zuther (eds.), *Plant Cold Acclimation: Methods and Protocols*, Methods in Molecular Biology, vol. 1166, DOI 10.1007/978-1-4939-0844-8_4, © Springer Science+Business Media New York 2014

display a strong vernalization requirement and sensitivity to short days; and intermediate or alternative (also called facultative) types which flower quickly in long days but in which floral induction is more or less inhibited by short days. In the light of the knowledge achieved by molecular genetics three major vernalization loci, *Vrn-1*, *Vrn-2*, and *Vrn-3*, have been identified as the determinants of the vernalization response [2]. Since no allelic variation at the *Vrn-3* locus was observed within the cultivated germplasm, a two-gene epistatic model was proposed in barley [3] as well as in wheat [4]. A higher level of frost tolerance is generally associated with the winter growth habit; nevertheless, some studies have also reported a high level of frost tolerance in facultative genotypes without vernalization response [5, 6]. Plants with facultative growth habit are more ready to react to changes in environmental factors (light intensity, temperature) assuring flowering under a wide range of climatic conditions, a trait associated with a high adaptation capacity.

The general association between winter habit and frost tolerance is explained by the genetic linkage between the *Vrn-1* locus and the two loci controlling frost tolerance, *Frost Resistance-1* (*Fr-1*) and *Fr-2*, all located on the long arm of chromosome 5A in wheat [7, 8] and 5H in barley [9]. *Fr-1* is a pleiotropic effect of *Vrn-1* (or it co-segregates with *Vrn-1*), while *Fr-2* maps about 30 cM proximal from *Vrn-1/Fr-1*. *Fr-2* contains a cluster of *CBF* genes, a family of cold inducible transcription factors known to control the expression of a large part of the cold-regulated genes. Allelic variations at the *Fr-1/CBF* locus are known to modify cold acclimation capacity and, in turn, frost tolerance [10].

Frost tolerance per se is only a part of the mechanisms that allow the plants to survive during winter and to synchronize their life cycle with the seasonal cycle. From an agricultural (and economical) point of view, winterhardenes (or winter survival) is a more relevant and broad concept than frost tolerance, although frost tolerance often represents the main factor for winter survival. Winterhardiness consider the plants within the whole ecosystem where other organisms (e.g., pathogens specifically adapted to low temperature) or physical conditions (e.g., anoxia, limited soil fertility) can challenge the plants, thus limiting their survival rate [11]. Winterhardiness can be determined only with field evaluations.

Barley and wheat crops can also experience frost damage at the reproductive stage. Post-heading frost damage is a main problem in southern Australia where barley and wheat are planted in autumn with the majority of the growing season over winter. Winter is usually mild, and winter frost damage is virtually absent. The predominant frost damage occurs from radiation frost events in spring during the reproductive stage. Radiation frosts occur under clear night skies, where more heat is radiated away from the crop canopy than it receives. The loss of radiant energy causes the temperature to fall, which can damage sensitive reproductive tissues at sub-zero temperatures. These frost events can cause floret and spike abortion as well as damage to the developing grain, which can have a significant impact on yield and quality [12, 13].

2 Methods

2.1 Winterhardiness

The intensity and the frequency of frost events during winter are unpredictable and vary significantly depending on locations and years limiting the effectiveness of winterhardiness field trials (*see* **Note 1**). A reliable evaluation of winterhardiness is carried out with field trials replicated across years and locations. In each location, a randomized design with three or more replicates is recommended, and small plots (2–3 m^2) are usually sufficient to evaluate the winterhardiness. The sowing date, being determinant for growth stage, plays a decisive role in frost tolerance and winterhardiness [14]. Late sowing limits the plant development before frost events and does not allow the complete deployment of the plant acclimation potential. Therefore, the application of different sowing dates might reveal differences in the frost tolerance capability of the genotypes under evaluation. Overall, a complete design to test winterhardiness considers three or more locations in different climatic regions, in 2 or more years and, if possible, two sowing dates with a replicated field design. A weather station in the proximity of the field trial is used to record the temperature throughout the winter season.

The extent of winter damage is usually assessed at the end of the winter by visual scoring. A frequently used scoring system is based on a 0–9 scale [14, 15] with

0: no damage

1: slightly yellowed leaf tips

2: half-yellowed basal leaves

3: fully yellowed basal leaves

4: whole plants slightly yellowed

5: whole plants yellowed and some plants withered

6: whole plants yellowed and 10 % plant mortality

7: whole plants yellowed and 20 % plant mortality

8: whole plants yellowed and 50 % plant mortality

9: all plants killed

Some examples of field experiments for the evaluation of winterhardiness and of winter damage are given in Figs. 1, 2, 3, and 4.

In environments where snow is frequent, a long-lasting snow cover can interfere significantly with the evaluation of winterhardiness (*see* **Note 2**). In these conditions, the field evaluation can be

Fig. 1 Barley field trial after a cold winter with a long snow cover period (Fiorenzuola, Northern Italy). The cultivar on the *left* has no damage (score 0), while the cultivar on the *right* shows all plants completely yellowed but with no dead plants (score 5)

Fig. 2 A close-up of a barley plant with clear symptoms of winter damage on the older leaves, but with the crown still alive

integrated with a parallel experiment where plants are grown in boxes in an open air space and protected from snow with a shelter [5]. Under the shelter the plants are exposed to natural temperature variations, but without snow cover. A comparison between shelter-protected and field-grown plants allows the estimation of the snow- and other winter-related stress factors on winterhardiness.

Fig. 3 Barley field trial after a cold winter with a long snow cover period (Fiorenzuola, Northern Italy). Some cultivars are completely killed by winter conditions (an example is given in Fig. 4), while other cultivars show extended leaf damage but limited plant mortality. Border plots (*dark green*) are sown with bread wheat, a cereal species with a higher frost tolerance than barley (Color figure online)

Fig. 4 A barley genotype showing a high level of winter damage, where only a few plants are still alive after winter (score 9)

The progress of the cold acclimation during winter in field-grown plants can be monitored taking leaf or crown samples from some representative plants and using them for standard frost evaluation tests such as the assessment of electrolyte leakage after freezing in controlled conditions.

2.2 Post-heading Frost Damage

As for winterhardiness, the evaluation of post-heading frost damage requires a multi-location field trial design to overcome the unpredictability of late frost events. Furthermore, the evaluation of post-heading frost damage has a high risk of escape due to difference in plant phenology and morphology and spatial temperature variations. To control for plant phenology two main strategies are used: (1) test genotypes with similar heading time, and (2) use different sowing dates in order to compare plants at similar phenological stage deriving from different sowings (*see* **Note 3** for additional techniques). Typically an experimental design for the evaluation of post-heading frost damage is run on several locations and with two to four sowing dates. Usually, early sowing encourages early flowering during the period of highest frost risk. Different sowing dates are required to allow for maturity differences between genotypes and to collect data from multiple frost events during the same season. Small plots are used to give the maximum number of genotypes in a small area to reduce the effect of spatial temperature variation. Thermometers distributed in the field are used to monitor spatial temperature differences. After each frost event, a number of tillers at the same developmental stage are tagged to allow a comparison at similar developmental stages. Frost-induced sterility is assessed 10–20 days later on each spike and expressed as percentage of total florets [12]. Additional frost-induced grain damage is scored at maturity.

3 Notes

1. Field evaluation of winterhardiness has been the first and simplest method used to select for frost tolerance. Quite often, the irregular occurrence of natural conditions which satisfactorily differentiate genotypes results in large experimental errors and complicates the detection of small but meaningful differences among cultivars. This limitation has prompted the development of a number of methods for the assessment of frost tolerance under controlled conditions. Nevertheless, field evaluation offers the unique opportunity to assess the overall winterhardiness capacity that is more than a simple evaluation of frost tolerance measured under controlled temperature conditions in a growth chamber. To overcome the intrinsic limitations of a field evaluation of winterhardiness, replication of the experiments

across locations and years must always be considered. A fundamental question concerns what kind of frost-tolerant plants we will need in the future, when the foreseen global climate changes will generally increase the temperature, reducing the frequency and the extent of harsh winters [11]. At a first impression the global warming might reduce frost damage in crops, but this is not likely to happen. A consequence of the warming is the fluctuation of winter temperatures; winter warm spells are becoming more frequent than in the past [16, 17], and this has a strong impact on the frost tolerance both in crops and natural flora. Vernalization is generally saturated or partially saturated during the first month of the winter. As a consequence, during a winter warm spell plants start active growth, thereby loosing most of their freezing tolerance. This condition leads to the exposition of not hardened plants to subsequent frost events. Since frost tolerance is intimately associated with a reduction in plant growth [18], it is unlikely that actively growing plants retain their hardening capacity. Therefore, the ability to rapidly change the physiological conditions in response to a fluctuation of temperature might become more important than absolute frost tolerance capacity. Testing the adaptation of cereal species and genotypes to the new winter climate under natural conditions will be an essential aspect of cereal breeding for temperate and cold regions in the coming years. Therefore, although in the last decades the evaluation of winterhardiness has often been substituted with tests for frost tolerance in controlled conditions, the field evaluation will remain and will acquire even more relevance in the future since there will be a need to breed new varieties for the new climatic conditions.

2. In deep-snow regions, the plants survive winter under a long-lasting snow cover. Snow provides a protection from deep frost keeping soil temperature at crown level between 0 and −10 °C despite the very low air temperature. Prolonged snow cover prevents photosynthesis, reduces plant metabolism, and exposes the plants to psychrophilic pathogenic fungi known as snow mould fungi [19] as well as to some risks of anoxic stress when the snow melts [20]. The most relevant snow mould fungi are snow scald (*Sclerotinia borealis*), speckled snow mould (*Typhula ishikariensis*), and pink snow mould (*Microdochium nivale*) [19]. When snow mould infections occur during an experiment for the evaluation of winterhardiness, it should be considered that the performance of the plants will be mainly determined by their susceptibility to snow mould, which, to some extent, is determined by the cold acclimation state of the plants [21].

3. A new method for the evaluation of post-heading frost damage has recently been proposed. The application of supplementary artificial light in the field, aiming to modify the natural photoperiod, together with a specific field design allows the generation of a "photoperiod gradient." Thereby, same genotypes, depending on the light conditions, can be stimulated to flower at different times. This will allow the screening of genotypes with different phenologies under natural field frost conditions at matched developmental stages [22].

References

1. Fowler DB, Carles RJ (1979) Growth, development, and cold tolerance of fall-acclimated cereal grains. Crop Sci 19:915–922
2. Trevaskis B, Hemming MN, Dennis ES, Peacock WJ (2007) The molecular basis of vernalization-induced flowering in cereals. Trends Plant Sci 12:352–357
3. von Zitzewitz J, Szucs P, Dubcovsky J, Yan LL, Francia E, Pecchioni N, Casas A, Chen THH, Hayes PM, Skinner JS (2005) Molecular and structural characterization of barley vernalization genes. Plant Mol Biol 59:449–467
4. Yan L, Loukoianov A, Tranquilli G, Helguera M, Fahima T, Dubcovsky J (2003) Positional cloning of the wheat vernalization gene VRN1. Proc Natl Acad Sci U S A 100:6263–6268
5. Rizza F, Pagani D, Gut M, Prasil IT, Lago C, Tondelli A, Orrù L, Mazzucotelli E, Francia E, Badeck FW, Crosatti C, Terzi V, Cattivelli L, Stanca AM (2011) Diversity in the response to low temperature in a set of representative barley genotypes cultivated in Europe. Crop Sci 51:2759–2779
6. von Zitzewitz J, Cuesta-Marcos A, Condon F, Castro AJ, Chao S, Corey A, Filichkin T, Fisk SP, Gutierrez L, Haggard K, Karsai I, Muehlbauer GJ, Smith KP, Veisz O, Hayes PM (2011) The genetics of winter-hardiness in barley: perspectives from genome-wide association mapping. Plant Genome 4:76–91
7. Vágújfalvi A, Galiba G, Cattivelli L, Dubcovsky J (2003) The cold regulated transcriptional activator *Cbf3* is linked to the frost-tolerance gene *Fr-A2* on wheat chromosome 5A. Mol Genet Genomics 269:60–67
8. Vágújfalvi A, Aprile A, Miller A, Dubcovsky J, Delogu G, Galiba G, Cattivelli L (2005) The expression of several *Cbf* genes at the *Fr-A2* locus is linked to frost resistance in wheat. Mol Genet Genomics 274:506–514
9. Francia E, Rizza F, Cattivelli L, Stanca AM, Galiba G, Tóth B, Hayes PM, Skinner JS, Pecchioni N (2004) Two loci on chromosome 5H determine low temperature tolerance in the new 'winter' × 'spring' ('Nure' × 'Tremois') barley map. Theor Appl Genet 108: 670–680
10. Tondelli A, Barabaschi D, Francia E, Pecchioni N (2011) Inside the CBF locus in Gramineae. Plant Sci 180:39–45
11. Cattivelli L (2011) More cold tolerant plants in a warmer world. Plant Sci 180:1–2
12. Reinheimer JL, Barr AR, Eglinton JK (2004) QTL mapping of chromosomal regions conferring reproductive frost tolerance in barley (*Hordeum vulgare* L.). Theor Appl Genet 109:1267–1274
13. Fuller MP, Fuller AM, Kaniouras S, Christopher J, Fredericks T (2007) The freezing characteristics of wheat at ear emergence. Eur J Agron 26:435–441
14. Crosatti C, Pagani D, Cattivelli L, Stanca AM, Rizza F (2008) Effects of the growth stage and hardening conditions on the association between frost resistance and the expression of the cold induced protein COR14b in barley. Environ Exp Bot 62:93–100
15. Rizza F, Crosatti C, Stanca AM, Cattivelli L (1994) Studies for assessing the influence of hardening on cold tolerance of barley genotypes. Euphytica 75:131–138
16. Beniston M (2005) Warm winter spells in the Swiss Alps: strong heat waves in a cold season? A study focusing on climate observations at the Saentis high mountain site. Geophys Res Lett 32, L01812
17. Shabbar A, Bonasal B (2003) An assessment of changes in winter cold and warm spells over Canada. Nat Hazards 29:173–188
18. Achard P, Gong F, Cheminant S, Alioua M, Hedden P, Genschik P (2008) The cold-inducible *CBF1* factor-dependent signaling pathway modulates the accumulation of the growth-repressing DELLA proteins via its effect on gibberellin metabolism. Plant Cell 20:2117–2129

19. Gaudet DA (1994) Progress towards understanding interaction between cold hardiness and snow mould resistance and development of resistant cultivars. Can J Plant Pathol 16:241–246
20. Andrews CJ, Pomeroy MK (1979) Toxicity of anaerobic metabolites accumulating in winter wheat seedlings during ice encasement. Plant Physiol 64:120–125
21. Gaudet DA, Wang Y, Frick M, Puchalski B, Penniket C, Ouellet T, Robert L, Singh J, Laroche A (2011) Low temperature induced defence gene expression in winter wheat in relation to resistance to snow moulds and other wheat diseases. Plant Sci 180: 99–110
22. Frederiks TM, Christopher JT, Harvey GL, Sutherland MW, Borrell AK (2012) Current and emerging screening methods to identify posthead-emergence frost adaptation in wheat and barley. J Exp Bot 63:5405–5416

Chapter 5

A Whole-Plant Screening Test to Identify Genotypes with Superior Freezing Tolerance

Annick Bertrand, Yves Castonguay, and Josée Bourassa

Abstract

Freezing tolerance is a determinant factor of persistence of perennials grown in northern climate. Selection for winterhardiness in field nurseries is difficult because of the unpredictability of occurrence of test winters allowing the identification of hardy genotypes. Here we describe a whole-plant assay entirely performed indoor in growth chambers and walk-in freezers to identify genotypes with superior tolerance to freezing within populations of open pollinated species. Three successive freezing stresses are applied to progressively eliminate 90 % of the population and to retain only the 10 % best performing genotypes. This approach can be used to generate recurrently selected populations more tolerant to freezing in different species.

Key words Freezing tolerance, Recurrent selection, Controlled conditions, Freezing stress, Perennial crops

1 Introduction

Winter survival is determined by the plant capacity to withstand numerous abiotic and biotic aggressions during the overwintering period. Cumulative evidence indicates that tolerance to low freezing temperatures plays a central role with regard to adaptation to winter [1]. Freezing tolerance is a quantitative trait with low-to-moderate heritability. Success in breeding freezing-tolerant plants using conventional plant breeding methodologies has been limited in spite of the presence of large genetic diversity for this trait within populations of open-pollinated species [2]. Improvement of plant winterhardiness has historically been based on field selection of genotypes that survive winters [3]. However, the unpredictability of test winters due to large variations between and within locations and the environmental conditions to which the plants are exposed severely limits the predictability of this approach [4]. As a result, costly assessment of the genetic material at multiple locations over many years is often used to minimize environmental effects and to increase the likelihood to accurately discriminate plants with regard to their winterhardiness potential. New approaches are needed by

Dirk K. Hincha and Ellen Zuther (eds.), *Plant Cold Acclimation: Methods and Protocols*, Methods in Molecular Biology, vol. 1166, DOI 10.1007/978-1-4939-0844-8_5, © Springer Science+Business Media New York 2014

breeding programs to accelerate and reduce the cost of assessment of freezing tolerance. In addition, the efficiency of gene discovery studies is highly determined by the availability of assays for high-throughput and reliable screening of plant phenotypes.

To address issues associated with the unpredictability and resource-intensive assessment of winterhardiness in the field, we devised a method of selection entirely performed indoor under controlled conditions [5]. Using that approach, large numbers of genotypes from initial genetic backgrounds are subjected to successive freezing stresses to progressively eliminate freezing-sensitive plants and ascertain the phenotype of the most hardy plants. At the completion of the selection process, plants comprising the 10 % superior genotypes can be intercrossed to generate a new population putatively improved for its tolerance to freezing (TF populations). Using this approach, several cycles of selection (depending on the species and initial gene pool) can be performed in order to progressively increase the frequency of adaptive alleles and to promote epistatic interactions in a recurrent selection process. Up to now, we successfully applied this approach in a number of open-pollinated species including alfalfa [5], red clover [6], and perennial ryegrass [7].

2 Materials

Start with open-pollinated species with characterized extensive genetic variability to avoid genetic bottlenecks eventually leading to lower plasticity for adaptation and restricted fitness.

2.1 Plant Material

1. Seeds: 1,500 viable seeds of an initial genetic background ready for germination (*see* **Notes 1** and **2**).
2. Soil: Potting soil plus slow-release fertilizer (14-14-14, 300 mL per 100 L of soil; Osmocote, Scotts, Marysville, OH, USA).
3. Fertilizer: 20-20-20 + micronutrients (1 g/L).
4. Pots: The pot system consists of 1,500 individual cells arranged in trays: 164 mL volume Ray Leach Cone-tainers (SC10 Super Cell, low density) and RL98 trays as well as IPL Rigi-pots IP110 (Stuewe and Sons, Inc., Tangent, OR, USA).
5. Plastic labels.

2.2 Controlled Environment Chambers and Programmable Freezer

1. Controlled environment chambers: Growth chambers with a temperature range from 2 to 25 °C and irradiance from 150 to 600 μmol photons/m^2/s PPFD. A surface of 6.69 m^2 is required for seeding one population.
2. Large walk-in programmable freezer with a temperature range from −2 to −30 °C (*see* **Note 3**).

3 Methods

3.1 Seeding and Growth

1. Fill the pots (Cone-tainers), which have been previously placed in IP110 Rigi-Pots, with uniformly humid soil. Slightly compact the soil up to 1.5 cm below the top.
2. Place two seeds per pot.
3. Cover the seeds with 0.5 cm of sieved soil, and water the pots uniformly by hand.
4. Identify each IP110 tray with a plastic label.
5. Place trays in growth chambers under the following environmental conditions: 16-h photoperiod with an irradiance of 400–600 μmol/m²/s PPFD and a day:night temperature regime of 22:17 °C.
6. After 1 week, thin to one seedling per pot (*see* **Note 4**).
7. Water the plants when needed, and fertilize twice a week for the first 3 weeks of growth (*see* **Note 5**) with a 1 g/L solution of a commercial fertilizer (20-20-20 + micronutrients).
8. After 4 weeks of growth, transfer the plants to cold acclimation conditions.

3.2 Cold Acclimation

1. Place trays in growth chambers set to the following environmental conditions: 8-h photoperiod (200–250 μmol/m²/s PPFD) under a constant temperature of 2 °C.
2. Water the plants when needed. Do not fertilize.
3. After 2 weeks of acclimation at 2 °C, transfer the tubes containing the plants in RL98 racks leaving a free row between each plant row. This pattern will facilitate even temperature distribution in each pot (*see* **Note 6**).
4. Transfer the plants to a programmable freezer set at −2 °C after an adequate irrigation of the pots (*see* **Note 7**).
5. After the soil is well frozen (1–2 days), cover the racks with tarpaulins to avoid plant desiccation.
6. Acclimate the plants at nonlethal freezing temperature (−2 °C) in the dark for 2 weeks (*see* **Note 8**).
7. Remove the tarpaulin the day before application of freezing stress.

3.3 Application of Freezing Stresses (Fig. 1)

1. After 2 weeks at −2 °C, progressively decrease the temperature in the freezer to the expected lethal temperature for 50 % of the plants (LT_{50}) using the following stepwise decrease: decremental steps of 2 °C during a 30-min period followed by a 90-min plateau at each temperature (*see* **Note 9**).
2. When the expected LT_{50} is reached, withdraw the plants from the freezer after a 90-min exposure.

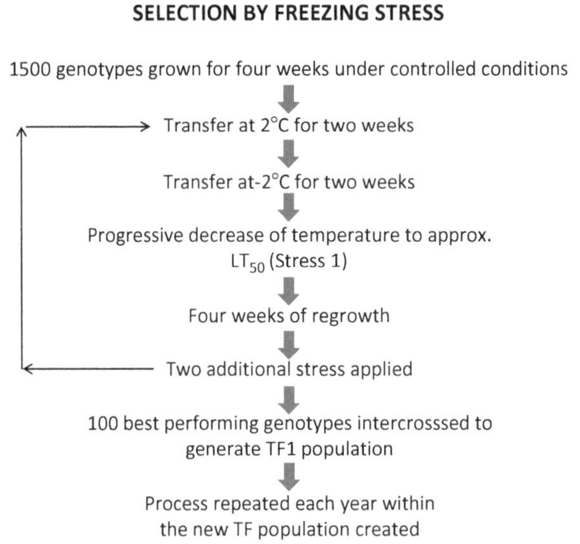

SELECTION BY FREEZING STRESS

1500 genotypes grown for four weeks under controlled conditions

↓

Transfer at 2°C for two weeks

↓

Transfer at-2°C for two weeks

↓

Progressive decrease of temperature to approx. LT_{50} (Stress 1)

↓

Four weeks of regrowth

↓

Two additional stress applied

↓

100 best performing genotypes intercrosssed to generate TF1 population

↓

Process repeated each year within the new TF population created

Fig. 1 Schematic illustration of the procedure used for the selection of populations tolerant to freezing. *Source*: Castonguay et al. (2009) [5], reprinted with permission from ASA, CSSA, SSSA

Fig. 2 Red clover regrowth following selection after a first freezing stress

3. Let the plants slowly thaw overnight at 4 °C in the dark.
4. Cut the plants to 3–4 cm height.
5. Transfer the plants to initial growth conditions (16-h photoperiod under 400–600 μmol/m^2/s PPFD; day:night temperature of 22:17 °C) for 4 weeks (Fig. 2).

6. Select genotypes that survived the freezing stress (*see* **Note 10**).
7. Expose the selected genotypes to a second round of cold acclimation followed by exposure to a freezing stress down to a temperature slightly below the initial test temperature to eliminate another group of genotypes (*see* **Note 11**).
8. Transfer the plants to initial growth conditions (16-h photoperiod under 400–600 μmol/m^2/s PPFD; day:night temperature of 22:17 °C) for 4 weeks.
9. Repeat the entire cycle of acclimation-freezing stress-regrowth a third time.
10. Among the plants remaining after the third cycle of selection, retain the 100 most vigorously growing genotypes (*see* **Note 12**).
11. Transplant these genotypes into 15 cm diameter pots, and transfer the plants to optimal growth conditions until flowering.

3.4 Intercrossing

Intercross the selected plants randomly using hand pollination or confined enclosure, depending on plant species and pollination type (Fig. 3).

3.5 Harvesting Seeds

1. Harvest the seeds of each genotype separately when they are ripened.
2. Pool seeds into a bulked sample using an equal representation of each genotype (*see* **Note 13**).

Fig. 3 Crossing 100 superior genotypes of alfalfa in a greenhouse

4 Notes

1. Depending on the plant species tested, seeds could need a pretreatment such as scarification or imbibition to optimize germination.
2. Proceed with a germination test before seeding to ensure having sufficient genetic material to complete three freezing stresses and end up with a sufficient number of genotypes to reduce the risk to create a genetic bottleneck during the recurrent selection process.
3. The lowest temperature of the freezer has to be inferior or equal to the LT_{50} of the population under selection. In cases when the expected LT_{50} is lower than the lowest temperature of the freezer, stress intensity can be varied by increasing exposure time at the last plateau.
4. Check the seeding regularly, and remove late-sprouting genotypes.
5. Fertilization should be stopped 1 week before transferring the plants to cold acclimation because it could interfere with the cold acclimation process.
6. Thermocouples measuring the actual temperature reached in the soil could be installed 1 cm deep in the soil.
7. The day before transferring the plants to –2 °C, it is important that the plants have been uniformly well watered and well drained. Soil water content has a major impact on the process of plant cold acclimation and freezing tolerance [8].
8. Acclimation at –2 °C in the dark is very important to reach the maximum cold acclimation of the plants [9].
9. Preliminary assays with a subset of plants are highly recommended to accurately target the LT_{50} and avoid insufficient or too intensive selection pressure that will markedly affect the screening process. LT_{50} varies with species, germplasm, and the number of cycles of selection.
10. Each cycle of acclimation-freezing stress-regrowth should eliminate ≈50 % of the plants. If you start with 1,500, around 750 plants should remain after a first cycle, 375 after two cycles, and 175 after three cycles. After the third cycle, select the 100 most vigorous among the 175 surviving plants.
11. The level of temperature of each freezing stress could be adjusted according to the number of surviving plants that are obtained after each cycle. If too many plants are killed after one cycle, set the second stress 1–2 °C above the temperature of the previous stress. If not enough plants are killed, set the following stress 1–2 °C below the previous temperature.

If after three stresses too many undamaged plants are obtained, choose the 100 most vigorous individuals of the population to undergo the next cycle of selection.

12. For a better estimate of the vigor of the genotypes, it is recommended to make this last screening after 2–3 weeks of regrowth.
13. This pool of seeds represents the next TF population that will be seeded to undergo another cycle of selection.

References

1. Volenec JJ, Cunningham SM, Haagenson DM, Berg WK, Joern BC, Wiersma DW (2002) Physiological genetics of alfalfa improvement: past failures, future prospects. Field Crop Res 75:97–110
2. Castonguay Y, Dubé M-P, Cloutier J, Bertrand A, Michaud R, Laberge S (2013) Molecular physiology and breeding at the crossroads of cold hardiness improvement. Physiol Plant 147:64–74
3. Castonguay Y, Laberge S, Brummer EC, Volenec JJ (2006) Alfalfa winter hardiness: a research retrospective and integrated perspective. Adv Agron 90:203–265
4. Limin AE, Fowler DB (1991) Breeding for cold hardiness in winter wheat: problems, progress and alien gene expression. Field Crops Res 16: 190–197
5. Castonguay Y, Michaud R, Nadeau P, Bertrand A (2009) An indoor screening method for improvement of freezing tolerance in alfalfa. Crop Sci 49:809–818
6. Bertrand A, Castonguay Y (2013) Molecular changes in recurrently selected populations of forage legumes. In: Imai R et al (eds) Plant and microbe adaptations to cold in a changing world. Springer, New York, pp 209–217
7. Iraba A, Castonguay Y, Bertrand A, Floyd D, Cloutier J, Belzile F (2013) Characterization of populations of turf-type perennial ryegrass recurrently selected for superior freezing tolerance. Crop Sci 53:2225–2238
8. Bélanger G, Castonguay Y, Bertrand A, Dhont C, Rochette P, Couture L, Drapeau R, Mongrain D, Chalifour F-P, Michaud R (2006) Winter damage to perennial forage crops in eastern Canada: causes, mitigation, and prediction. Can J Plant Sci 86:33–47
9. Dionne J, Castonguay Y, Nadeau P, Desjardins Y (2001) Freezing tolerance and carbohydrate changes during cold acclimation of green-type annual bluegrass (*Poa annua* L.) ecotypes. Crop Sci 41:443–451

Chapter 6

Mapping of Quantitative Trait Loci (QTL) Associated with Plant Freezing Tolerance and Cold Acclimation

Evelyne Téoulé and Carine Géry

Abstract

Most agronomic traits are determined by quantitative trait loci (QTL) and exhibit continuous distribution in segregating populations. The genetic architecture and the hereditary characteristics of these traits are much more complicated than those of oligogenic traits and need adapted strategies for deciphering. The model plant *Arabidopsis thaliana* is widely studied for quantitative traits, especially via the utilization of natural genetic diversity. Here we describe a QTL-mapping protocol for analyzing freezing tolerance after cold acclimation in this species based on its specific genetic tools. Nevertheless, this approach can also be applied for the elucidation of complex traits in other plant species.

Key words QTL-mapping, *Arabidopsis thaliana*, Freezing tolerance, Natural variability

1 Introduction

Low temperatures, besides drought and salt stress, are among the most important abiotic environmental factors affecting the geographical distribution of plant species as well as growth and yield of crop plants. The ability of plants to survive freezing temperatures is largely dependent on their ability to cold acclimate. The acclimation process triggers an increase of freezing resistance after an exposition to low but nonfreezing temperatures. It induces wide modifications in the physiology and metabolism of the plant, as changes in lipid composition of membranes, increase in cryoprotectant molecules such as sugars and proline for example. These changes are largely associated with variations in gene expression. Due to its major agronomic importance, this trait has been extensively studied not only in crops but also in model plants, such as *Arabidopsis thaliana*, and numerous cold-induced genes have been identified [1, 2]. Although these approaches, based on monogenic mutant analyses, have been very successful, freezing tolerance is a trait, quantitative in nature, and other genetic approaches, for example studying complex traits via quantitative trait loci (QTL)

Dirk K. Hincha and Ellen Zuther (eds.), *Plant Cold Acclimation: Methods and Protocols*, Methods in Molecular Biology, vol. 1166, DOI 10.1007/978-1-4939-0844-8_6, © Springer Science+Business Media New York 2014

mapping [3], have appeared as powerful tools to identify novel genes involved in freezing tolerance. These strategies have been applied successfully to crop plants [4–6] and more recently to model species.

The model species *A. thaliana* is a chilling-tolerant plant capable of cold acclimation. It is widely distributed all over the Northern hemisphere, and therefore natural accessions are exposed to various climates. Natural variation for freezing tolerance after acclimation has been demonstrated in several studies [7], and this trait appears to be correlated with habitat winter temperatures [8–10]. This suggests that this character could be under natural selection and that natural genetic variability could be larger than previously predicted. Deciphering this complex trait via QTL mapping in *A. thaliana*, where numerous genetic tools are available, appears particularly pertinent (Fig. 1). These approaches could increase our understanding of the gene networks and molecular mechanisms underlying the response to cold in *A. thaliana* and identify new targets for breeding in crop plants. Several fruitful studies confirmed this working hypothesis [3, 7, 11].

The key idea in QTL analysis and mapping is the identification of linkage disequilibrium or significant associations between genetic markers and quantitative phenotypic data in a segregating population, generally issued from a biparental cross: this implies the availability of dense genetic maps and efficient and reproducible phenotypic tests. Phenotyping has proved to be a crucial part of the work. Considering freezing tolerance, this trait is continuous and can be assessed through freezing tests (in fields or under

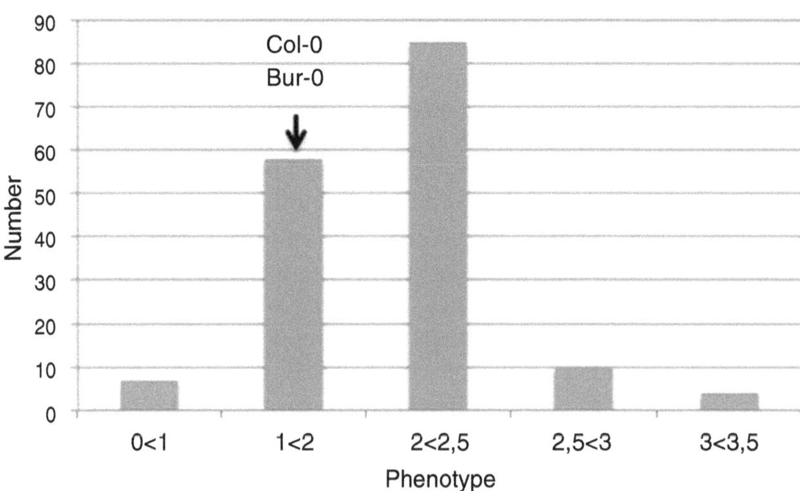

Fig. 1 Distribution of the score for the freezing tolerance phenotype of the core population of RILs derived from the Bur-0 × Col-0 cross. Both parental phenotypes are indicated by *arrows*. Despite a small difference in freezing tolerance phenotype between Bur-0 and Col-0, a strong variability, due to transgression, is observed in the RIL population

controlled conditions). Integrative measurements consist of the evaluation of freezing injuries by means of notes on a standard scale and/or evaluation of recovery by estimating the percentage of viability [7]. Electrolyte leakage tests, based on the evaluation of plasmalemma alterations [12], also allow the evaluation of freezing tolerance and are used in various species. Especially the method of LT_{50} temperature corresponding to 50 % lethality measured via electrolyte leakage measurements is very reliable and has been used in several studies [7, 11, 13].

In this chapter, we describe protocols that we currently use for the mapping of QTL associated with plant freezing tolerance after cold acclimation in *A. thaliana*. To summarize, protocols are developed for three main steps: (1) primary QTL detection in classical segregating populations and localization of wide candidate regions, (2) validation of the QTL by genetic approaches using more specific material, and then (3) fine mapping of positive regions. A strategy based on mutants and genetic complementation is described for the step of validation of potential candidate gene(s). Yet, the context could be largely different for other species, as for example crop species with fewer available genetic tools or generally less developed resources: the researcher will find it advantageous to take all the underlined points into account before initiating an experimental strategy for QTL detection and mapping.

2 Materials

2.1 QTL Mapping Populations

1. Seed stocks of *A. thaliana* accessions and mapping populations as Recombinant Inbred Lines (RILS), F2, Backcross (BC), Heterogeneous Inbred Lines (HIFs) (*see* **Notes 1–3**) are obtained from the *A. thaliana* Resources Center (BRC) in IJPB in INRA/Versailles. All the seed stocks are described and can be ordered at http://publiclines.versailles.inra.fr/ (*see* **Note 4**).
2. Available RILs are all derived from the same type of cross: the female parent is one accession of the core collection (maximizing the diversity of the whole collection), and the male parent is Col-0. For each RIL population are available a minimal set, a core population of 164 RILs, and the complete set. The minimal set includes 20 lines representative of the whole population that allows checking phenotypic diversity in the selected RIL population easily and rapidly. The core population is optimized for QTL detection. The complete set includes all the RILs issued from the initial cross, derived from the 500 starting F2 plants. All genotyping data are also available and can be downloaded as ".raw," ".maps," and ".xls" files. ".raw" and ".maps" are necessary and used for primary QTL detection. ".xls" files are presented in a format optimized to research for regions of residual heterozygosity in the different RILs.

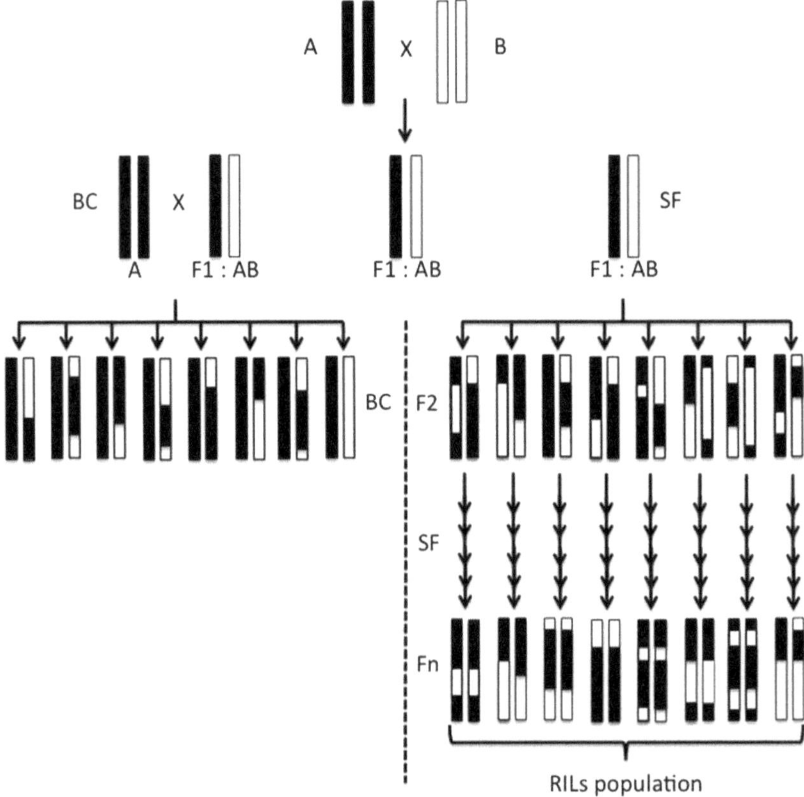

Fig. 2 Examples of mapping populations usable for QTL detection: Crossing homozygous parental accessions produces fully heterozygous F1 plants. The parents are homozygous due to the self-pollination regime of *A. thaliana*. F1 plants can be backcrossed with one parent, generating BC populations, or self-pollinated to produce F2 populations. The F2 population itself can be used as a mapping population or lines can be selfed through single-seed descent (SSD) until F6 generation without selection. Then one plant per line is chosen again for selfing to obtain F7 seeds, which are used as a bulk for genotyping. For easier reading, one chromosome pair only is represented on the scheme. *BC* backcross, *SF* self-fertilization

3. RILs are homozygous with low residual heterozygosity (around 3 %). RILs are obtained as follows (Fig. 2; *see* **Note 1**): F1 seeds are produced by crossing the selected parental accessions; plants are grown and self-fertilized. 500 F2 lines are generated through single-seed descent until F6 generation without selection. Then one plant per line is chosen again for selfing to obtain F7 seeds, which are used as a bulk for genotyping.

2.2 DNA Extraction for Genotyping of Plant Material

1. Extraction buffer: 200 mM Tris–HCl (pH 7.5), 250 mM NaCl, 25 mM EDTA, 0.5 % SDS.
2. Isopropanol.
3. 75 % ethanol.
4. TE buffer: 10 mM Tris (pH 8.0), 1 mM EDTA.

5. Ball mill MM30 (Retsch, Haan, Germany).
6. Eight-strip 1.2-ml collection tubes.
7. Polypropylene 96-well microplates.

2.3 Evaluation of Freezing Tolerance

1. For plant growth and freezing test: Greenhouse with cooling in summer and heating and light supplementation in winter, growth chambers for acclimation and freezing periods (chambers under 12-h photoperiod, 70 μE/m^2/s light intensity, and 70 % relative humidity), able to stay stable at 4 °C for acclimation and at negative temperatures from −5 to −8 °C for freezing tests.
2. 0.1 % agarose in water.

2.4 Software for QTL Analysis

1. MAPMAKER 3.0 [14] to establish genetic maps.
2. The Unix version of QTL CARTOGRAPHER 1.14 [15] to perform QTL analyses.

3 Methods

3.1 Generation of QTL Mapping Populations

1. The first step consists of choosing parental accessions of the mapping population. Both parental lines must exhibit significant genetic distance. Sometimes they are chosen for their phenotypic contrast, but this is not absolutely necessary. Effectively, despite a low difference in phenotype between accessions, large variability could be observed in the mapping populations derived from the cross (Fig. 1).
2. Several mapping populations are already available in BRC: F2 and RILs. RILs are often preferred to other populations when preexisting and available. If not, F2 populations are more rapid to generate and could be used in primary detection.
3. Mapping populations are genotyped with molecular markers: SNPs, microsatellites, and/or indels chosen to get regular spacing on the genome. The distance between markers can vary depending upon plant material (*see* **Note 5**).
4. MAPMAKER 3.0 is used to establish the genetic map, and the Kosambi [16] mapping function is used to convert recombination data into map distances, as it seems optimal for *A. thaliana* [17].

3.2 Evaluation of the Freezing Tolerance Phenotype

1. Seeds of the 164 RILs (*see* **Note 6**) are put in 0.1 % agarose at 4 °C in the dark for 48 h to ensure homogenous germination.
2. They are then sown with a pipette in square pots containing organic substrate and irrigated with mineral nutrient solution once a week and with water. Plants are sown in little bunches,

12 lines per pot in a random design allowing blind notation. Plants are grown in a greenhouse for 14 days, when they have reached the six- to eight-leaf stage.

3. They are then transferred for cold acclimation to a growth chamber at 5 °C under 12-h photoperiod, 70 μE/m^2/s light intensity, and 70 % relative humidity for 7 days.
4. Acclimated plants are then exposed to freezing temperatures varying from −4 to −8 °C, depending upon germplasm, for 48 h.
5. At this time, plants are removed from freezing conditions and put back in the greenhouse.
6. These experimental conditions have to be optimized in a first round of experiments to maximize the variation in the response.
7. A week later, tolerance to freezing is determined by evaluating leaf damage and capacity for continued growth. Using a method favored by agronomists [18], damage to leaves is evaluated by noting on a scale, ranging from 0 (no damage) to 6 (dead plants) [19].
8. Each line is sown in five replicates, and a mean score is calculated.

3.3 QTL Analysis

1. The phenotype scores are collected in an Excel file, and the mean and standard deviation are calculated for each RIL to check the quality of scoring. Aberrant values can be discarded at this step.
2. Matrix data for QTL detection is built by combining the mean phenotypic values with genotype data for each line from the RIL population (Fig. 3).
3. QTL analyses are performed with a Unix version of QTL Cartographer [15], which is a suite of programs for mapping QTL onto a genetic linkage map, but other software are available for this purpose as well (*see* **Note 7**).
4. First, interval mapping (IM) [20] is used to determine putative QTL involved in the variation of the trait. However, the localization and genetic effects of such detected QTL can be confused because of the presence of other linked QTL or because of nonrandom segregation of other QTL in the population.
5. Composite interval mapping (CIM) allows addressing these limitations: The significant markers in the population, detected by regression analysis, are chosen as cofactors to estimate maximum likelihood for QTL. Effects of possible other QTL are then taken into account.
6. Model 6 of QTL Cartographer is therefore performed on the same data: The closest marker to each local logarithms of odds (LOD) score peak, a putative QTL, is used as a cofactor to control the genetic background while testing at a position of the genome [15]. When a cofactor is also a flanking marker of

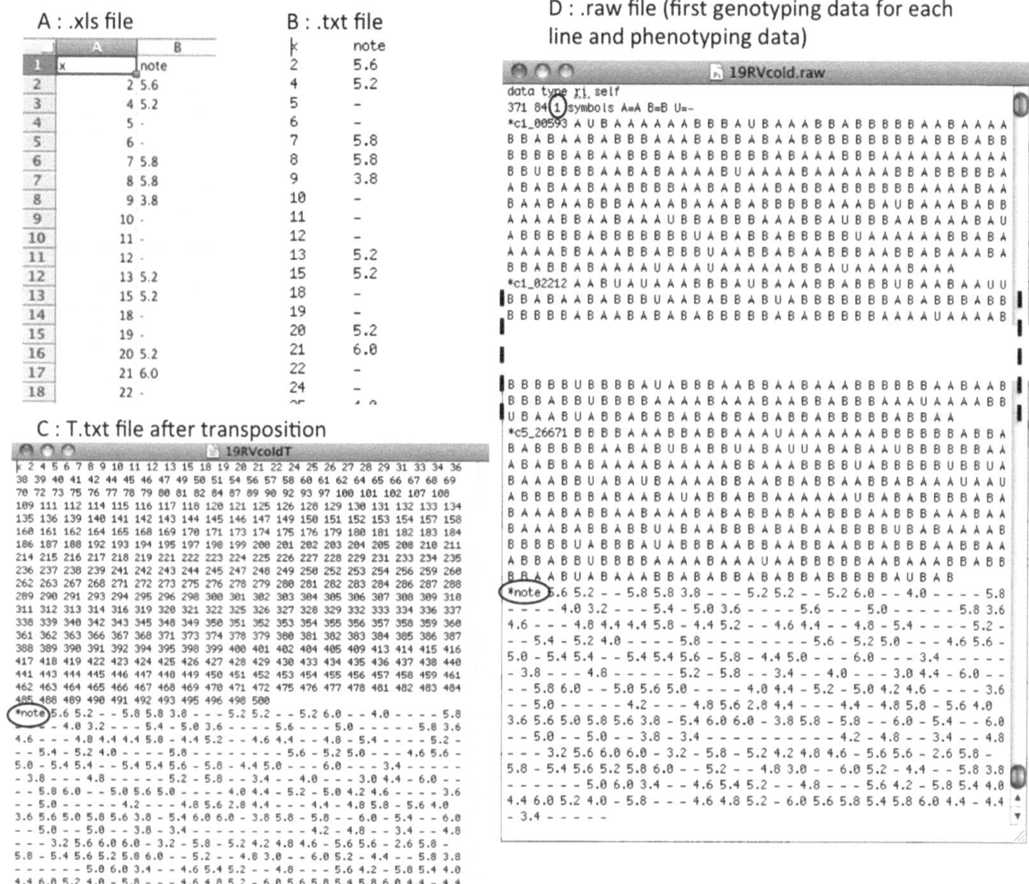

Fig. 3 Example of freezing tolerance data treatment for QTL detection: The successive transformations of raw data during the QTL detection process with QTL Cartographer. (**a**) Primary data file: All the RILs available in BRC must appear in the column X even if the work is done on a subset of RILs, missing data are noted with "-." (**b**) Transformed file in txt format: Data are organized in columns. (**c**) Transposed data file exhibiting the number of the whole set of RILs. (**d**) Treated data file ready to use in QTL Cartographer. Transposed phenotypic data are pasted at the end of the genotypic data

the tested region, it is excluded from the model. The number of cofactors involved in our models varied between one and six for a full RIL set (400–500 lines) and a maximum of four for a core population (164 lines).

7. The LOD significance threshold (2.3 LOD) is estimated from 1,000 permutation test analyses [21].

8. To better illustrate the process, a concrete example of the procedure is depicted below. When using the successive programs proposed in QTL Cartographer, some values or names are proposed by default. Here we show an example of how to manage the procedure (Fig. 4) (*see* **Note 8**). Then, the parameters used for our mapping populations to detect QTL

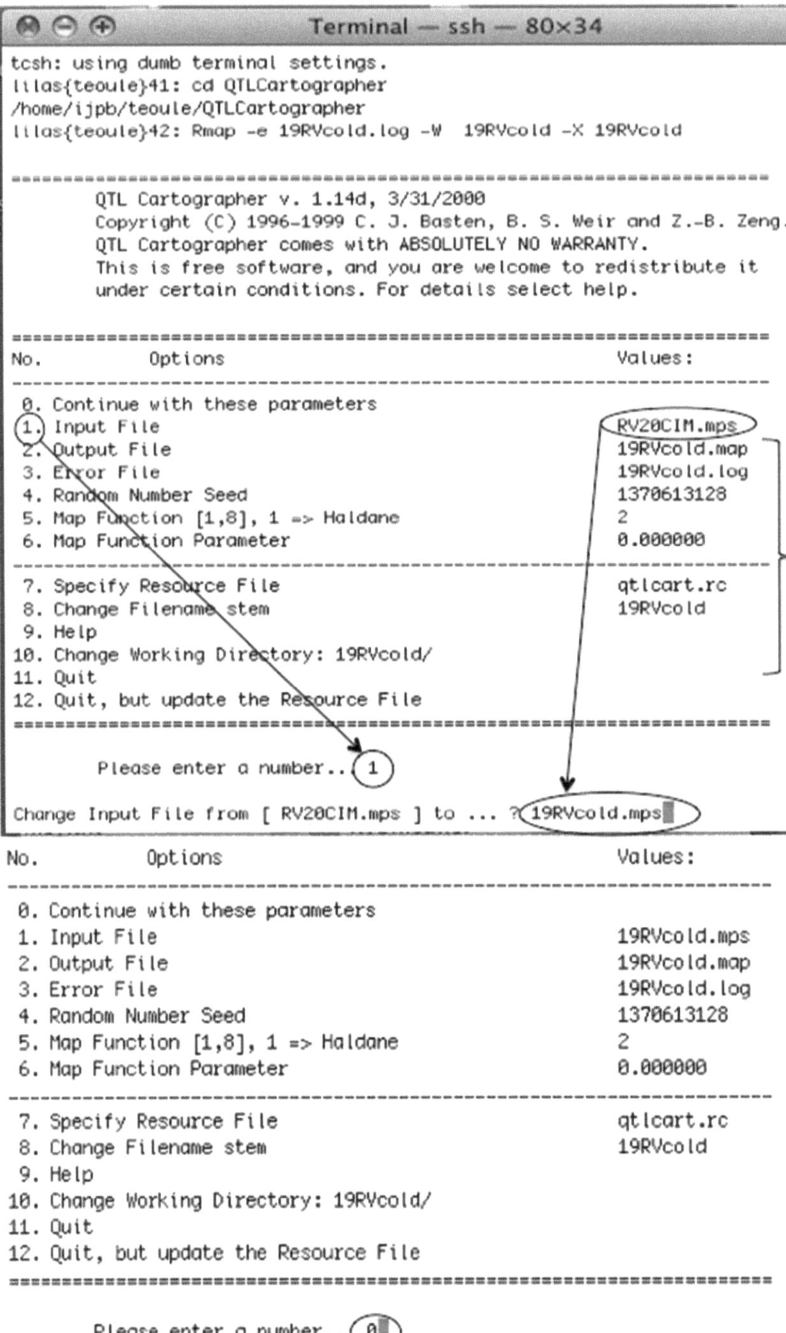

Fig. 4 Example of program usage: When opening QTL Cartographer, in the successive screens of the program, values are proposed by default and some parameters need to be changed. In this example, the created directory to work with is 19RVcold and all names of files must be identical. The proposed name for input file is RV20CIM.mps. To modify it, select line 1 and change the wrong name by the correct one 19RVcold.mps. When all parameters are correct, enter 0 to execute the program. This procedure can be applied on all programs of QTL Cartographer (*see* text for details)

associated with plant freezing tolerance are listed. The chosen population is the RIL core population of 19RV, derived from a cross between Can-0, an accession issued from Canary Islands, with a low potential of acclimation (damage score = 6 after freezing), and Col-0.

3.4 QTL Cartographer Process: Example of the RIL Population 19RV

1. In the original Excel file with all the phenotype data values put one trait per column, use "." instead of "," for numbers, use "-" for missing data, use « X » as title of the first column corresponding to RIL line numbers, and use the name of the phenotypic trait corresponding to the measures as title. Here, we have used "note" for the freezing damage score (Fig. 3a).
2. Save this file as a text file (Windows). Name it populationnumbertrait.txt (here: 19RVcold.txt).
3. Transpose lines/columns using functions of Splus (Fig. 3b) (http://www.insightful.com/products/splus/).
4. In the transposed file, add a "*" before the name of the trait without space and save (Fig. 3c).
5. In the QTL Cartographer directory, create a new directory for each population and trait studied (19RVcold in this example).
6. In this directory, add the ".raw" and ".maps" files, issued from MAPMAKER and downloaded from the *Arabidopsis* BRC website. Change the ".maps" by ".mps."
7. In the ".raw" file, change 0 (at the beginning of line 2) by the number of characters you have to analyze (here 1).
8. After the genotyping data, paste the phenotyping values from the transposed file with "*" before the trait name (Fig. 3d). Add all trait values in the same file if there is more than one trait.
9. In the software, use the following commands:
10. cd QTL Cartographer

```
Rmap -e 19RVcold.log -W 19RVcold -X 19RVcold
1 -> 19RVcold.mps
5 -> 2 (Kosambi)
Do not change the resource file (qtl.cart).
0 (= execute)
```

11. Rcross

```
1 -> 19RVcold.raw
0
```

12. Qstats

```
0
```

13. LRmapqtl (-> more affected markers by trait)

```
7 -> 99 (number higher than traits number -> all the traits)
0
```

14. SRmapqtl
```
6 -> 2 (FB model)
7 -> 99
0
```
15. Zmapqtl—Do this for each trait, and then take only one Eqtl for all.
```
8 -> 3
9 -> 1 then 2, 3…
0
```
16. Eqtl to extract data
```
6 -> 3
10 -> 1 (LOD score)
0
```
17. Preplot
```
8 -> postscript (file.ps)
10 -> 1
0
```
18. Then go back to the directory, in file 19RVcold.plt:
    (a) At line 10, replace set output's name "19RVcold/ 19RVcold.ps" by set output's name "19RVcold.ps."
    (b) Adjust Y limits (LOD score): Set Y range, and replace 100 by 30.
19. Go back to the application.
```
cd 19RVcold (directory name)
gnuplot        (postscript file is created)
     load "19RVcold.plt."
quit.
```
20. The most useful files created in the directory to represent the detected QTL are the following:
    (a) ".ps," a pdf file with the graphs (Fig. 5).
    (b) ".eqt" will give you marker positions for each significant LOD score, LOD score value, and the associated additive value (Fig. 6).
    (c) ".z3" will give you LOD score values all along the chromosome. Use this data if you want to make a graph with Excel.
21. It is then necessary to choose cofactors to run CIM.
    (a) Use the markers closest to the higher peak, even if this is not the peak with the higher LOD score value.
    (b) Avoid markers at the extremity of the chromosomes.
22. Modify the ".sr" file to contain only markers used as cofactors. If there are only three cofactors, keep rank 1, 2, and 3 (Fig. 6).
23. Go back to the application.

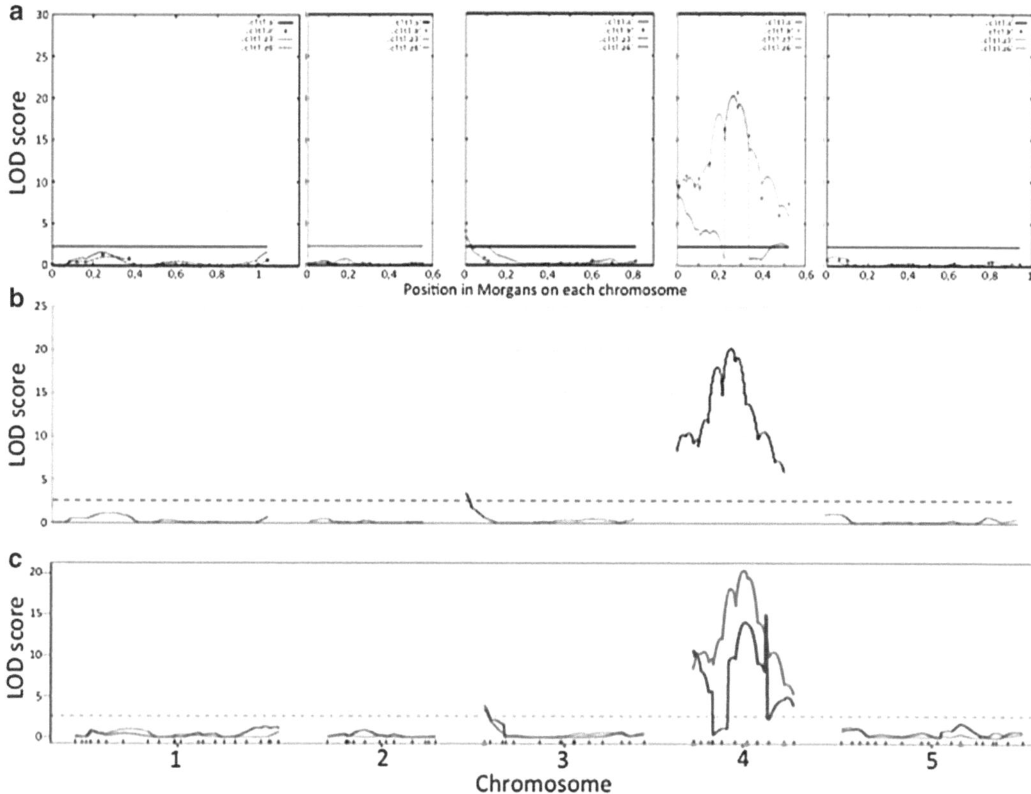

Fig. 5 Graphic localization of detected QTL in the 19RV population for the freezing tolerance trait. The five chromosomes of *A. thaliana* are represented individually: positions of the markers are on the abscissa, and the LOD score is on the ordinate. (**a**) Graphs issued from QTL Cartographer in ".ps" files, IM and CIM curves are superimposed. (**b**) Graphs issued from QTL Cartographer using ".z3" data, only the IM curve is represented. (**c**) Graphs obtained with treatment of the same data with R/qtl, as in 5.1, IM and CIM curves are superimposed

24. Zmapqtl—To be done for each trait, and then apply only one Eqtl for all.

```
17 -> use the good stem 19RVcold
19 -> use the good working directory 19RVcold
8 -> 6 (CIM)
9 -> 1 then 2, 3…
12 -> maximum number of cofactors retained
13 -> Ws=5
0
```

25. Preplot

```
8 -> postscript (file.ps)
10 -> 1
0
```

26. Go back to the directory, in file 19RVcold.plt:
    (a) At line 10, replace set output name "19RVcoldREP/19RVcold.ps" by set output name "19RVcold.ps."
    (b) Adjust Y limits (LOD score): Set Y range, and replace 100 by 30.

    Go back to the application.

.eqt file

#	..Chrom.	.Markr.	.Position.	.LOD Score.	.Additive.	.Dominance.	.R2.	.TR2.	.S.
1	3	1	0.0100	3.4879	-0.2612	0.0000	0.0940	0.1001	12.0755
2	4	2	0.4200	8.8871	-0.3983	0.0000	0.2195	0.2256	5.8502
3	4	4	3.0400	10.2632	-0.4368	0.0000	0.2609	0.2670	3.9779
4	4	5	5.9300	10.4923	-0.4460	0.0000	0.2710	0.2771	5.2434
5	4	6	10.1400	9.7713	-0.4174	0.0000	0.2393	0.2454	8.4984
6	4	8	14.4000	12.0408	-0.4711	0.0000	0.3106	0.3167	11.9498
7	4	9	19.7300	18.1203	-0.5737	0.0000	0.4620	0.4681	12.0174
8	4	10	26.2800	20.2539	-0.5800	0.0000	0.4720	0.4781	7.1729
9	4	11	29.2900	19.2764	-0.5651	0.0000	0.4478	0.4539	5.0278
10	4	12	34.4000	13.8913	-0.4950	0.0000	0.3434	0.3495	4.9159
11	4	13	42.1500	10.6036	-0.4642	0.0000	0.3029	0.3090	5.0435
12	4	14	48.7300	7.1847	-0.3762	0.0000	0.1994	0.2055	7.2097

.sr file : keep only marker(s) you want to use as cofactor(s)

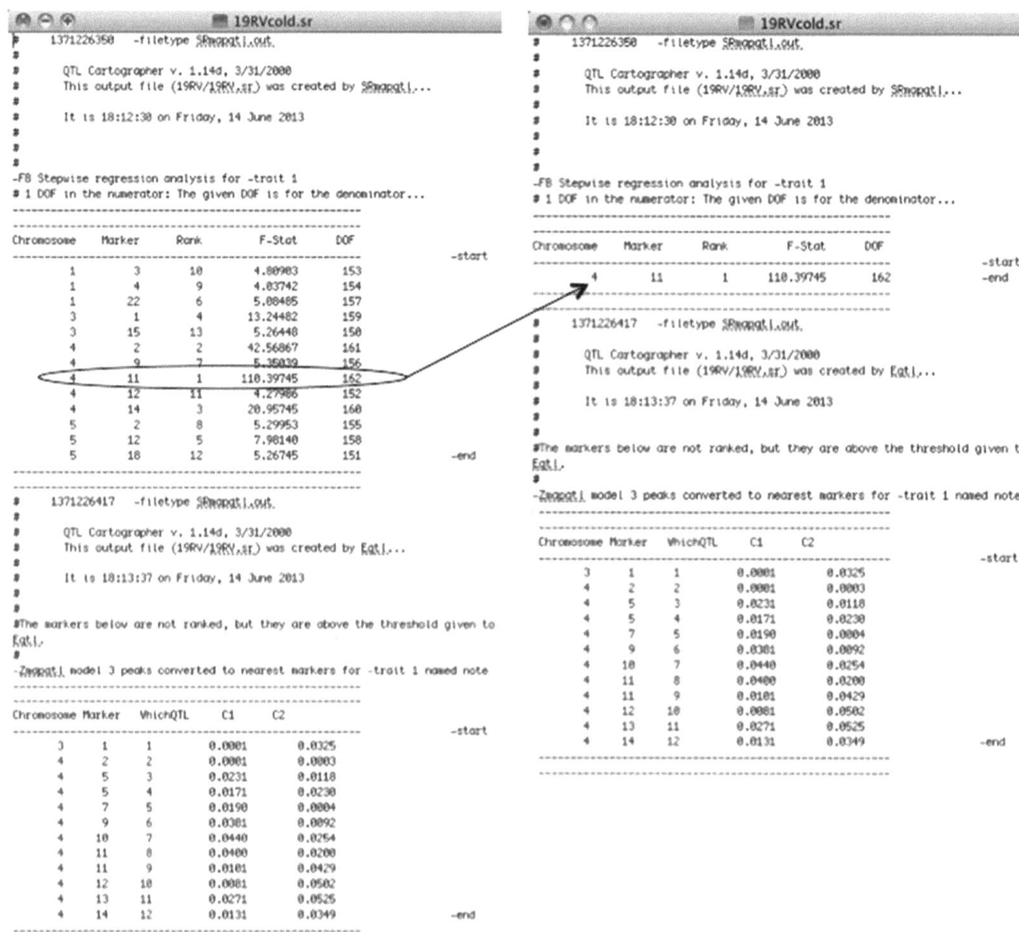

```
19RVcold.sr
#     1371226350   -filetype SRmapqtl.out.
#
#       QTL Cartographer v. 1.14d, 3/31/2000
#       This output file (19RV/19RV.sr) was created by SRmapqtl...
#
#       It is 18:12:30 on Friday, 14 June 2013
#
#
#
-FB Stepwise regression analysis for -trait 1
# 1 DOF in the numerator: The given DOF is for the denominator...
-------------------------------------------------------
-------------------------------------------------------
Chromosome    Marker     Rank        F-Stat      DOF
-------------------------------------------------------          -start
         1        3       10        4.80903      153
         1        4        9        4.03742      154
         1       22        6        5.08485      157
         3        1        4       13.24482      159
         3       15       13        5.26448      150
         4        2        2       42.56867      161
         4        9        7        5.35039      156
         4       11        1      110.39745      162
         4       12       11        4.27986      152
         4       14        3       20.95745      160
         5        2        8        5.29953      155
         5       12        5        7.98140      158
         5       18       12        5.26745      151          -end
-------------------------------------------------------
-------------------------------------------------------
#     1371226417   -filetype SRmapqtl.out.
#
#       QTL Cartographer v. 1.14d, 3/31/2000
#       This output file (19RV/19RV.sr) was created by Eqtl...
#
#       It is 18:13:37 on Friday, 14 June 2013
#
#
#The markers below are not ranked, but they are above the threshold given to
Eqtl.
#
-Zmapqtl model 3 peaks converted to nearest markers for -trait 1 named note
-------------------------------------------------------
-------------------------------------------------------
Chromosome Marker   WhichQTL     C1        C2
-------------------------------------------------------          -start
      3      1       1        0.0001      0.0325
      4      2       2        0.0001      0.0003
      4      5       3        0.0231      0.0118
      4      5       4        0.0171      0.0230
      4      7       5        0.0190      0.0004
      4      9       6        0.0301      0.0092
      4     10       7        0.0440      0.0254
      4     11       8        0.0400      0.0200
      4     11       9        0.0101      0.0429
      4     12      10        0.0081      0.0502
      4     13      11        0.0271      0.0525
      4     14      12        0.0131      0.0349          -end
-------------------------------------------------------
```

```
19RVcold.sr
#     1371226350   -filetype SRmapqtl.out.
#
#       QTL Cartographer v. 1.14d, 3/31/2000
#       This output file (19RV/19RV.sr) was created by SRmapqtl...
#
#       It is 18:12:30 on Friday, 14 June 2013
#
#
#
-FB Stepwise regression analysis for -trait 1
# 1 DOF in the numerator: The given DOF is for the denominator...
-------------------------------------------------------
-------------------------------------------------------
Chromosome    Marker     Rank        F-Stat      DOF
-------------------------------------------------------          -start
         4       11        1      110.39745      162          -end
-------------------------------------------------------
-------------------------------------------------------
#     1371226417   -filetype SRmapqtl.out.
#
#       QTL Cartographer v. 1.14d, 3/31/2000
#       This output file (19RV/19RV.sr) was created by Eqtl...
#
#       It is 18:13:37 on Friday, 14 June 2013
#
#
#The markers below are not ranked, but they are above the threshold given to
Eqtl.
#
-Zmapqtl model 3 peaks converted to nearest markers for -trait 1 named note
-------------------------------------------------------
-------------------------------------------------------
Chromosome Marker   WhichQTL     C1        C2
-------------------------------------------------------          -start
      3      1       1        0.0001      0.0325
      4      2       2        0.0001      0.0003
      4      5       3        0.0231      0.0118
      4      5       4        0.0171      0.0230
      4      7       5        0.0190      0.0004
      4      9       6        0.0301      0.0092
      4     10       7        0.0440      0.0254
      4     11       8        0.0400      0.0200
      4     11       9        0.0101      0.0429
      4     12      10        0.0081      0.0502
      4     13      11        0.0271      0.0525
      4     14      12        0.0131      0.0349          -end
-------------------------------------------------------
-------------------------------------------------------
```

Fig. 6 Characteristics of the detected QTL explaining freezing tolerance variation in the Can-0 × Col-0 population (19RV). (**a**) Raw output Eqt file: Raw file can be converted in .xls file to work with the data. *Position*: Position of the QTL is expressed in cM from the first marker of the chromosome. *Additive*: Represents the mean effect of the replacement of the non-Col-0 alleles by Col-0 at the locus. R^2: Represents the contribution of an identified QTL (or interaction QTL × QTL when significant) to the total phenotypic variation. The marker 11, retained as cofactor for CIM, is *underlined* with an oval. (**b**) The .sr file is modified to process by CIM. Statistical analysis classifies the markers by rank: here the number 11 is the first and will be kept as a cofactor allowing detection of intervals to localize QTL more precisely

27. cd 19RVcold

```
gnuplot      (postscript file is created)
    load "19RVcold.plt"
cd (executive command line which allows to shut open
files and go back to QTL Cartographer)
```

28. Eqtl to extract data

```
6 -> 6
10 -> 1 (LOD score)
0
```

29. In the 19RVcold.eqt file, the values under $R^2 \times 100$ explain *x*% of the phenotype variation and additive × 2 correspond to 2a., representing the proper additive values of alleles. This parameter is used to estimate sensu stricto heritability as additive variance/ phenotypic variance.
    (a) ".z6" will give you LOD score values all along the chromosome. Use these data if you want to make a graph with Excel.
30. After running the programs, detected QTL are represented on graphs (".ps" files or converted ".z3" in Excel files). Fig. 5 shows an example of the 19RV population analysis.
31. All the characteristics of detected QTL are summarized in the Eqt.out files (Fig. 6). For each detected QTL, the position on the chromosome and relative to the closest marker is indicated. Additive effects represent the mean effect of the replacement of the non-Col-0 alleles by Col-0 alleles at the locus. The R^2 value represents the contribution of each QTL to the total phenotypic variation for the trait (*see* **Note 9**).

3.5 Validation of QTL

1. Residual heterozygosity in RILs is used to confirm the existence of a QTL in the candidate region via HIFs. This material is produced by self-fertilization of selected RILs exhibiting residual heterozygosity in the detected candidate region.
2. Twenty-four seeds are sown and plants grown individually. DNA of each plant is extracted, and genotyping is performed with markers localized on the borders of the heterozygous region (and one in the middle if the region is wide). Five plants that are homozygous for the allele of the respective parent and five heterozygous plants are kept, and their selfed progenies are collected (*see* **Note 10**).
3. Three of these fixed progenies are phenotyped, and interpretation is based on the difference in the level of freezing tolerance between individuals with fixed A or B genotype: if their phenotype is significantly different, the HIF is retained (Fig. 7; *see* **Note 11**).
4. The global analysis consists of searching for a correlation between genotype and phenotype. Due to the characteristics

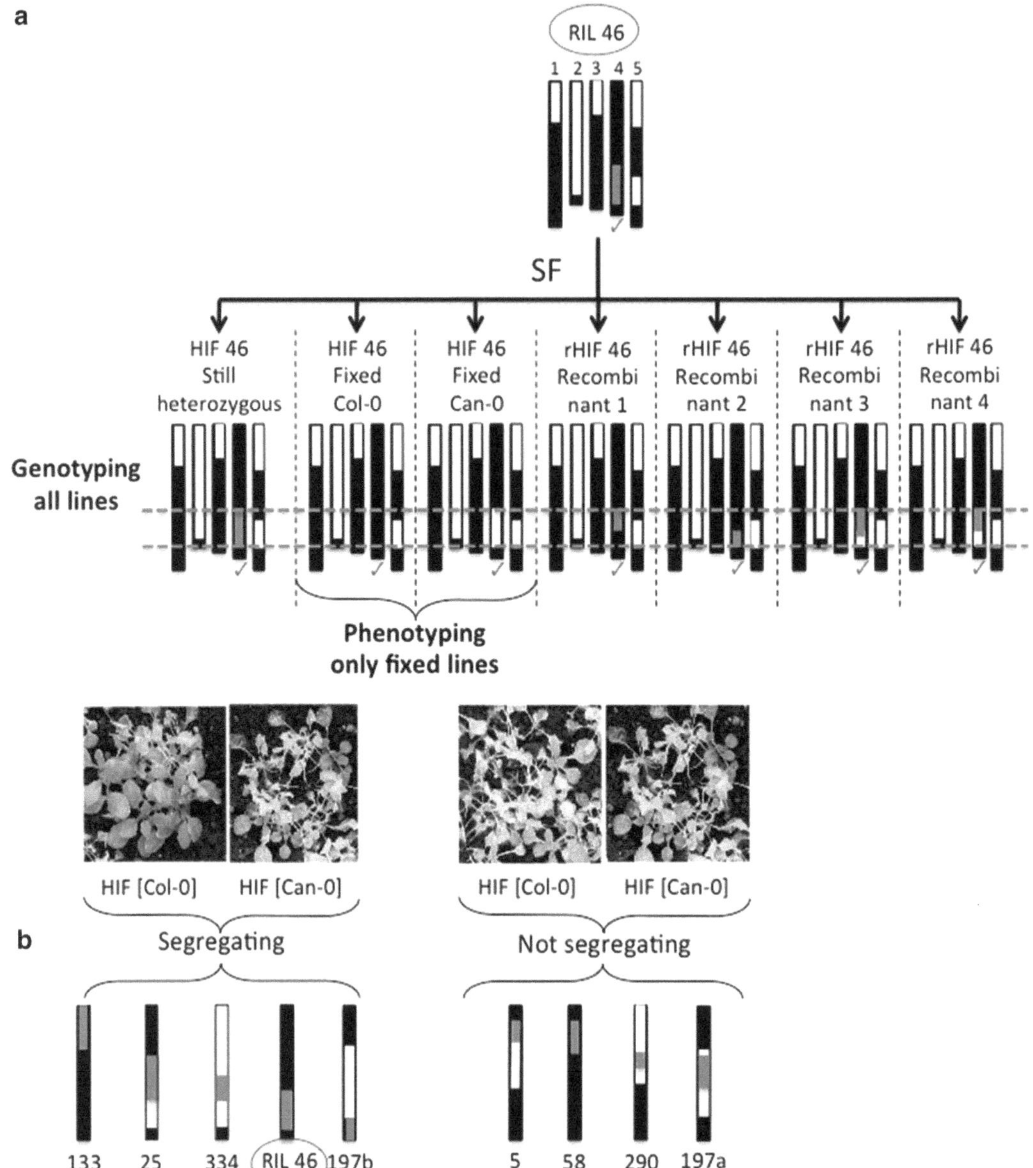

Fig. 7 QTL validation with HIFs. The Can-0 × Col-0 population (19RV) is shown as an example. (**a**) The example RIL (46) is self-fertilized, and the progeny is genotyped. For easier reading, chromosome pairs are represented as one block only with colors depending upon the genotype: *black* is homozygous for Col-0, *white* is homozygous for Can-0, and *grey* is a heterozygous region. In the RIL 46, the residual heterozygous region is located on chromosome 4. The fixed progenies are phenotyped, and if there is a significant difference between HIF fixed Col-0 and HIF fixed Can-0, the QTL is validated (*left illustration*). If not (*right illustration*) the RIL is excluded from further analysis. (**b**) Localization of residual heterozygous regions along chromosome 4 in several RILs. Five RILs, including RIL 46, exhibiting a segregating phenotype in HIF, are retained for further analysis. The other four are excluded

of phenotyping tests using small numbers of plants, fixed progeny testing (described here) is only appropriate for freezing tests and is used for this step. However, another technique is available (*see* **Note 12**).

3.6 Fine Mapping

1. The suitable HIFs are then submitted to a process to reduce the size of the initial QTL regions, which are often large (sometimes several Mb) to allow further progress in identifying causal polymorphisms. This step is based on production and analysis of recombinants. Recombinant HIF (rHIF) plants are identified by genotyping the progeny of self-fertilized HIF. To be efficient at this step, the number of plants that have to be screened has to be estimated. For this purpose, the following formula can be used:

$$\text{Number of expected recombinants} = \left[\frac{[\text{Size of the QTL region}]}{[\text{Size of the chromosome}]}\right] \times 2 \times [\text{number of genotyped plants}].$$

The result constitutes an approximation, and this step could be submitted to practical constraints (*see* **Note 13**). It necessitates a rapid method to extract DNA because several thousand plants have to be analyzed.

2. "Quick" SDS 96-well format DNA preparation from *Arabidopsis* to perform genotyping: Collect samples (one or two leaves from 2-week-old plants) in 1.2 ml 8-strip collection tubes. Include a metal ball in each tube for ball milling.
3. Add 120 μl of extraction buffer, and then cap tubes.
4. Disrupt tissues for 5 min at maximum speed (30 vibrations/s). Turn racks over, and disrupt again for 5 min.
5. Centrifuge for 10 min at maximum speed (5,200 × *g*).
6. In a 96-well plate, add 100 μl of isopropanol to each well and then transfer 100 μl of supernatant (pipet slowly). Mix by pipetting. Cover with a flexible assay plate cover.
7. Incubate samples at room temperature for at least 10 min.
8. Centrifuge at maximum speed (5,200 × *g*).
9. Decant supernatant: Turn plate upside down carefully over the sink, and then tap gently on a paper towel.
10. Add 100 μl of 75 % ethanol. Cover samples with a flexible assay plate cover.
11. Centrifuge for 5 min at maximum speed (5,200 × *g*).
12. Decant supernatant as described above, and allow drying.
13. Add 50–100 μl (depending on the amount of starting material) of TE buffer, and let DNA suspend.

14. Use 1 μl of DNA for genotyping PCR.
15. As in the first round of QTL detection, fixed progeny is the only suitable strategy for our criteria of damage scores. The analysis of phenotyping results at this step allows including or excluding the tested regions: if the fixed rHIF differs significantly in its damage score from the surrounding regions, one of the genes in this QTL region is probably causing the phenotypic difference. If there is no difference in phenotype, this zone can be excluded from further analysis.
16. This phenotyping/genotyping loop is repeated until the region of interest is reduced enough to only cover a small number of genes.
17. Below a certain size (around 10 kb in *Arabidopsis*), it is useless to go on with this procedure: the number of plants to test becomes too large (several thousands), and the results become unreliable. The better strategy is to use specific rHIF to cross them and get advanced rHIFs (arHIF, [22]) (Fig. 8).

3.7 Validation of Candidate Gene(s)

1. When the candidate region is reduced sufficiently to include a manageable number of genes, go to appropriate databases to identify candidate genes from genome sequences and check if knockout (KO, such as T-DNA insertion or chemical mutagenesis) or knockdown mutants (such as RNA interference or artificial microRNA) exist.
2. Do the phenotyping of the mutants and compare to the RIL phenotypes. If the phenotype of the mutant is similar to one of the fixed HIF progenies, the gene is a serious candidate.
3. Do complementation of the mutant(s) via classical transgenesis, using the allele of each parent: if a functional, even partial, complementation is observed, the gene is confirmed as being (at least partially) responsible for the observed QTL.
4. After this step, numerous analyses such as comparison of transcriptomes [11] or quantitative complementation by crosses with other mutants [22] can be performed to reinforce the identified candidate.

3.8 Conclusion

Cold stress is a major abiotic stress affecting crop productivity all over the world. Increasing the knowledge in the detailed mechanisms controlling freezing tolerance would be of great help for agronomic purposes. The QTL strategy is powerful because it allows the identification of new genes involved in molecular pathways without a priori knowledge of possible targets since it is based on natural variability. All the steps, described here, for mapping and cloning, are summarized in Fig. 9. This protocol is directly usable on *Arabidopsis* and is adaptable on other species.

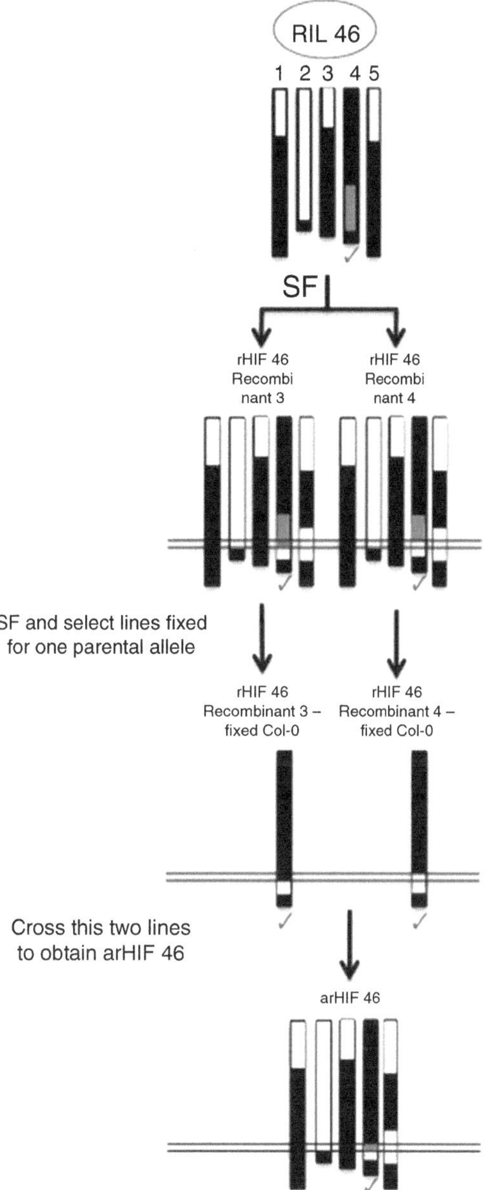

Fig. 8 Creation of advanced rHIFs (arHIFs): Two complementary rHIFs are chosen in the restricted candidate zone. They are fixed by self-fertilization to focus on a small candidate zone: one fixed for one parent and the other for the second parent. Then, they are crossed, constituting an arHIF, heterozygous only on a small part of the candidate region. Validation is done by progeny or fixed progeny testing. This material is free of interaction with borders and is therefore very valuable for further analysis

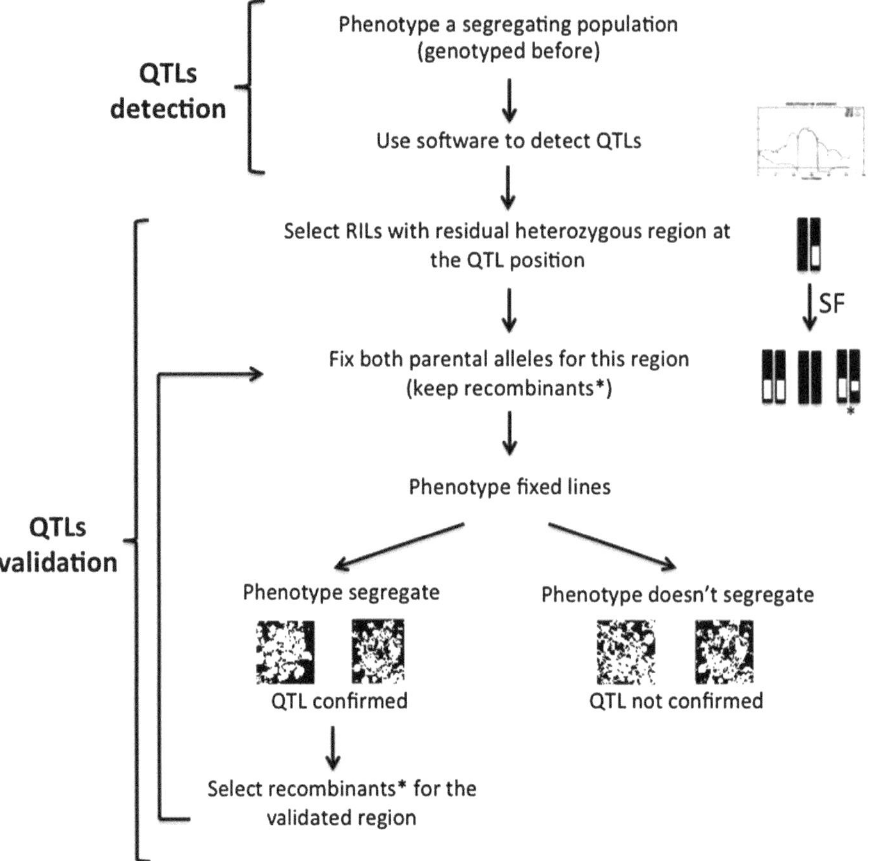

Fig. 9 General chart outlining the procedure to identify QTL associated with freezing tolerance in *Arabidopsis*. Main steps are summarized on this figure, and the repeated loop to reduce the initial candidate zone is shown by an *arrow*

4 Notes

1. Each line in the population is a unique mosaic of the two parental genomes of the cross. It has retained specific recombination events along the cycles of selfing. A great advantage of RIL populations is that the genotypes of the lines are fixed, and then, once genotypes for a population are identified, the population can be used for any number of replications or analyzed and measured under specific conditions for any number of traits. The main weakness of this material is that, in some species, there are no preexisting RILs and that building such populations is time consuming and expensive. In this case, backcrosses or F2 can be preferred. The genetic complexity of this kind of material could be bypassed by increasing the number of tested lines to keep a sufficient power of QTL detection.

2. F2 populations are developed by selfing F1 plants. Each individual in the F2 generation receives recombinant chromosome from each parent, so that at each locus the genotype is AA, AB, or BB, which is powerful for QTL detection, as all genotypes are represented. This structure also allows analyzing additive and dominant effects of the detected QTL. The main weakness of F2 material is the impossibility of replication in scoring the phenotype, so that some traits cannot be analyzed with these populations. For example, in the *Arabidopsis* freezing tolerance project, integrative scores for freezing damage after acclimation cannot be evaluated on individual plants, so F2 populations are unusable.
3. Backcrosses progenies are intermediate in their genetic structure: two inbred lines, denoted A and B, are crossed to obtain the first filial generation (F1) which is fully heterozygous AB. These F1 individuals are then crossed with one of the parental lines producing the backcross progeny. This strategy is easy to set up, but at each locus one homozygous parental genotype is not present, so only one side of the cross can be exploited for QTL detection, and detection of dominant traits is excluded. Moreover, this kind of population is very weak for epistasis detection.
4. The Versailles *Arabidopsis* Stock Center provides seeds useful to the international research community. Numerous seed stocks are available: more than 600 natural accessions, 132 F2 populations, 16 RIL populations, and 3 near-isogenic line sets (HIF: heterogeneous inbred family).
5. SNP, microsatellite, and indel markers are used for genotyping. The selected markers must be evenly distributed along the chromosomes; a mean distance of 5–10 cM is efficient enough, but distance must not exceed 20 cM. In our experiments, 80–90 markers are used on 200–300 individuals. The proportion of each parental allele has to be around 50 % in the RIL population to avoid biased analysis. At last, due to the large amount of work involved in genotyping, the selected markers have to be easy to use: high specificity and easy identification of heterozygotes. SNPs are generally typed with multiplex system.
6. The definition of the number of lines to test is a complex task because the question of the best size of the population has no easy answers: the researcher must find a consensus between theoretical rules for power of QTL detection and practical constraints combining genotyping of individuals and collecting phenotypic data. The general theory is "the more RILs the better" [23–25]. This could be particularly important when small-effect QTL are supposed to interfere in genetic architecture of a complex trait. So, if the effort to collect phenotypic data is a limiting constraint, it appears as a good choice to find the best

equilibrium between the number of lines to test and the number of repeats. Our phenotypic analysis is based on an integrative parameter, easy to measure and nondestructive for the plant. It is an efficient way to analyze in a first approach a great number of RIL populations to evaluate natural variability for the character. When using long or destructive tests, or using very expensive analysis such as metabolomics, it could be better to adapt the sample collection strategy. It has been shown that studying more RILs could be more powerful than increasing the number of repetitions [17]. Nevertheless, in order to reduce the phenotypic task it is possible to define "core populations" of selected RILs: the core population is an optimal subset of 164 lines allowing the user to phenotype only a reduced number of lines without loosing QTL detection power.

7. Besides the new versions of QTL Cartographer, R software is also usable for QTL mapping with the package (R/qtl).
8. Figure 4 illustrates the practical utilization of the software: what the purpose is when opening and how to change the parameters. It is very important to check that all the files have the right name: here 19RVcold has been chosen to create the directory, and all the files must have exactly this name; otherwise, the program would stop running.
9. In quantitative trait analysis, the model includes values for the effect of dominance and additive effects of alleles. Depending upon the genetic structure of the mapping population it is possible to calculate the additive value: here, and mentioned as additive in the raw file columns of Eqtl, the additive values are calculated as the effect of replacement of the allele from the Can-0 parent by the allele of Col-0. Negative values are due to the score scale: the higher the score is, the more damage had occurred. On the other hand, the effects of dominance can be estimated only if the analysis has been run on F2 population data. In all other cases, values for dominance effect are equal to zero.
10. Five plants of each type will be kept to be able to replace unexpected losses (dead plants, sterility, delayed flowering, etc.) and recover enough (three individuals) useful plants of each type without the need to go through an additional generation.
11. At this step, only positive HIFs can be selected. Effectively, as the genome of an RIL is a unique mosaic of parental genomes, the absence of validation can be due to epistatic interactions between other non-identified regions. That is the reason why, when possible, it is better to test several HIFs covering the candidate region to avoid local phenomena or specific genetic interactions. At last, easily "workable lines" must be chosen with normal development and fertility, and low sensitivity to pathogens, to increase the efficiency of the next steps.

12. When screening and phenotyping are possible on the same individual plant, progeny testing can be used. It consists of screening large numbers of plants to correlate individual phenotype with genotype. Practically, around 100 seeds of an HIF are sown individually, grown in phenotyping conditions, and genotyped with markers located at the extremities of the heterozygote zone. The interpretation is based on the same scheme as that used for the fixed progeny.
13. For example, when genotyping 1,000 plants on a zone of around 50 kb on chromosome 4 (which is about 19,000 kb), around five recombinants are expected. This can vary a lot between localization of the regions, screens, and chance. The general rule is to divide the candidate region into regular intervals with molecular markers. It is not useful to analyze a large population of recombinants at the same time because on this material the genetic background is globally homogenous between lines and phenotypic fixed progenies allow excluding regions at each step. So, it is better to search for regularly spaced markers to be able to refine the candidate zone as quickly as possible.

References

1. Thomashow M (1999) Plant cold acclimation: freezing tolerance genes and regulatory mechanisms. Annu Rev Plant Physiol Plant Mol Biol 50:571–599
2. Warren G, McKown R, Marin A, Teutonico R (1996) Isolation of mutations affecting the development of freezing tolerance in *Arabidopsis thaliana (L.) Heynh.* Plant Physiol 111:1011–1019
3. Alonso-Blanco C, Gomez-Mena C, Llorente F et al (2005) Genetic and molecular analyses of natural variation indicate CBF2 as a candidate gene for underlying a freezing tolerance quantitative trait locus in *Arabidopsis*. Plant Physiol 139:1304–1312
4. Vágújfalvi A, Galiba G, Cattivelli L, Dubcovsky J (2003) The cold-regulated transcriptional activator Cbf3 is linked to the frost-tolerance locus Fr-A2 on wheat chromosome 5A. Mol Genet Genomics 269:60–67
5. Francia E, Barabaschi D, Tondelli A et al (2007) Fine mapping of a HvCBF gene cluster at the frost resistance locus Fr-H2 in barley. Theor Appl Genet 115:1083–1091
6. Shirasawa S, Endo T, Nakagomi K et al (2012) Delimitation of a QTL region controlling cold tolerance at booting stage of a cultivar, 'Lijiangxintuanheigu', in rice *Oryza sativa* L. Theor Appl Genet 124:937–946
7. Géry C, Zuther E, Schulz E et al (2011) Natural variation in the freezing tolerance of *Arabidopsis thaliana*: effects of RNAi-induced CBF depletion and QTL localisation vary among accessions. Plant Sci 180:12–23
8. Hannah MA, Wiese D, Freund S et al (2006) Natural genetic variation of freezing tolerance in *Arabidopsis*. Plant Physiol 142:98–112
9. Zhen Y, Ungerer MC (2008) Clinal variation in freezing tolerance among natural accessions of *Arabidopsis thaliana*. New Phytol 177: 419–427
10. Zuther E, Schulz E, Childs LH, Hincha DK (2012) Clinal variation in the non-acclimated and cold-acclimated freezing tolerance of *Arabidopsis thaliana* accessions. Plant Cell Environ 35:1860–1878
11. Meissner M, Orsini E, Ruschhaupt M et al (2013) Mapping quantitative trait loci for freezing tolerance in a recombinant inbred line population of *Arabidopsis thaliana* accessions Tenela and C24 reveals REVEILLE1 as negative regulator of cold acclimation. Plant Cell Environ 36:1256–1267
12. Whitlow TH, Bassuk NL, Ranney TG, Reichert DL (1992) An improved method for using electrolyte leakage to assess membrane competence in plant tissues. Plant Physiol 98: 198–205

13. Burr KE, Tinus RW, Wallner SJ, King RM (1990) Comparison of three cold hardiness tests for conifer seedlings. Tree Physiol 6: 351–369
14. Lander ES, Green P, Abrahamson J et al (1987) MAPMAKER: an interactive computer package for constructing primary genetic linkage maps of experimental and natural populations. Genomics 1:174–181
15. Basten CJ, Weir BS, Zeng ZB (2000) QTL CARTOGRAPHER Version 1.14. North Carolina State University, Raleigh, NC
16. Kosambi DD (1944) The estimation of map distances from recombinant values. Ann Eugen 12:172–175
17. Loudet O, Chaillou S, Camilleri C et al (2002) Bay x Shadara recombinant inbred line population: a powerful tool for the genetic dissection of complex traits in *Arabidopsis*. Theor Appl Genet 104:1173–1184
18. Andaya VC, Mackill DJ (2003) Mapping of QTLs associated with cold tolerance during the vegetative stage in rice. J Exp Bot 54: 2579–2585
19. McKhann HI, Gery C, Bérard A et al (2008) Natural variation in CBF gene sequence, gene expression and freezing tolerance in the Versailles core collection of *Arabidopsis thaliana*. BMC Plant Biol 8:105
20. Lander ES, Botstein D (1989) Mapping Mendelian factors underlying quantitative traits using RFLP linkage maps. Genetics 121: 185–199
21. Churchill GA, Doerge RW (1994) Empirical threshold values for quantitative trait mapping. Genetics 138:963–971
22. Vlad D, Rappaport F, Simon M, Loudet O (2010) Gene transposition causing natural variation for growth in *Arabidopsis thaliana*. PLoS Genet 6:e1000945
23. Charcosset A, Gallais A (1996) Estimation of the contribution of quantitative trait loci (QTLs) to the variance of a quantitative trait by means of genetic markers. Theor Appl Genet 93:1193–1201
24. Borevitz JO, Chory J (2004) Genomics tools for QTL analysis and gene discovery. Curr Opin Plant Biol 7:132–136
25. Simon M, Loudet O, Durand S et al (2008) Quantitative trait loci mapping in five new large recombinant inbred line populations of Arabidopsis thaliana genotyped with consensus single-nucleotide polymorphism markers. Genetics 178:2253–2264

Chapter 7

Common Garden Experiments to Characterize Cold Acclimation Responses in Plants from Different Climatic Regions

Andrey V. Malyshev, Hugh A.L. Henry, and Juergen Kreyling

Abstract

Cold acclimation is a crucial factor to consider in the context of ongoing climate change. Maladaptation with regard to frost damage and use of the growing season may occur depending on cold acclimation cues. Importance of photoperiod and preceding temperatures as cues needs therefore to be evaluated within (ecotypes) and among species. Common garden designs, in particular the (1) establishment of multiple common gardens along latitudinal/altitudinal gradients, (2) with in situ additional climate manipulations and (3) with manipulations in climate chambers are proposed as tools for the detection of local adaptations and relative importance of temperature and photoperiod as cues for cold adaptation. Here, we discuss issues in species and ecotype selection, establishment of common gardens including manipulations of temperature and photoperiod, and quantification of cold adaptation.

Key words Common garden experiments, Provenance trials, Cold acclimation, Frost tolerance, Experimental design, Within-species variability, Intraspecific variability

1 Introduction

Cold acclimation (also known as hardening) is a suite of changes in gene expression and physiology that increases plant tolerance to cold temperatures [1, 2]. In the fall a reduced photoperiod and declining temperatures trigger cold acclimation in perennial plants [3, 4]. The topic of plant winter acclimation has received increasing attention in the context of global climate change, because the latter is expected to increase temperature variability. Highly fluctuating temperatures can disrupt plant cold acclimation, making plants less hardy to withstand unexpected frosts, which are predicted to occur with prior magnitude despite decreased frequency [5].

The relative importance of temperature and photoperiod may be critical for determining how effectively plants acclimate to prevent frost damage in a changing climate. Plants that use temperature as the main acclimation cue can benefit from a longer growing

Dirk K. Hincha and Ellen Zuther (eds.), *Plant Cold Acclimation: Methods and Protocols*, Methods in Molecular Biology, vol. 1166, DOI 10.1007/978-1-4939-0844-8_7, © Springer Science+Business Media New York 2014

season but can also be more susceptible to unexpected frosts. On the other hand, plants that respond strongly to photoperiod may be less susceptible to unexpected frosts. Cold acclimation cues have been studied most intensively in trees, for which there is high interspecific variation in the dominance of temperature versus photoperiod in driving cold acclimation responses [6]. In temperate regions, woody plants cold acclimate to a greater extent than herbaceous species, likely because the latter can be insulated from cold air temperatures by snow cover [7]. Snow cover is decreasing substantially in some regions [8, 9], which can increase the exposure of herbaceous plants to cold temperatures over winter, despite an overall increase in mean air temperatures. Therefore, herbaceous plants are an important contrasting functional group to woody plants in the study of cold acclimation responses to climate change.

Frost susceptibility also varies substantially among ecotypes within species. An ecotype is a general term used to describe plants within a species that display genetic differentiation to particular environmental parameters within the ecotype's habitat [10, 11]. Ecotypes at lower latitudes are generally less cold tolerant than ecotypes at higher latitudes [12]. Species distributions can span across continents and along great elevational gradients. Local adaptation creates ecotypes that may differ as much in their acclimation and freezing tolerance as different plant species [13]. Thus, knowledge of local adaptation is needed to more precisely predict species' climate change responses. In addition, because of the potential for long-distance dispersal and population range shifts, ecotypes of different species from different climatic regions need to be incorporated into cold acclimation studies in order to predict regional changes in species composition. Relating cold hardening responses to local environmental conditions will also improve the understanding of cold hardening cues in general.

Common garden experiments involve the transplantation of experimental subjects into a common growth environment before taking measurements [14] and allowing plant responses to be compared under standard environment conditions, with control over factors such as ontogenetic stages. Measurement of ecotypes in their natural environment along a spatial or a temporal climatic gradient is an alternative approach that can be used to study cold acclimation responses [14], but it includes plant–plant interactions which are minimized in common garden experiments and are beyond the scope of this chapter. Although common garden experiments can be performed under naturally variable weather conditions, the addition of controlled environmental conditions (e.g., variation in photoperiod/frost intensity) allows for a better mechanistic understanding of plant responses. Here we describe the setup of common garden experiments designed to test cold acclimation responses, performed with and without additional manipulations. First, experimental guidelines are presented that apply to common garden experiments without additional environmental manipulation. Implementation of additional environmental manipulations is then discussed, and finally potential response parameters for such experiments are presented. An overview of the stepwise procedure involved in the setup and implementation of cold acclimation experiments on plants from different climatic regions is presented in Fig. 1.

The guidelines are intended to be general and to apply to most vascular plants. The actual experimental design needs will likely require modification based on the specific species used, taking factors such as plant size and seed production into consideration.

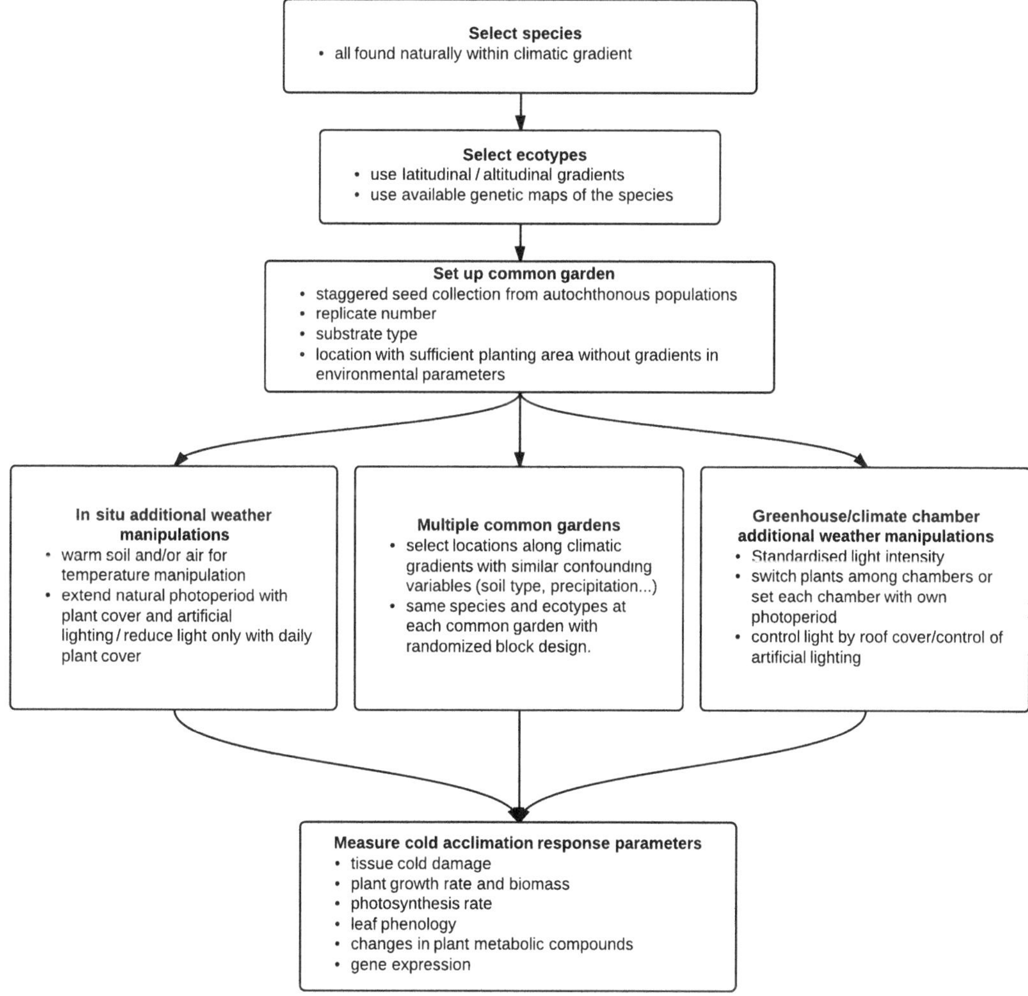

Fig. 1 Steps required to carry out cold acclimation experiments on plants from different climatic regions (different ecotypes) in one or more common gardens, with or without additional weather manipulations (*see* Table 1 for respective treatment comparisons). *Bullet points* highlight factors to consider in each step

2 Materials

2.1 Ecotype Selection

After choosing the target plant species, ecotypes from each species need to be carefully selected in the context of the research question. Altitudinal and latitudinal gradients are typically chosen to select ecotypes [14, 15]. To aid with ecotype selection, a species distribution map can be overlaid on a climatic map. For example, minimum winter temperature and annual photoperiod changes within the species range can be used as proxies for climate adaptation and ecotype selection. Most commonly, latitude has been taken as a proxy for climate, because it correlates with many biologically relevant environmental gradients, including photoperiod [16]. Existing databases such as Worldclim database (http://www.worldclim.org) [17], as well as photoperiod [18], can be used to acquire such data.

Altitudinal gradients allow an isolation of temperature adaptation, with greater changes in temperature per distance than latitudinal gradients [19], while latitudinal gradients allow the combined effects of photoperiod and temperature to be considered. If the latitudinal species distribution effect on acclimation is of prime interest, the elevation of the ecotype origins should be as equal as possible to avoid confounding environmental factors. In both cases, care should be exercised in selecting ecotypes that stem from locations differing as little as possible in confounding factors not under study, such as precipitation and soil type [15, 20]. There also exist statistical methods to control for confounding factors (*see* Subheading 3.2). Cold tolerance is commonly used to describe the extent of the achieved cold acclimation. To maximize the likelihood of selecting ecotypes that differ most with respect to cold tolerance, the longest possible gradient along the distribution of the species range should be used. It is advisable to select at least two distinct ecotypes (separated enough to limit cross-pollination) in relatively close proximity to each other for each latitudinal/altitudinal point of origin in order to increase the confidence that the variation in differences is due to the latitudinal position, rather than to specific local conditions.

Differences in geographically influenced acclimation trends result from genetic adaptation or from epigenetic modifications (plastic plant responses) [21]. Climatic influence on the genetic adaptation of ecotypes is well known [22]. Available genetic data of plant populations across the distribution range of a species can thus be considered for selecting the most genetically distinct ecotypes, and genetic differences can be compared to differences in the acclimation responses upon completion of the experiment (for an example of genetic analysis *see* ref. 23, and for an example of climatic origin vs. cold hardiness comparison *see* ref. 24).

2.2 Seed Collection and Plant Numbers

Ecotypes originating from different latitudes flower and produce seed at different times, which necessitates staggered timing of seed collection along the latitudinal gradient. Ideally, single clones or mother plants need to be sampled individually from autochthonous populations and kept separated throughout germination trials. If variation among individuals within each ecotype is to be included in the analysis, at least five different clones or mother plants per ecotype should be used as seed sources, with at least five replicate plants being grown per mother. Especially for clonal plants it needs to be assured that the mother plants do not stem from the same clone, by taking into account the distance typically reached by rhizome growth and dispersal potential of detached propagules. Without analyzing mother plant response variability, the replicate number depends on the number of ecotypes being compared. Ideally, plant traits should be measured for the ecotype to be used in the experiment to gauge natural variability, which will help the right replicate number to be selected effectively. Examples below are given as a reference:

- Ten replicates were used for three ecotypes, both in determining the relationship of differences in allele frequencies at polymorphic traits to phenotypic plant traits [25] and in quantifying autumn cold tolerance [21].
- Five trees were sampled per species, using 14 species in total for bud burst as a function of photoperiod [6].
- Ten plants per ecotype with a total of ten grass ecotypes were used in a frost tolerance experiment [26].
- Four replicate raspberry plants per cultivar with a total of six cultivars were used to monitor growth at different cold acclimation temperatures [27].

2.3 Plant Propagation

After seed collection, germination and propagation methods should be standardized across all ecotypes, with seedlings being first potted or planted in a standard substrate in a common garden. Pot size depends on the duration of time the plants need to remain potted. If space is limited, deeper rather than wider pots are preferred to ensure that roots do not become pot bound. Once roots reach the bottom of the pot, acclimation temperature should be lowered to 5 °C or below, which causes roots to cease growth [28]. Although root growth is species specific, roots in 3-month-old juniper cuttings have been shown to grow 0.2 mm per day at 6 °C, as compared with 1 mm per day at 15 °C [29]. Roots can also be trimmed and repotted to promote regular growth [30].

Substrate type can influence the rate of root acclimation as a result of differences in heat-conducting ability and water-holding capacity, with higher water moisture content reducing the freezing and thawing rates [31]. A major deciding factor in substrate selection is whether the roots will need to be washed and analyzed after the conclusion of the experiment. Proportionally higher sand content allows easier root washing while minimizing root damage. The drawback of a high sand substrate is that it has low water and nutrient-holding capacities [31]. Frequent watering and nutrient addition (e.g., Hoagland's solution) can be used to offset this problem. Additionally, sandier soils (1) cause faster root development [32], which can effect cold hardiness by altering root distribution within the substrate, and (2) reduce insulation of the roots [33].

Fertilization should be done with care, because the effects of fall nitrogen application can increase or decrease cold tolerance depending on the stage of cold acclimation at which it is added as well as the particular species involved [34–38].

The chosen substrate at the experimental location should be standardized and characterized to the depth reached by the respective plant species. For long-term common gardens (spanning multiple years), the distance between trees should be at least 3 m for plants lower than 15 m in height and increased by a minimum of a meter for every additional 15 m of height. Standard planting measures based on the spacing of plants are specified by the International Phenological Gardens in Europe (http://www.arm.ac.uk/nci/docs/Instructions-IPG.pdf) [39].

3 Methods

3.1 Applying Cold Acclimation Treatments

For all types of treatments spatiotemporal replication of experiments is suggested to increase the confidence in results [14]. A comparison of the three ways in which cold acclimation treatments can be administered is presented in Table 1.

1. Establishment of multiple common gardens along a latitudinal/altitudinal gradient

   First, species should be used which all naturally occur within the altitudinal/latitudinal gradient of interest. Replicated common gardens are then established in two or more locations along the selected gradient. Two or more ecotypes from every species, typically stemming from two opposite climatic extremes, are planted in every common garden in a randomized block design, with species and ecotypes being randomly mixed within each block. Using reciprocal native soils is not advised due to uncontrolled bias (site × ecotype × soil source interaction) [40]. For experiments where plants are only to be observed for one to two seasons, potted seedlings (*see* Subheading 2.3 on preventing pot-bound roots) can be left in containers, placed in prepared nursery beds, and sand/soil can be used to fill the gaps among pots for insulation (as done for tree seedlings by [40] and for grass by [36]).

Table 1
Comparison of cold acclimation treatments in common garden experiments to evaluate cold acclimation differences among plant species and ecotypes

Methodology	Best suited for	Advantages	Disadvantages
1. Common garden experiments without additional climate manipulation - replicated common gardens along a latitudinal/altitudinal gradient	Disentangling photoperiodic and temperature cues for cold acclimation, replicating common gardens simultaneously along latitudinal and altitudinal gradients	• Simulates most probable future climate scenario by incorporating most environmental variables of the new location • "Hands-off" experimental approach	• Presence of confounding factors (e.g., wind patterns/precipitation sunlight) makes attributing differences to specific factors (photoperiod/temperature) difficult = "black box" experimental design • Results depend on actual climatic conditions during the experiment at the single sites while sites were chosen due to their long-term climatic differences
Common garden experiments with additional climate manipulation treatments			
2. In situ additional weather manipulations	Mechanistic exploration of photoperiod/temperature effects on cold acclimation in a natural environment	• More realistic than # 1 • Less space/logistical constraints (larger/more plants possible)	• Hard to replicate experimental setup and compare results to previous experiments due to high inter- and intra-annual temperature variability • Limited control of temperature and photoperiod adjustments
3. Manipulations in climate chambers	Mechanistic exploration of threshold cold acclimation responses to photoperiod/temperature changes General exploration of plant acclimation cues (e.g., comparing the rates of temperature and photoperiod decreases on plant acclimation)	• High level of control over treatments • Can be replicated with high precision	• Not as realistic as in situ photoperiod and temperature manipulations • Space/logistical constraints (smaller/fewer plants possible)

Additional weather manipulations can offer an increased number of cold acclimation scenarios within the same time frame, aiming at a more mechanistic understanding of temperature and photoperiod vs. ecotype-specific acclimation responses. Additional climate simulations can be administered either in the field (in situ) or in enclosed climate-controlled chambers/greenhouses.

2. Additional climate manipulations in situ

   A variety of fall light and temperature manipulations can be applied to the plants in the study of cold acclimation.

   Examples of temperature manipulation:

   Overhead heating lamps to warm the plants with infrared radiation [13, 41].

   Buried heating wires to increase soil temperature [42].

   Open top passive warming chambers that increase temperature via decreased airflow and greenhouse effect [43].

   Examples of light manipulation:

   Light-tight aluminum boxes in the field that close daily via a remote control [44].

   Opaque plastic wrapped around plants with additional incandescent light inside the covered units [45].

3. Manipulations in climate chambers

   Greenhouses and small climate chambers can also be used to control temperature and photoperiod. In addition to computer-controlled daily photoperiodic cycles, extension of natural photoperiod in greenhouses with artificial light can be used [46] or exposure to natural light can be decreased by automatically closing greenhouse roof panels [30]. Two strategies can be used to impose different photoperiods: (1) different photoperiod cycles can be run in different chambers while keeping temperature constant or (2) the same temperature can be maintained in two chambers with light being always on in one and off in another, switching plants between the chambers twice daily.

   The first method requires less maintenance while the experiment is running, but it suffers from pseudoreplication, because no two chambers are alike, and potential chamber effects are confounded with the desired treatment differences. To circumvent this shortcoming, chamber settings can be switched at set time intervals to ensure that plants spend equal time periods in all chambers and to ensure that the chamber effect is as equal as possible for all plants. This method is advised when multiple combinations of temperature × photoperiod factors are used and when there is little mean variability in temperature and photoperiod among chambers. Treatments that differ in acclimating temperature will cause plants to lose water at different rates, and the bigger the temperature difference among treatments, the more often the plants should be watered to minimize soil moisture differences.

In the second method, in order to impose progressively lower photoperiod treatments, sets of plants are transferred twice daily at staggered intervals from one chamber to another to shorten or lengthen the effective photoperiod. This method allows for several photoperiod manipulations using only two climate chambers, while minimizing the chamber effect by all plants sharing time in both chambers, as compared to using a separate chamber for every different photoperiod. Periodically the no-light and light chambers can be reversed to ensure that the same time period is spent by all plants in both chambers.

4. Response parameters

   Table 2 presents plant parameters that can be used to quantify plant cold acclimation responses. Ultimately, the nature of the study will dictate the types of parameters that are to be measured. Functional responses attributable to responses of a plant as a whole have high ecological importance. Yet, they usually require natural or artificial frost events after acclimation in order to test for differences in cold tolerance. Ideally, a gradient of minimum temperatures can be tested. Imposing controlled frost events is even possible in remote field locations [47]. Assessment of physiological pathways or plant responses at a molecular level improves mechanistic understanding of the processes involved and does not depend on actual frost events after acclimation.

3.2 Data Analysis

Differences among climate treatments and among ecotypes/species, and in particular interactions between both factors, are tested by applying two-factor analysis of variance (ANOVA). In the case of nested replicates, mixed models with the blocking factor assigned as a random effect can be applied [48]. Preferable designs are replicated common garden experiments in conditions similar to the home site of all ecotypes [49]. Local adaptation of populations can be assessed using a regression between the relative performances of the ecotypes in the common garden experiment and their home-site climate (e.g., winter minimum temperature) [24]. Generally, adequate statistical power in the regression analysis is achieved when the number of ecotypes exceeds ten. Statistical techniques such as multiple regression, hierarchical or mixed-effect models, variation partitioning, or structural equation modeling [20, 50–52] can be used to correct for confounding factors.

Table 2
An overview of parameters that can be used to quantify plant cold acclimation responses. Whole-plant functional responses are advised to be measured before more mechanistic response parameters

Plant functional traits	Commonly measured parameter/s	Rationale	References for detailed explanation and methods	Level at which response parameter is measured
Fall phenology	• Leaf color • Leaf fall	Plant phenology is one of the most sensitive parameters to changes in photoperiod and temperature, making it an ideal cold acclimation parameter	Phenological guide of the international phenological gardens: [53]	Whole plant, functional long term responses to cold acclimation
Plant growth	• Growth rate • Biomass • Seed production after acclimation and exposure to frost	Growth rate may be faster following cold stress and equalize among treatments with time. Reproductive output may be independent from growth performance	Reproductive output: [36]	
Functional effect of root frost damage	• Root N uptake after acclimation and exposure to frost	Roots are exposed to a solution of isotopically labelled N solution and then washed, dried, and ground before tissue ^{15}N content is determined with a mass spectrometer	[54]	
Extent of plant issue cold damage	• Relative electrolyte leakage after acclimation	When plant cell membranes are cold damaged, they release ions into the apoplast. The conductivity of the solution containing the ions is then measured and is proportional to cold damage	[21, 55]	
Plant photo-synthetic activity	• Chlorophyll fluorescence after acclimation	Acclimation stage as well as amount of cold damage sustained can be assessed from the amount of photoinhibition of plant photosystems as well as rates of CO_2 production/O_2 consumption	[56–58]	Mechanistic immediate responses to cold acclimation

(continued)

Table 2 (continued)

Plant functional traits	Commonly measured parameter/s	Rationale	References for detailed explanation and methods	Level at which response parameter is measured
Plant acclimation stage/stress level	• Compounds that change their concentration with cold acclimation to prepare plants to withstand sub-zero temperatures	Increase in soluble sugar content and abscisic acid and the formation of antifreeze proteins as well as structural changes in cell membrane retard ice formation and keep the cell membrane fluid, allowing plants to maintain normal metabolism. Specific genes are also only expressed when plants cold acclimate	Proteome changes: [59–61] Sugar concentration: [62] Gene expression: [63] Abscisic acid: [64] Cell membrane structural changes: [45]	

References

1. Kalberer SR, Wisniewski M, Arora R (2006) Deacclimation and reacclimation of cold-hardy plants: current understanding and emerging concepts. Plant Sci 171:3–16
2. Thomashow MF (1999) Plant cold acclimation: freezing tolerance genes and regulatory mechanisms. Annu Rev Plant Physiol Plant Mol Biol 50:571–599
3. McKenzie JS, Pacquin R, Duke SH (1988) Cold and heat tolerance. In: Hanson AA, Barnes DK, Hill RR (eds) Alfalfa and alfalfa improvement. ASA, CSSA and SSSA, Madison, WI
4. Stout DG, Hall JW (1989) Fall growth and winter survival of alfalfa in interior British-Columbia. Can J Plant Sci 69:491–499
5. Kodra E, Steinhaeuser K, Ganguly AR (2011) Persisting cold extremes under 21st-century warming scenarios. Geophys Res Lett 38, L08705
6. Basler D, Koerner C (2012) Photoperiod sensitivity of bud burst in 14 temperate forest tree species. Agric For Meteorol 165:73–81
7. Welling A, Moritz T, Palva ET, Junttila O (2002) Independent activation of cold acclimation by low temperature and short photoperiod in hybrid aspen. Plant Physiol 129: 1633–1641
8. Belanger G, Rochette P, Castonguay Y, Bootsma A, Mongrain D, Ryan DAJ (2002) Climate change and winter survival of perennial forage crops in eastern Canada. Agron J 94:1120–1130
9. Kreyling J, Henry HAL (2011) Vanishing winters in Germany: soil frost dynamics and snow cover trends, and ecological implications. Climate Res 46:269–276
10. Linhart YB, Grant MC (1996) Evolutionary significance of local genetic differentiation in plants. Annu Rev Ecol Syst 27:237–277
11. Turesson G (1922) The genotypical response of the plant species to the habitat. Hereditas 3:211–350
12. Kozlowski TT, Pallardy SG (2002) Acclimation and adaptive responses of woody plants to environmental stresses. Botanical Rev 68:270–334
13. Kreyling J, Thiel D, Simmnacher K, Willner E, Jentsch A, Beierkuhnlein C (2012) Geographic origin and past climatic experience influence the response to late spring frost in four common grass species in central Europe. Ecography 35:268–275
14. De Frenne P, Graae BJ, Rodriguez-Sanchez F, Kolb A, Chabrerie O, Decocq G, De Kort H, De Schrijver A, Diekmann M, Eriksson O,

Gruwez R, Hermy M, Lenoir J, Plue J, Coomes DA, Verheyen K (2013) Latitudinal gradients as natural laboratories to infer species' responses to temperature. J Ecol 101:784–795

15. Korner C (2007) The use of 'altitude' in ecological research. Trends Ecol Evol 22:569–574
16. Vasemagi A, Primmer CR (2005) Challenges for identifying functionally important genetic variation: the promise of combining complementary research strategies. Mol Ecol 14:3623–3642
17. Hijmans RJ, Cameron SE, Parra JL, Jones PG, Jarvis A (2005) Very high resolution interpolated climate surfaces for global land areas. Int J Climatol 25:1965–1978
18. Naval Meteorology and Oceanography Command (2013) U.S. Navy Naval oceonography portal. vol 2013, 1100 Balch Blvd, Stennis Space Center, MS 39529
19. Jump AS, Matyas C, Penuelas J (2009) The altitude-for-latitude disparity in the range retractions of woody species. Trends Ecol Evol 24:694–701
20. De Frenne P, Kolb A, Verheyen K, Brunet J, Chabrerie O, Decocq G, Diekmann M, Eriksson O, Heinken T, Hermy M, Jogar U, Stanton S, Quataert P, Zindel R, Zobel M, Graae BJ (2009) Unravelling the effects of temperature, latitude and local environment on the reproduction of forest herbs. Glob Ecol Biogeogr 18:641–651
21. Gomory D, Foffova E, Kmet J, Longauer R, Romsakova I (2010) Norway spruce (*Picea abies* l. karst.) provenance variation in autumn cold hardiness: adaptation or acclimation? Acta Biol Cracov Ser Bot 52:42–49
22. Franks SJ, Hoffmann AA (2012) Genetics of climate change adaptation. Annu Rev Genet 46:185–208
23. Michalski SG, Durka W, Jentsch A, Kreyling J, Pompe S, Schweiger O, Willner E, Beierkuhnlein C (2010) Evidence for genetic differentiation and divergent selection in an autotetraploid forage grass (*Arrhenatherum elatius*). Theor Appl Genet 120:1151–1162
24. Kreyling J, Wiesenberg GLB, Thiel D, Wohlfart C, Huber G, Walter J, Jentsch A, Konnert M, Beierkuhnlein C (2012) Cold hardiness of *Pinus nigra* Arnold as influenced by geographic origin, warming, and extreme summer drought. Environ Exp Bot 78:99–108
25. Romsakova I, Foffova E, Kmet J, Longauer R, Pacalaj M, Gomory D (2012) Nucleotide polymorphisms related to altitude and physiological traits in contrasting provenances of Norway spruce (*Picea abies*). Biologia 67: 909–916
26. Bykova O, Sage RF (2012) Winter cold tolerance and the geographic range separation of *Bromus tectorum* and *Bromus rubens*, two severe invasive species in North America. Glob Chang Biol 18:3654–3663
27. Palonen P (2006) Vegetative growth, cold acclimation, and dormancy as affected by temperature and photoperiod in six red raspberry (*Rubus idaeus* L.) cultivars. Eur J Hortic Sci 71:1–6
28. Lyr H, Hoffmann G (1967) Growth rates and growth periodicity of tree roots. Int Rev For 2:181–206
29. Bigras FJ, Rioux JA, Therrien HP, Paquin R (1989) Influence de la photoperiode et de la temperature sur l'evolution de la tolerance au gel, de la croissance et de la teneur en eau, sucres, amidon et proline des rameaux et des racines de genevrier (*Juniperus chinensis* L. 'Pfitzerana'). Can J Plant Sci 69:305–316
30. Garris A, Clark L, Owens C, McKay S, Luby J, Mathiason K, Fennell A (2009) Mapping of photoperiod-induced growth cessation in the wild grape *Vitis riparia*. J Am Soc Hortic Sci 134:261–272
31. Hanninen H (1991) Does climatic warming increase the risk of frost damage in Northern trees. Plant Cell Environ 14:449–454
32. Gill JS, Sivasithamparam K, Smettem KRJ (2000) Soil types with different texture affects development of Rhizoctonia root rot of wheat seedlings. Plant Soil 221:113–120
33. Masaka K, Sato H, Kon H, Torita H (2010) Mortality of planted *Pinus thunbergii* Parlat. saplings subject to coldness during winter and soil types in region of seasonal soil frost. J For Res 15:374–383
34. Dalen LS, Johnsen O (2004) CO(2) enrichment, nitrogen fertilization and development of freezing tolerance in Norway spruce. Trees Struct Funct 18:10–18
35. Junttila O, Svenning MM, Rosnes K (1995) Influence of mineral nitrogen source on growth and frost resistance of white clover (*Trifolium repens* L.) and timothy (*Phleum pratense* L.) seedlings. Acta Agric Scand B 45:261–267
36. Malyshev AV, Henry HAL (2012) Frost damage and winter nitrogen uptake by the grass *Poa pratensis* L.: consequences for vegetative versus reproductive growth. Plant Ecol 213:1739–1747
37. Ishikawa M, Robertson AJ, Gusta LV (1990) Effect of temperature, light, nutrients and dehardening on abscisic-acid induced cold

hardiness in *Bromus inermis* Leyss suspension cultured-cells. Plant Cell Physiol 31:51–59
38. Webster DE, Ebdon JS (2005) Effects of nitrogen and potassium fertilization on perennial ryegrass cold tolerance during deacclimation in late winter and early spring. HortScience 40:842–849
39. Humboldt University of Berlin, Faculty of Agriculture and Horticulture, Institute of Crop Sciences Subdivision of Agricultural Meteorology. http://www.arm.ac.uk/nci/docs/Instructions-IPG.pdf
40. Vitasse Y, Hoch G, Randin CF, Lenz A, Kollas C, Scheepens JF, Korner C (2013) Elevational adaptation and plasticity in seedling phenology of temperate deciduous tree species. Oecologia 171:663–678
41. Hutchison JS, Henry HAL (2010) Additive effects of warming and increased nitrogen deposition in a temperate old field: plant productivity and the importance of winter. Ecosystems 13:661–672
42. Kreyling J, Beierkuhnlein C, Pritsch K, Schloter M, Jentsch A (2008) Recurrent soil freeze-thaw cycles enhance grassland productivity. New Phytol 177:938–945
43. Bokhorst S, Huiskes A, Aerts R, Convey P, Cooper EJ, Dalen L, Erschbamer B, Gudmundsson J, Hofgaard A, Hollister RD, Johnstone J, Jonsdottir IS, Lebouvier M, Van De Vijver B, Wahren CH, Dorrepaal E (2013) Variable temperature effects of Open Top Chambers at polar and alpine sites explained by irradiance and snow depth. Glob Chang Biol 19:64–74
44. Fennell A, Hoover E (1991) Photoperiod influences growth, bud dormancy, and cold-acclimation in *Vitis labruscana* and *V. riparia*. J Am Soc Hortic Sci 116:270–273
45. Vogg G, Heim R, Hansen J, Schafer C, Beck E (1998) Frost hardening and photosynthetic performance of Scots pine (*Pinus sylvestris* L.) needles. I. Seasonal changes in the photosynthetic apparatus and its function. Planta 204:193–200
46. Pietsch GM, Anderson NO, Li PH (2009) Cold tolerance and short day acclimation in perennial *Gaura coccinea* and *G. drummondii*. Sci Hortic 120:418–425
47. Thorpe PC, Macgillivray CW, Priestman GH (1993) A portable device for the simulation of air frosts at remote field locations. Funct Ecol 7:503–505
48. Thiel D, Nagy L, Beierkuhnlein C, Huber G, Jentsch A, Konnert M, Kreyling J (2012) Uniform drought and warming responses in *Pinus nigra* provenances despite specific overall performances. For Ecol Manage 270: 200–208
49. O'Neill GA, Hamann A, Wang T (2008) Accounting for population variation improves estimates of the impact of climate change on species' growth and distribution. J Appl Ecol 45:1040–1049
50. Milla R, Escudero A, Iriondo JM (2009) Inherited variability in multiple traits determines fitness in populations of an annual legume from contrasting latitudinal origins. Ann Bot 103:1279–1289
51. Figueiredo M, Anderson AJ (2009) Digestive enzyme spectra in crustacean decapods (Paleomonidae, Portunidae and Penaeidae) feeding in the natural habitat. Aquac Res 40: 282–291
52. Stomp M, Huisman J, Mittelbach GG, Litchman E, Klausmeier CA (2011) Large-scale biodiversity patterns in freshwater phytoplankton. Ecology 92:2096–2107
53. Phenological observational guide of the international phenological gardens. http://www.agrar.hu-berlin.de/fakultaet/departments/dntw/agrarmet/phaenologie/ipg
54. Malyshev AV, Henry HAL (2012) N uptake and growth responses to sub-lethal freezing in the grass *Poa pratensis* L. Plant Soil 360:175–185
55. Repo T, Ryyppo A (2008) The electrolyte leakage method can be misleading for assessing the frost hardiness of roots. Plant Biosyst 142: 298–301
56. Rizza F, Pagani D, Stanca AM, Cattivelli L (2001) Use of chlorophyll fluorescence to evaluate the cold acclimation and freezing tolerance of winter and spring oats. Plant Breed 120:389–396
57. Huner NPA, Oquist G, Hurry VM, Krol M, Falk S, Griffith M (1993) Photosynthesis, photoinhibition and low-temperature acclimation in cold tolerant plants. Photosynth Res 37: 19–39
58. Huner NPA, Oquist G, Sarhan F (1998) Energy balance and acclimation to light and cold. Trends Plant Sci 3:224–230
59. Antikainen M, Griffith M (1997) Antifreeze protein accumulation in freezing-tolerant cereals. Physiol Plantarum 99:423–432
60. Kosova K, Vitamvas P, Prasil IT, Renaut J (2011) Plant proteome changes under abiotic stress—contribution of proteomics studies to understanding plant stress response. J Proteomics 74:1301–1322

61. Guy C, Kaplan F, Kopka J, Selbig J, Hincha DK (2008) Metabolomics of temperature stress. Physiol Plantarum 132:220–235
62. Jaszberenyi C, Lukacs L, Inotai K, Nemeth E (2012) Soluble sugar content in poppy (*Papaver somniferum* L.) and its relationship to winter hardiness. Zeitschrift für Arznei- & Gewürzpflanzen 17:169–174
63. Knight MR, Knight H (2012) Low-temperature perception leading to gene expression and cold tolerance in higher plants. New Phytol 195:737–751
64. Bravo LA, Zuniga GE, Alberdi M, Corcuera LJ (1998) The role of ABA in freezing tolerance and cold acclimation in barley. Physiol Plantarum 103:17–23

Chapter 8

Identification of *Arabidopsis* Mutants with Altered Freezing Tolerance

Carlos Perea-Resa and Julio Salinas

Abstract

Low temperature is an important determinant in the configuration of natural plant communities and defines the range of distribution and growth of important crops. Some plants, including *Arabidopsis*, have evolved sophisticated adaptive mechanisms to tolerate low and freezing temperatures. Central to this adaptation is the process of cold acclimation. By means of this process, many plants from temperate regions are able to develop or increase their freezing tolerance in response to low, nonfreezing temperatures. The identification and characterization of factors involved in freezing tolerance are crucial to understand the molecular mechanisms underlying the cold acclimation response and have a potential interest to improve crop tolerance to freezing temperatures. Many genes implicated in cold acclimation have been identified in numerous plant species by using molecular approaches followed by reverse genetic analysis. Remarkably, however, direct genetic analyses have not been conveniently exploited in their capacity for identifying genes with pivotal roles in that adaptive response. In this chapter, we describe a protocol for evaluating the freezing tolerance of both non-acclimated and cold-acclimated *Arabidopsis* plants. This protocol allows the accurate and simple screening of mutant collections for the identification of novel factors involved in freezing tolerance and cold acclimation.

Key words Low temperature, Mutant screening, Freezing-tolerant mutants, Freezing-sensitive mutants, Cold acclimation, Constitutive freezing tolerance, *Arabidopsis*

1 Introduction

Plants are sessile organisms continuously adapting to the environmental changes to ensure an appropriate development. Low temperatures are one of the most important environmental constraints that limit the development and survival of plants and determine their geographical distribution [1]. The stress induced by low temperatures also produces important economic losses, reducing the yield of agricultural crops every year. It is known that modest increases in the freezing tolerance of crop species would positively affect agricultural production [2]. Plants from temperate regions have evolved an adaptive response, known as a cold acclimation [1, 3], whereby they develop or increase their freezing tolerance after being

Dirk K. Hincha and Ellen Zuther (eds.), *Plant Cold Acclimation: Methods and Protocols*, Methods in Molecular Biology, vol. 1166, DOI 10.1007/978-1-4939-0844-8_8,

exposed during several days to low, nonfreezing temperatures (0–10 °C). Understanding the molecular mechanisms underlying this response is essential to conceive how plants grow and develop under adverse conditions originated by abiotic stresses and to generate new biotechnological strategies to improve crop tolerance to freezing temperatures and other related stresses such as drought and high salt.

Genetic analysis is a classical and powerful tool for identifying genes implicated in a given physiological process. In the case of freezing tolerance, the identification and characterization of mutant plants with altered freezing tolerance before and/or after cold acclimation have been carried out essentially in *Arabidopsis*, a model plant that is able to acclimate to low temperature increasing its constitutive freezing tolerance. Its small genome, the first to be sequenced in plants, together with its physiological characteristics, facilitates the subsequent molecular identification and characterization of the mutated genes. The most commonly used mutants in the screenings were generated by *e*thyl *me*thane*s*ulfonate (EMS), an organic compound that randomly produces nucleotide substitutions in the DNA [4–6]. For instance, Warren et al. [7] identified several *Arabidopsis* EMS mutants, termed *sensitivity to freezing* (*sfr*), that showed reduced freezing tolerance compared with wild-type (WT) plants. Consistent with the expectations that *sfr* should be loss-of-function mutations, most of them were recessive. Seven *sfr* mutants were nonallelic and only acquired partial freezing tolerance after cold acclimation. A preliminary study revealed that four *sfr* mutations, *sfr3*, *sfr4*, *sfr6*, and *sfr7*, reduced or blocked anthocyanin accumulation during this adaptive response. *sfr4* mutant was also impaired in cold-induced accumulation of sucrose and glucose levels, and both *sfr4* and *sfr7* mutants showed abnormal fatty acid composition when cold acclimated [8]. In another study [9], Xin and Browse identified several *Arabidopsis* EMS mutants with increased freezing tolerance. One of them, *eskimo1 (esk1)*, presented an increase in both constitutive freezing tolerance and cold acclimation capacity. *esk1* was originated by a single recessive mutation in the *AT3G55990* locus that produced elevated proline levels but did not generate constitutive expression of cold-regulated genes. Finally, Llorente and colleagues [10] identified *freezing sensitive 1* (*frs1*), an *Arabidopsis* EMS mutant that exhibited decreased constitutive freezing tolerance and capacity to cold acclimate. Complementation analysis revealed that *frs1* mutation was a new allele of *ABA3*, supporting that ABA is essential for full development of cold acclimation and for constitutive freezing tolerance in *Arabidopsis*.

In this chapter, we describe a simple and precise protocol for evaluating the freezing tolerance of both non-acclimated and cold-acclimated *Arabidopsis* plants. The protocol is suitable for screening mutants generated by EMS, fast neutron (FN), or T-DNA insertions and can be carried out with plants grown on media, in Petri dishes, or on soil, in pots. Important aspects, depending on searching for mutants with increased (tolerant) or decreased (sensitive) freezing tolerance, are also detailed. In addition, this protocol can also be used in reverse genetic studies to determine the involvement of a gene of interest in freezing tolerance and to assess the effect that different treatments may produce on the tolerance of *Arabidopsis* to freezing temperatures.

2 Materials

2.1 Plant Material

WT seeds of the appropriate ecotype, mutagenized M_2 families or M_2 pools, depending if screening for sensitive or tolerant mutants, respectively (*see* **Note 1**), and seeds of previously reported tolerant and/or sensitive freezing mutants to be used as controls in the screenings [7, 9].

2.2 Plate Assay

1. 1.5 mL Eppendorf or 50 mL Falcon tubes.
2. MS growth media (0.5× Murashige and Skoog basal salt mix; 2.5 mM *m*orpholino *e*thane*s*ulfonic acid (MES), pH 5.7; 0.8 % agar).
3. Amphotericin B (final concentration 2.5 mg/L).
4. Round plates (Ø 15 cm).
5. 3 M Micropore tape.
6. Filter paper or nylon mesh.
7. Bell jar.
8. Bleach.
9. HCl.
10. Forceps.
11. Liquid nitrogen.
12. Mortar and pestle.
13. Spoon.

2.3 Soil Assay

1. Peat substrate.
2. Vermiculite.
3. Clay pots (Ø 10 cm).
4. Trays.
5. Plastic film.

2.4 Growth Chambers and Other Equipment

1. Plant growth chamber set at 20–22 °C with cool-white light (100 μE/m^2/s).
2. Plant growth chamber set at 4 °C with cool-white light (50 μE/m^2/s).

3. Plant growth chamber with a range of programmable temperatures from 4 °C to –14 °C with lights off.
4. Cold room.
5. Autoclave.
6. Fume hood.
7. Sterile hood.

3 Methods

3.1 Seed Mutagenesis

Protocols to obtain EMS or FN mutagenized seeds have been previously described [5, 11]. T-DNA mutant collections can be generated as reported [12]. EMS and FN mutagenized seeds from different ecotypes are also commercially available at Lehle seeds (http://www.arabidopsis.com), while T-DNA mutant collections can be ordered at the European Arabidopsis Stock Centre (NASC). Generation of M_2 families and M_2 pools from M_1 mutagenized seeds has already been communicated [4] (*see* **Note 1**).

3.2 Screening Using Plates

We strongly recommend vapor-phase seed sterilization using chlorine gas (*see* **Note 2**).

3.2.1 Seed Sterilization and Plating

1. Put WT, control, and mutagenized seeds into appropriate tubes depending on number (Eppendorf or Falcon tubes are suitable).
2. Open the tubes into a hermetic bell jar placed in a fume hood.
3. Generate chlorine gas by combining 100 mL of bleach and 3 mL of HCl in a 200 mL glass placed inside the jar.
4. Close the jar, and let the seeds be exposed to chlorine gas for 3 h.
5. Open the jar inside the fume hood, and air-ventilate the seeds for 15 min before closing the tubes.
6. Cut filter paper pieces according to the plate size, and autoclave (*see* **Note 3**).
7. Place sterile filter papers on MS plates by using sterilized forceps.
8. Distribute sterilized seeds over the sterile papers (*see* **Note 4**).
9. For seed stratification seal plates with 3M Micropore tape and transfer them to a cold room (4 °C) under darkness for 2 days.
10. Transfer plates to a growth chamber, and let seeds germinate and develop for 12 days at 20–22 °C under long-day conditions (16-h light/8-h darkness).

3.2.2 Freezing Assay

1. Transfer plates to the growth chamber for freezing assay (*see* **Notes 5** and **6**), and expose seedlings to freezing temperatures under dark conditions. Appropriate temperatures should be empirically established depending on the accessions employed, the type of seedlings used in the screening (non-acclimated or cold-acclimated), and the searched mutants (tolerant or sensitive to freezing) (*see* **Notes 7** and **8**) (Fig. 1).
2. Prepare fresh ice chips by vaporizing sterile distilled water in a mortar containing liquid nitrogen. Grind the ice to a fine powder, and gently apply over the seedlings homogeneously when temperature decreases to −2 °C. Close the plates, and let the program finish (*see* **Note 9**).
3. After the freezing assay, when seedlings are exposed to 4 °C and media is still frozen, move plates to a sterile hood and,

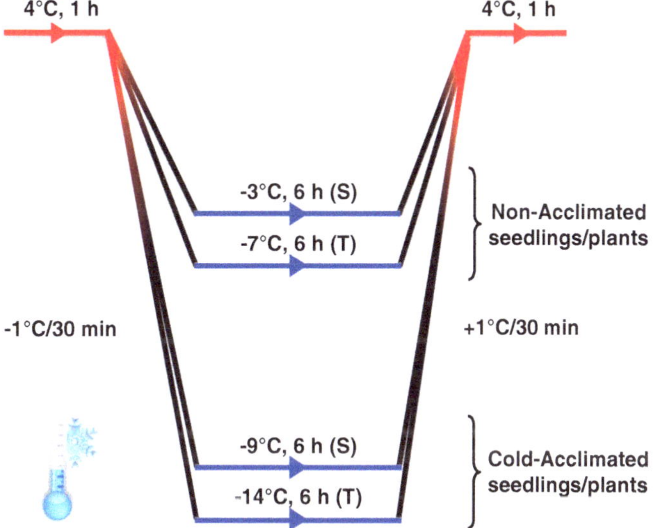

Fig. 1 Schematic representation of the freezing program used for the screenings. In all cases, before being subjected to freezing temperatures, seedlings and plants are exposed for 1 h to 4 °C in the freezing chamber. Then, temperature is progressively decreased (−1 °C/30 min) until reaching the desired freezing temperature. As an example, the two different temperatures we generally use to screen for non-acclimated seedlings or plants (−3 and −7 °C) and the two temperatures we generally use to screen for cold-acclimated seedlings or plants (−9 and −14 °C) are shown. −3 and −9 °C are employed when looking for sensitive mutants (S), while −7 and −14 °C when looking for tolerant ones (T). After exposing plants to the appropriate freezing temperature for 6 h, temperature is gradually increased to 4 °C (+1 °C/30 min). One hour later, plants are transferred to 20 °C under long-day light regime for recovering and subsequent survival evaluation

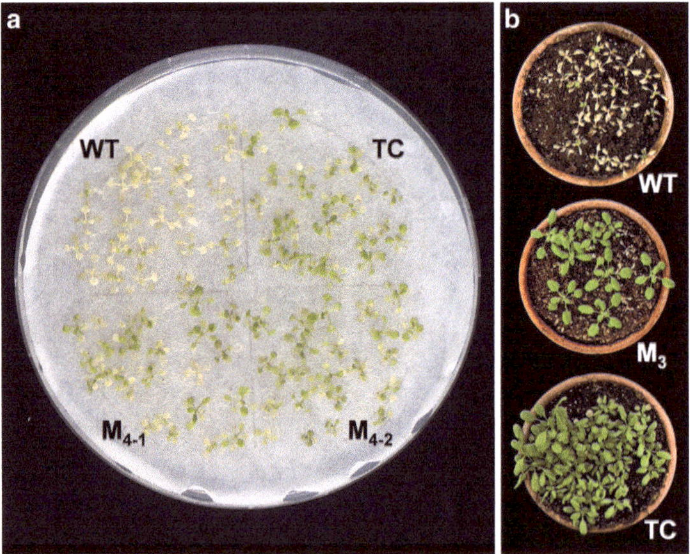

Fig. 2 Identification of *Arabidopsis* mutants with increased freezing tolerance before and after cold acclimation. (**a**) Non-acclimated seedlings 1 week after being frozen on plate at −7 °C for 6 h. Col-0 (WT), tolerant control (TC), and two M_4 homozygous families for two tolerant mutations are shown. (**b**) Cold acclimated plants 1 week after being frozen on pot at −14 °C for 6 h. Col-0 (WT), tolerant control (TC), and an M_3 family segregating for a tolerant mutation are shown

using sterilized forceps, transfer carefully the filter papers with the frozen seedlings to new plates containing MS media supplemented with amphotericin B (*see* **Notes 10** and **11**).

4. Move seedlings to a growth chamber, and let them to recover at 20–22 °C for 1 week under long-day light regime. An example of recovered seedlings is shown in Fig. 2a (*see* **Note 12**).
5. When screening for tolerant mutants, transfer surviving M_2 seedlings to soil and allow them to reproduce for phenotype confirmation in a secondary screening with the corresponding M_3 families (*see* Subheading 3.4).
6. If screening for sensitive mutants, repeat the freezing assays with additional seeds from the selected M_2 lines (*see* **Note 13**), confirm the survival rate, and proceed to phenotype confirmation (*see* Subheading 3.4).

3.3 Freezing Using Pots

3.3.1 Pot Preparation and Growing Conditions

1. Mix peat substrate with vermiculite in a 3:1 ratio, add one volume of water per 3 volumes of mix, and sterilize at 120 °C for 20 min.
2. Fill pots homogeneously with sterile soil avoiding leaving air bubbles (*see* **Note 14**).
3. Distribute the seeds on the soil as separated as possible (*see* **Note 15**).

4. Place the pots into trays, cover it with plastic film for humidity maintenance during the first days after germination, and transfer them to a cold room (4 °C) under darkness for 2 days for stratification.
5. Move the trays containing the pots to a growth chamber set at 20–22 °C with a long-day light regime, and allow seeds to germinate and develop for 2 weeks.

3.3.2 Freezing Assay

1. Transfer individual pots to the growth chamber for freezing assay (*see* **Notes 6** and **16**), and expose plants to freezing temperatures.
2. Screening conditions are essentially the same as those described for seedlings on plates (*see* **Notes 7**, **8**, and **17**) (Fig. 1).
3. After the freezing assay, move plants to a growth chamber and let them to recover at 20–22 °C for 1 week under long-day light regime (Fig. 2b).
4. When screening for tolerant mutants, allow surviving M_2 plants to reproduce for phenotype confirmation in a secondary screening with the corresponding M_3 families (*see* below).
5. If screening for sensitive mutants, repeat the freezing assays with additional seeds from the selected M_2 lines (*see* **Note 13**), confirm the survival rate, and proceed to phenotype confirmation in a secondary screening (*see* below).

3.4 Confirmation of Mutant Phenotypes

1. When screening for freezing-tolerant mutants, screen about 100 seedlings or plants from each generated M_3 family for their freezing tolerance, and calculate the survival rate.
2. M_3 families from freezing tolerant M_2 seedlings or plants showing survival rates of 100 % are likely produced by a single mutation in homozygosis. Families showing a 3:1 or 1:3 tolerant/sensitive ratio would be generated by a single dominant or recessive mutation in heterozygosis, respectively (*see* **Note 18**). In these cases, select and reproduce at least ten freezing-tolerant seedlings or plants to obtain the corresponding M_4 families. The screening of these will allow the identification of the homozygous mutant lines (*see* **Note 19**).
3. If screening for freezing-sensitive mutants, collect seeds from at least ten M_2 plants from each selected M_2 family to obtain the corresponding M_3 ones.
4. The screening of around 100 seedlings or plants from each generated M_3 family will allow the identification of homozygous mutant lines for the phenotype selected (those showing survival rates of 0 %).

3.5 Molecular Identification of Selected Mutations

The number of mutations that a selected homozygous mutant line contains in its genome varies depending on the method employed for seed mutagenesis. Before proceeding to the molecular identification of the mutations, mutant lines should be backcrossed several times in order to clean up their genetic backgrounds of mutations not related to the freezing phenotype of interest.

Until 2 or 3 years ago, the molecular identification of EMS- or FN-produced mutations causing a phenotype of interest in *Arabidopsis* was performed through a map-based cloning approach. This method is based on crossing a mutant plant of interest with a WT of a different accession. Subsequent polymorphism analysis determines the chromosome region where the mutation is located and, finally, by complementation experiments, the mutagenized locus is identified [13]. Unfortunately, the enormous amount of time required to identify a mutation causing a phenotype of interest by this method constitutes an important obstacle when planning a forward genetic screening. During the last 2 years, next-generation genomic sequencing technologies have been applied for the rapid and precise molecular identification of mutations in different *Arabidopsis* ecotypes. For instance, Austin and colleagues [14] have described a rapid and robust method for mapping mutations, independently of their chromosomal location, by sequencing only a small pooled M_2 population. The protocol was able to identify a highly restricted region containing very few SNP candidates that can be easily validated using standard reverse genetic techniques. On the other hand, Uchida et al. [15] have been able to identify a mutated locus in a non-reference *Arabidopsis* accession, i.e., whose genome is not publicly available, by only one round of genome sequencing.

When screening mutant collections generated by T-DNA insertions, identifying the site of insertion in the genome is commonly performed using an adapter ligation-mediated PCR protocol [16]. This method consists of three steps and takes about 3 weeks to be completed. First, an adapter is ligated to genomic DNA after digestion with a restriction enzyme. Then, by using specific primers to the adapter and T-DNA, the T-DNA/genomic DNA junction is amplified by PCR. Finally, sequencing the T-DNA/genomic junction allows mapping the T-DNA location in the genome.

4 Notes

1. M_2 families collected by pedigreeing are the recommended material to screen for sensitive mutants. Since sensitive mutants will not survive the screening, you must ensure a high (>200) number of mutant seeds for each family. When screening for tolerant mutants, M_2 pools are the material of choice. Generation of M_2 families and M_2 pools has been reported in detail [4].

2. Seeds can also be sterilized with bleach as described [17].
3. Wrap paper pieces with aluminum paper. For a correct sterilization, do not autoclave more than ten paper pieces together. Other supports, such as a nylon mesh, can also be used.
4. The seed number per plate depends on the germination rate and on the type of screening that is going to be performed. When screening for freezing-tolerant mutants, a high number of pooled M_2 seeds can be plated (~300 seeds/Ø15 cm plate). When the screening is performed to identify freezing-sensitive mutants, the number of seeds plated from each M_2 family should allow establishing a significant sensitive/tolerant segregation (~100 seeds/Ø15 cm plate).
5. For evaluation of freezing tolerance after cold acclimation, before freezing, plates should be transferred to a growth chamber set at 4 °C for 5 days under long-day conditions to ensure the full development of the adaptive response.
6. We strongly recommend transferring seedlings or plants to the appropriate growth chamber for the freezing assay always at the same time of the day, since the expression of several cold-regulated genes involved in cold acclimation is subjected to circadian regulation [18, 19].
7. Appropriate freezing temperatures to evaluate the tolerance of non-acclimated or cold-acclimated seedlings depend on the type of screening to be performed (searching for freezing-tolerant or -sensitive mutants) and should be previously established in each case using WT seedlings and previously reported tolerant and/or sensitive mutants that will act as positive controls. If searching freezing-tolerant mutants, the highest temperature that produces 0 % surviving seedlings should be used. On the contrary, when screening for freezing-sensitive mutants, the lowest temperature that allows 100 % seedling survival is the convenient one. The time that seedlings should be exposed to the appropriate freezing temperatures must be determined at the same time as freezing temperatures. Different temperatures are used for the evaluation of non-acclimated and cold-acclimated seedlings, the latter always requiring lower temperatures (~2–3 °C). Optimally, freezing temperatures must be gradually reached (−1 °C/30 min) starting from 4 °C (Fig. 1).
8. During the freezing assay, it is critical that the temperature inside the chamber is homogeneous in such a way that all seedlings being screened are exposed to the same conditions.
9. Ice chips are ice nucleation sites that favor freezing homogeneity in all seedlings of the plate.
10. Seedlings should be transferred to new plates since freezing temperatures depolymerize the growth media.

11. Amphotericin B is a polyene antifungal drug that helps to minimize plate contaminations.
12. In our experience, 1 week of recovery is enough to establish if seedlings have survived to the freezing treatment or not. Fungal contaminations usually appear when longer recovering times are allowed.
13. We recommend selecting M_2 lines with a 3:1 or a 1:3 sensitive/tolerant segregation, indicating that the sensitive phenotype is produced by a single dominant or recessive mutation, respectively. Phenotypes produced by single mutations are preferred because these mutations are easy to map and molecularly identify.
14. In our hands, clay pots allow water transpiration and work better than plastic pots.
15. The seed number per pot depends on the germination rate and on the type of screening that is going to be performed. When screening for freezing-tolerant mutants, a high number of pooled M_2 seeds can be sown (~60 seeds/Ø 10 cm pot). When the screening is performed to identify freezing-sensitive mutants, the number of seeds sown from each M_2 family should allow to establish a significant sensitive/tolerant segregation (~40 seeds/Ø 10 cm pot).
16. When screening for mutants with altered freezing tolerance after cold acclimation, plants should be previously exposed to 4 °C during 1 week to ensure the adaptive response.
17. When freezing plants grown on soil, the addition of ice chips is not necessary because freezing occurs very homogeneously on the surface of the pot.
18. Other segregations will suggest that the freezing mutant phenotype is originated by more than one mutation.
19. When screening for freezing-tolerant mutants by pooling, mutant seedlings or plants containing the same mutation may be selected. Allelism tests should then be performed between the identified mutant lines.

References

1. Levitt J (1980) Responses of plants to environmental stresses: chilling, freezing and high temperatures stresses, 2nd edn. Academic, New York
2. Steponkus P, Uemura M, Joseph RA et al (1998) Mode of action of the *COR15a* gene on the freezing tolerance of *Arabidopsis thaliana*. Proc Natl Acad Sci U S A 95: 14570–14575
3. Guy CL (1990) Cold acclimation and freezing stress tolerance: role of protein metabolism. Annu Rev Plant Physiol Plant Mol Biol 41: 187–223
4. Lightner J, Caspar T (1998) Seed mutagenesis of *Arabidopsis*. In: Martínez-Zapater JM, Salinas J (eds) Methods in molecular biology, vol 82. Humana Press, Totowa, NJ, pp 91–103

5. Kim YS, Schumaker KS, Zhu JK (2006) EMS mutagenesis of *Arabidopsis*. In: Salinas J, Sanchez-Serrano JJ (eds) Methods in molecular biology, vol 323. Humana Press, Totowa, NJ, pp 101–103
6. Weigel D, Glazebrook J (2006) EMS mutagenesis of *Arabidopsis* seed. CSH Protoc 28. doi:10.1101/pdb.prot4621
7. Warren G, McKown R, Marin AL et al (1996) Isolation of mutations affecting the development of freezing tolerance in *Arabidopsis thaliana* (L.) Heynh. Plant Physiol 111: 1011–1019
8. McKown R, Kuroki G, Warren G (1996) Cold responses of *Arabidopsis* mutants impaired in freezing tolerance. J Exp Bot 47:1919–1925
9. Xin Z, Browse J (1998) *eskimo1* mutants of *Arabidopsis* are constitutively freezing-tolerant. Proc Natl Acad Sci U S A 95:7799–7804
10. Llorente F, Oliveros JC, Martínez-Zapater JM et al (2000) A freezing-sensitive mutant of *Arabidopsis*, *frs1*, is a new *aba3* allele. Planta 211:648–655
11. Koornneef M, Dellaert LWM, van der Veen JH (1982) EMS- and radiation-induced mutation frequencies at individual loci in *Arabidopsis thaliana* (L.) Heynh. Mutat Res 93:109–123
12. Alonso JM, Stepanova AN (2003) T-DNA mutagenesis in *Arabidopsis*. In: Grotewold E (ed) Methods in molecular biology, vol 236. Humana Press, Totowa, NJ, pp 177–188
13. Jander G (2006) Gene identification and cloning by molecular marker mapping. In: Salinas J, Sanchez-Serrano JJ (eds) Methods in molecular biology, vol 323. Humana Press, Totowa, NJ, pp 115–126
14. Austin RS, Vidaurre D, Stamatiou G et al (2011) Next-generation mapping of *Arabidopsis* genes. Plant J 67:715–725
15. Uchida N, Sakamoto T, Kurata T et al (2011) Identification of EMS-induced causal mutations in a non-reference *Arabidopsis thaliana* accession by whole genome sequencing. Plant Cell Physiol 52:716–722
16. O'Malley RC, Alonso JM, Kim CJ et al (2007) An adapter ligation-mediated PCR method for high-throughput mapping of T-DNA inserts in the *Arabidopsis* genome. Nat Protoc 2: 2910–2917
17. McCourt P, Keith K (1998) Sterile techniques in *Arabidopsis*. In: Martínez-Zapater JM, Salinas J (eds) Methods in molecular biology, vol 82. Humana Press, Totowa, NJ, pp 13–17
18. Mikkelsen MD, Thomashow MF (2009) A role for circadian evening elements in cold-regulated gene expression in *Arabidopsis*. Plant J 60: 328–339
19. Dong MA, Farré EM, Thomashow MF (2011) Circadian clock-associated 1 and late elongated hypocotyl regulate expression of the c-repeat binding factor (CBF) pathway in *Arabidopsis*. Proc Natl Acad Sci U S A 108:7241–7246

Chapter 9

Infrared Thermal Analysis of Plant Freezing Processes

Gilbert Neuner and Edith Kuprian

Abstract

Infrared thermal analysis is an invaluable technique to study the plant freezing process. In the differential mode infrared thermal analysis allows to localize ice nucleation and ice propagation in whole plants or plant samples at the tissue level. Ice barriers can be visualized, and supercooling of cells, tissues, and organs can be monitored. Places where ice masses are accommodated in the apoplast can be identified. Here, we describe an experimental setting developed in the laboratory in Innsbruck, give detailed information on the practical procedure and preconditions, and give additionally an idea of the problems that can be encountered and how they by special precautions may be overcome.

Key words Freezing stress, Ice barriers, Ice nucleation, Ice propagation, Frost, Deep supercooling

1 Introduction

While symplastic (intracellular) ice formation has always lethal consequences for plant cells, apoplastic (extracellular) ice formation can safely be survived down to a certain freezing temperature. Where ice forms and how ice propagates and is accommodated within the apoplast but also how ice formation can be avoided in certain cells, tissues, and organs are essential aspects of plant cold hardiness and are all important factors that affect the ability of a plant to survive freezing. These traits may be even as important as the ability to withstand the dehydration stresses associated with ice formation [1]. As far as known the response to the presence of ice within a plant's tissue can be complex and quite divers [2]. A deeper understanding of the control of apoplastic ice formation processes will lead to new strategies and technologies for improving plant cold hardiness [2].

Below 0 °C without ice nucleation water remains in the supercooled state. Once ice has nucleated it readily propagates at high rates of up to 27 cm/s [3–5] throughout all plant parts that are not protected by an ice barrier. A precondition for ice nucleation is the formation of an active ice crystal nucleus. An ice crystal nucleus is a

Dirk K. Hincha and Ellen Zuther (eds.), *Plant Cold Acclimation: Methods and Protocols*, Methods in Molecular Biology, vol. 1166, DOI 10.1007/978-1-4939-0844-8_9, © Springer Science+Business Media New York 2014

cluster of regularly ordered water molecules. The cluster size is temperature dependent. With decreasing temperature the critical size decreases successively increasing the probability of ice nucleation [6, 7]. There are principally two ways how ice crystal nuclei can form: (1) autonomously, i.e., homogeneous ice nucleation, or (2) water molecules are forced to form an ice crystal nucleus by an ice nucleation-active (INA) substance (heterogeneous ice nucleation). Apoplastic ice nucleation is considered to occur heterogeneously, and symplastic ice nucleation very likely is of homogeneous nature as it can be close to the homogeneous ice nucleation temperature of water which is −38.5 °C [6, 7].

When below 0 °C crystallization of water occurs liberation of heat (enthalpy of fusion) produces the so-called freezing exotherm, i.e., a sudden rise in sample temperature. Depending on the amount of water freezing this can be several K (e.g., 6 K for *Rhododendron ferrugineum* leaf [8]) or can be close to the thermometric resolution limit of current instrumentation as for example observed during the so-called low-temperature freezing exotherms produced by symplastic freezing of deeply supercooled cells (e.g., xylem parenchyma cells [8]).

Measurement of freezing exotherms can principally be conducted punctually by the use of thermocouples. By differential thermal analysis (DTA [9]) very small freezing events get detectable at low noise, as the temperature of a dry and dead reference sample is subtracted from the measured alive plant sample. More sophisticated is the employment of high-resolution infrared thermography as a two-dimensional thermal image of the sample can be obtained which allows to localize ice nucleation sites and to monitor ice propagation throughout the plant [10–13]. Again by using the infrared camera in the differential imaging mode (infrared differential thermal analysis, IDTA) the resolution of the thermal images obtained during the freezing process can be significantly increased [3–5, 14–16]. IDTA is based on the subtraction of a reference image, captured just before the occurrence of freezing, from the sequence of images during freezing that then show only the changes in temperatures during freezing of water in the plant sample [3]. By this background correction temperature fluctuations and thermal gradients on the image are canceled out and smallest changes in temperature caused by freezing of water in the plant tissue can be visualized.

2 Materials

For measurement of IDTA during the freezing process a high-resolution infrared camera is necessary that is suitable for operation temperatures in the sub-zero temperature range. Additionally a temperature-controlled freezing chamber large enough to accommodate the infrared camera together with the plant sample is necessary. In the following the measurement procedure as developed in the laboratory in Innsbruck is described.

2.1 Infrared Camera

A digital infrared camera model ThermaCAM™ S60 (FLIR Systems AB, Danderyd, Sweden) is employed for measurement of sequences of two-dimensional infrared images of plant samples. The thermal resolution is 0.08 K. The camera is equipped with a close-up lens (LW64/150). By this a spatial resolution on the thermal images of 200 μm is achieved. The infrared camera is connected by a FireWire/IEEE1394 interface with a control computer. The Therma CAM™ S60 has a maximum time resolution of 25 images/s. Recording at maximum time resolution results in large amounts of data (3.8 MB/s, 13.7 GB/h). For the original infrared video records a powerful PC providing sufficiently high disk space is necessary. IDTA images can be extracted by subtraction of a reference image from the original infrared image which is performed during data analysis with the ThermaCAM™ Researcher software package (FLIR Systems AB, Danderyd, Sweden). Absolute tissue temperature is recorded with thermocouples (*see* **Note 1**) placed close to the surface of the analyzed plant samples. Thermocouples are connected to a data logger (CR10X, Campbell Scientific, Loughborough, UK). The lowest freezing temperature that can be studied is currently set by the minimum operation temperature of the ThermaCAM™ which is −25 °C.

2.2 Thermally Insulated Camera Housing

In order to protect the infrared camera from fast temperature changes and potential water condensations (*see* **Note 2**), we use a thermally insulated housing for the camera, when used inside the freezing compartment. The housing is made out of a 21 × 25 × 31 cm Plexiglas box which is thermally insulated inside with 3 cm thick Styrofoam plates. The box has a hole that fits to the close-up lens of the infrared camera and a lid on top to insert and remove the infrared camera. A second hole opposite to the opening for the close-up lens allows inserting FireWire connection and electric currency cables.

2.3 Temperature-Controlled Freezing Chamber

For freezing treatment of the plant samples together with the infrared camera a sufficiently spacious fully temperature-controlled freezing chamber is necessary (*see* **Note 3**). This is realized in Innsbruck by a computer-controlled commercial chest freezer [17]. The freezing compartment has a volume of 141.9 L (43 × 50 × 66 cm). The freezing device needs to be fully temperature controlled and should allow a precise setting of cooling and warming rates (*see* **Note 4**). The control technique should keep temperature oscillations at a minimum, preferentially less than 0.2 K (*see* **Note 5**).

2.4 Sample Holder

The entire plant sample investigated must be in a focusing plane to ensure that ice formation can be monitored with high resolution. The sample holder plate (approximately 15 × 15 cm) should be impermeable for infrared radiation and provide a homogeneous thermal background. Usually an ordinary plastic plate is sufficient. For convenient distance adjustments this plate may be placed on a laboratory lifting plate. The actual measurement area of the infrared camera equipped with the close-up lens is approximately 3.7 × 4.6 cm.

3 Methods

3.1 Sample Preparation

1. In case that detached plant parts have to be used, it must be considered that the sample size has strong effects on ice nucleation temperature. The size of detached plant samples should not be too small (*see* **Note 6**).
2. For identification of action sites of intrinsic INA substances plant samples should have a dry surface. Water on the plant surface can freeze extrinsically at first and can then potentially trigger consequent intrinsic freezing of the sample [18].
3. For control of ice nucleation temperature application of droplets of INA bacteria is recommended (*see* **Note 7**).
4. Plant samples must be mounted on the sample holder in such a way that all plant parts investigated are exposed in the focusing plane of the infrared camera, particularly when the infrared camera is used with the close-up lens. For small plant parts such as single leaves this is not problematic (Fig. 1a). For leafy twigs or whole herbaceous plants this can be done by fanning out the plant samples on the plain sample holder plate (*see* **Note 8**).

3.2 Setup of Equipment

1. For recording of the absolute temperature it is advisable to use a set of at least four thermocouples as a reference and additional temperature control. The thermocouples are fixed onto the sample holder close to the investigated plant samples (*see* **Note 1**) with the double-sided glue tape but additionally outside the measurement frame by further pieces of adhesive tape to securely hold them in place (*see* **Note 9**).
2. Altogether, the sample holder plate including the mounted plant samples and thermocouples on top of the laboratory lifting plate is then placed on the bottom of the freezing compartment of the freezing equipment.
3. The infrared camera is inserted into the camera housing, and the lid is closed. The infrared camera is then turned on to record the live image.
4. Altogether, camera plus housing is then put upside down into the freezing compartment of the freezing equipment and is

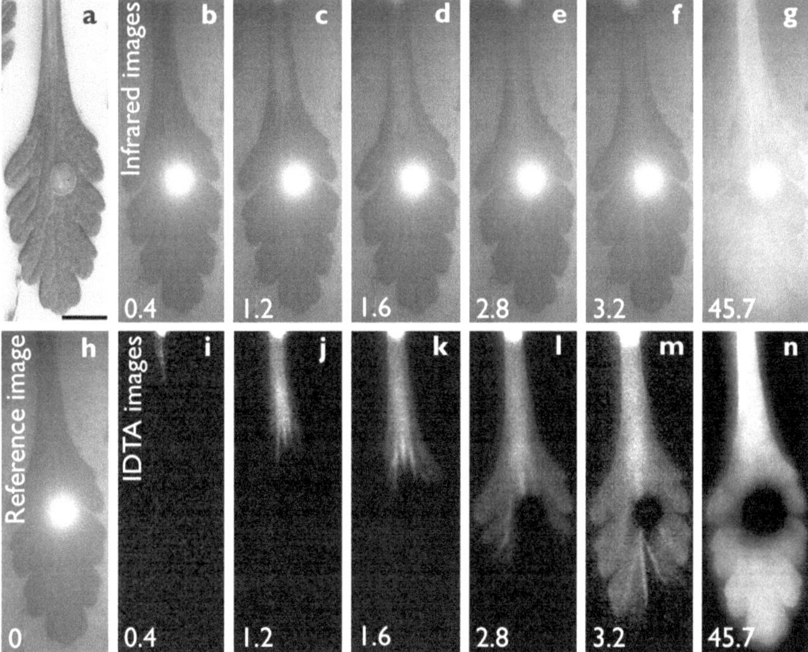

Fig. 1 (**a**) On the digital black and white image of a single, detached leaf of *Senecio incanus* a 50 μl droplet of INA bacteria suspension positioned on the upper leaf surface can be seen (*horizontal black bar* width is 0.5 cm). Ice nucleated initially in the droplet on the surface at −4.3 °C, but this surface ice did not enter the leaf. Freezing is visualized by a brightening, while unfrozen areas remain black. At −4.5 °C a second ice nucleation event in the petiole of the leaf initiated an ice wave via the vascular system into the leaf blade. The original sequence of infrared images during this second freezing event (**b**–**g**) gives only a blurred picture. When these infrared images are referenced to the image immediately before the leaf freezing event (**h**), IDTA images (**i**–**n**) are obtained that show more details of the freezing process. The numbers in the *bottom left corner* of each image indicate the time in seconds after ice nucleation in the petiole of the leaf

positioned in a distance of approximately 9 cm to the plant sample by the help of two additional lifting plates.

5. To bring the point of interest into focus fine adjustments of the distance between the close-up lens of the infrared camera and the plant sample can then be made by the use of the lifting plate supporting the sample holder plate.
6. A single digital image of the sample setting should be recorded by the infrared camera.
7. The infrared camera is switched into the infrared mode. The recording conditions (time resolution, storage space for recordings) are set (*see* **Note 10**). The lid of the freezing compartment is closed.

3.3 Recording of Ice Formation Processes

1. The parameters of the freezing treatment have to be set. The experimental settings largely affect ice nucleation temperature, particularly the cooling rate, and should be selected to come close to natural night frost conditions (*see* **Notes 3** and **4**). After setting of all parameters the freezing treatment can be started.

2. As soon as 0 °C is reached the recording mode of the infrared camera is manually started (*see* **Note 11**).
3. After the programmed freezing treatment is finished the infrared recording is stopped, the lid of the freezing compartment is opened, and the whole equipment is allowed to thaw to room temperature.
4. The recorded video sequences of infrared images are scanned by the use of the ThermaCAM Researcher software package to elucidate video sequences that show freezing events. On the original infrared images freezing processes can be detected but they are often blurred (Fig. 1b–g). Once identified IDTA images can be extracted. IDTA is based on the subtraction of a reference image, captured just before the occurrence of freezing (Fig. 1h), from the sequence of images during freezing that then show only the changes in temperatures during freezing of water in the plant sample revealing much more details of the ice formation processes (Fig. 1i–n).

4 Notes

1. Thermocouples themselves can be a source of ice nucleation [9]. In case that ice nucleation sites in the plant sample should be identified it is hence advisable to avoid close contact of thermocouple solder junctions to the sample.
2. The housing ensures that the infrared camera is kept at a higher temperature than the surroundings. This prevents water condensation on the close-up lens which could lead to erroneous measurements.
3. Currently commercially available freezing devices are all convective systems. Convective freezing devices cannot simulate radiant heat loss and do not produce temperature gradients in plants during freezing as in nature [19]. As plants are not colder than air in these devices dew and ice deposit on the plant surface is absent. These conditions favor supercooling which should be kept in mind.
4. Moderate cooling rates below 0 °C should be selected in order to simulate natural night frosts. In nature cooling rates below 0 °C often are not faster than −2 K/h [19]. Higher cooling rates tend to provoke supercooling in the sample. Additionally, if frost survival is investigated exposure times and thawing rates have to be controlled. For plant samples with unknown frost survival mechanisms it is advisable to use a freezing protocol that eventually extends down to freezing temperatures where after the initial apoplastic ice wave additional freezing events at lower temperatures may be recorded. These are either further

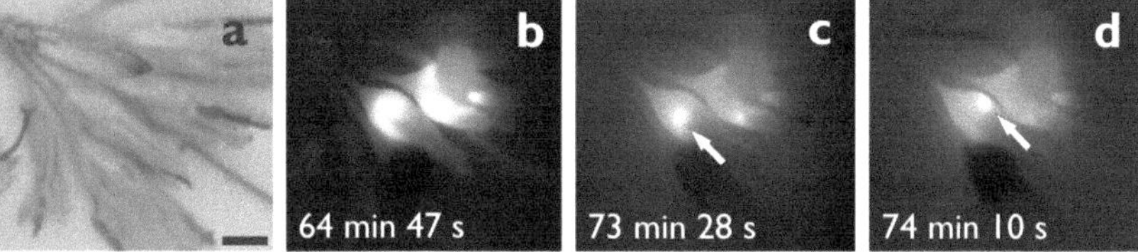

Fig. 2 (**a**) Digital black and white image of an infructescence of *Primula veris* before exposure to a controlled freezing treatment (*horizontal black bar* width is 0.5 cm). Initial ice nucleation in the supporting stem occurred at −3.2 °C. Single fruits and developing seeds inside the fruits froze separately at various lower freezing temperatures: IDTA images of the infructescence show freezing of (**b**) two single fruits at −10.7 °C and (**c**, **d**) separate freezing of developing seeds inside these fruits at around −12.0 °C (*white arrows*). Freezing is visualized by a brightening, while unfrozen areas remain black. The time span after ice nucleation in the supporting stem is indicated in the *bottom left corner* of each image

apoplastic freezing events occurring later due to ice barriers (Fig. 2) or freezing exotherms that originate from symplastic freezing.

5. Higher temperature oscillations can be buffered by placing the sample holder and the camera inside of a thermally insulated box inside the freezing compartment.
6. Sample size [16] and detachment [19] significantly influence the ice nucleation temperature. Generally spoken the smaller the sample size is down to lower freezing temperatures samples tend to supercool. For apple twigs only if sized 40 cm in length ice nucleation temperature recorded in the laboratory experiment on detached twigs was roughly equivalent to that measured in the field on intact trees [16].
   Several plants have structural ice barriers between different plant organs [8], and in some cases their role may only be studied in whole plants. Thermal ice barriers as found in alpine cushion plants [15] can only be measured on whole intact plants in experimental settings as natural as possible.
7. INA bacteria (e.g.: strain 5176 of *Pseudomonas syringae* has a high ice-nucleating activity, Deutsche Sammlung von Mikroorganismen und Zellkulturen GmbH (DSMZ), Braunschweig, Germany) can be applied as a bacterial suspension in 50 μl droplets (*see* Fig. 1). Depending on the question droplets can be deposited on intact, scratched, or cut plant surfaces.
8. For fanning out plant samples double-sided glue tape (5 cm width) has proved to be most convenient.
9. For documentation of thermocouple positioning it is advisable to make a photo of the experimental setting.
10. A measurement frequency of 10 images/s is mostly sufficient. When the maximum of 25 images s is recorded this results in a large amount of data (3.8 MB/s, 13.7 GB/h). Sufficient disk space on the computer must be provided.

11. During the recording mode it is advisable to regularly conduct an image matching, at least every half an hour. For an easier handling of the video material during subsequent analysis it is recommended to produce several consequent short videos rather than one single long video.

References

1. Wisniewski ME, Gusta LV, Fuller MP, Karlson D (2009) Ice nucleation, propagation and deep supercooling: the lost tribes of freezing studies. In: Gusta LV, Wisniewski ME, Tanino KK (eds) Plant cold hardiness: from the laboratory to the field. CAB International, Cambridge, pp 1–11
2. Gusta LV, Wisniewski ME, Trischuk RG (2009) Patterns of freezing in plants: the influence of species, environment and experimental procedures. In: Gusta LV, Wisniewski ME, Tanino KK (eds) Plant cold hardiness: from the laboratory to the field. CAB International, Cambridge, pp 214–223
3. Hacker J, Neuner G (2007) Ice propagation in plants visualized at the tissue level by infrared differential thermal analysis (IDTA). Tree Physiol 27:1661–1670
4. Hacker J, Neuner G (2008) Ice propagation in dehardened alpine plant species studied by infrared differential thermal analysis (IDTA). Arc Antarc Alp Res 40:660–670
5. Hacker J, Spindelböck J, Neuner G (2008) Mesophyll freezing and effects of freeze dehydration visualized by simultaneous measurement of IDTA and differential imaging chlorophyll fluorescence. Plant Cell Environ 31:1725–1733
6. Franks F (1985) Biophysics and biochemistry at low temperatures. Cambridge University Press, Cambridge
7. Chen S-H, Mallamace F, Mou C-Y, Broco M, Corsavo C, Faraone A, Liu L (2006) The violation of the Stokes-Einstein relation in supercooled water. Proc Natl Acad Sci U S A 103: 12974–12978
8. Sakai A, Larcher W (1987) Frost survival of plants. Responses and adaptation to freezing stress. In: Billings WD, Golley F, Lange OL, Olson JS, Remmert H (eds) Ecological studies, vol 62. Springer, New York
9. Burke MJ, Gusta LV, Quamme HA, Weiser CJ, Li PH (1976) Freezing and injury in plants. Annu Rev Plant Physiol Plant Mol Biol 27: 507–528
10. Wisniewski M, Lindow SE, Ashworth EN (1997) Observations of ice nucleation and propagation in plants using infrared video thermography. Plant Physiol 113:327–334
11. Lutze JL, Roden JS, Holly CJ, Wolfe J, Egerton JJG, Ball MC (1998) Elevated atmospheric [CO_2] promotes frost damage in evergreen tree seedlings. Plant Cell Environ 21:631–635
12. Ball MC, Wolfe J, Canny M, Hofmann M, Nicotra AB, Hughes D (2002) Space and time dependence of temperature and freezing in evergreen leaves. Funct Plant Biol 29: 1259–1272
13. Sekozawa Y, Sugaya S, Gemma H (2004) Observations of ice nucleation and propagation in flowers of Japanese Pear (*Pyrus pyrifolia* Nakai) using infrared video thermography. J Jpn Soc Hortic Sci 73:1–6
14. Neuner G, Xu BC, Hacker J (2010) Velocity and pattern of ice propagation and deep supercooling in woody stems of *Castanea sativa*, *Morus nigra* and *Quercus robur* measured by IDTA. Tree Physiol 30:1037–1045
15. Hacker J, Ladinig U, Wagner J, Neuner G (2011) Inflorescences of alpine cushion plants freeze autonomously and may survive subzero temperatures by supercooling. Plant Sci 180: 149–156
16. Pramsohler M, Hacker J, Neuner G (2012) Freezing pattern and frost killing temperature of apple (*Malus domestica*) wood under controlled conditions and in nature. Tree Physiol 32:819–828
17. Neuner G, Buchner O (1999) Assessment of foliar frost damage: a comparison of in vivo chlorophyll fluorescence with other viability tests. J Appl Bot 73:50–54
18. Pearce RS (2001) Plant freezing and damage. Ann Bot 87:417–424
19. Neuner G, Hacker J (2012) Ice formation and propagation in alpine plants. In: Lütz C (ed) Plants in alpine regions: cell physiology of adaptation and survival strategies. Springer, Wien, pp 163–174

Chapter 10

Cryo-Scanning Electron Microscopy to Study the Freezing Behavior of Plant Tissues

Seizo Fujikawa and Keita Endoh

Abstract

A cryo-scanning electron microscope (cryo-SEM) is a valuable tool for observing bulk frozen samples to monitor freezing responses of plant tissues and cells. Here, essential processes of a cryo-SEM to observe freezing behaviors of plant tissue cells are described.

Key words Cryo-scanning electron microscope (cryo-SEM), Ice crystal, Extracellular freezing, Intracellular freezing, Cryo-fixation, Recrystallization, Cooling rate

1 Introduction

A cryo-scanning electron microscope (cryo-SEM) or a low-temperature SEM has been used to observe bulk biological samples under a freezing condition. Since early reports on the development of a simple cryo-SEM, in which the SEM was equipped with only a cold stage in the SEM column [1], the instruments have been improved in step-by-step fashion [2–4]. Various cryo-SEMs with their own unique combination of features are now available.

A cryo-SEM has been used to observe hydrated structures of biological materials including plant tissues, in which water in samples is kept from conversion to ice by cryo-fixation using very rapid freezing (for recent review, *see* ref. 5). A cryo-SEM has also been used to observe the distribution of water in plant tissues [6–9] as well as the distribution of contents dissolved in water [10]. Moreover, a cryo-SEM has been used to analyze effects of the freeze-drying process in relation to morphological changes of plant tissues [11].

A cryo-SEM is particularly useful for observing responses in plant tissue cells to freezing. Studies using a cryo-SEM have shown interaction between extracellular ice crystals and cells in the fruit body of mushrooms [12, 13], in leaves of cereals [14, 15], in leaves

Dirk K. Hincha and Ellen Zuther (eds.), *Plant Cold Acclimation: Methods and Protocols*, Methods in Molecular Biology, vol. 1166, DOI 10.1007/978-1-4939-0844-8_10, © Springer Science+Business Media New York 2014

of *Arabidopsis* [16], in evergreen leaves of trees [17–19], in leaves of freezing-sensitive plant species [20, 21] and in bark tissues of fruit trees [22]. Direct observation of frozen tissue cells by a cryo-SEM has provided information on freezing responses of xylem parenchyma cells in several tree species which adapt to subfreezing temperatures by deep supercooling [23–31] and information on freezing behavior of tissue cells in dormant buds in trees that adapt to subfreezing temperatures by extraorgan freezing [32].

Here, we introduce our manuals in addition to general manuals for observing freezing responses of plant tissue cells by cryo-SEM. Additionally, we recommend the use of freeze-fracture replica electron microscopy in combination with cryo-SEM for understanding freezing responses of plant tissue cells in more detail (for the manual of diverse freeze-replica technique, *see* ref. 33).

2 Materials

2.1 Main Apparatus

1. Cryo-SEM: Although there are many different types of commercially available cryo-SEM, a cryo-SEM generally consists of a specimen preparation chamber and SEM column equipped with cold stages. A pre-evacuation chamber is connected to the specimen preparation chamber in order to transfer frozen samples from outside to the specimen preparation chamber. The specimen preparation chamber is equipped with not only a cold stage but also a cold knife for fracturing frozen samples and a metal-coating system for coating freeze-fracture faces to facilitate radiation of secondary electrons as well as to inhibit electric charging. The specimen preparation chamber is directly connected to the SEM column with another cold stage on which frozen samples are observed with cover by a cold trap for decontamination of samples. In different types of cryo-SEM, the cold stage is cooled by a connected copper braid cooled by liquid nitrogen (LN_2) [34], by a piped system for circulating LN_2 [3] or by a Joule–Thomson refrigerator principle [35]. Metal coating is done by resistance heating [2, 35, 36], sputter coating [3, 34], or using electron-beam guns [37, 38].
2. Specimen carrier (standard apparatus of cryo-SEM): The carrier is used for transferring frozen samples from outside to inside the cryo-SEM and within the cryo-SEM by a rod (Fig. 1). A frozen sample is firmly fixed to the specimen carrier under LN_2. The specimen carrier with the sample is transferred to the cold stage in the specimen preparation chamber through the pre-evacuation chamber, and, after treatment in the preparation chamber, the sample is transferred to a cold stage in the SEM column. The specimen carrier with the sample is temporarily fixed on both cold stages, and sophisticated temperature control of the samples is done through the carrier.

Fig. 1 A specimen carrier (*right*) and sample holders (*left*) with cavities of different sizes. Sample holders are made of aluminum in and are 5 mm in diameter and 5 mm in height with cavities of different sizes (depth and diameter) in the center in which samples are put. The specimen carrier has a hole in which a specimen holder is fixed by a screw under LN_2

3. Sample holders: Samples to be frozen and observed by the cryo-SEM are put in sample holders. The samples in holders can be firmly fixed, without physical stresses to frozen samples themselves, by using a screw to a hole of the specimen carrier under LN_2. Since samples have various sizes and shapes, diverse kinds of sample holders are convenient. We usually have more than 100 sample holders made of aluminum with different sizes of holes in which samples are put (Fig. 1). Different sizes of sample holders are also useful. When size of the sample holder is changed, the size of holes in the specimen carrier must also be changed.
4. LN_2 Dewar for fixing samples to the specimen carrier under LN_2: Proper size of Dewar filled with LN_2 is used to fix frozen samples into holes in the specimen carrier under LN_2. Most types of cryo-SEM are equipped with such a Dewar as a standard apparatus.
5. LN_2 container for stock frozen samples: A large container is filled with LN_2 to stock frozen samples until use. We use 20-L containers that have several partitions for keeping samples. When stocked samples are used for cryo-SEM observation, they are transferred into a small Dewar vessel (1-L) filled with LN_2.
6. Programmed freezer: In addition to cryo-SEM related apparatus, a programmed freezer is necessary to perform the described experiments. The cooling rate of samples is controlled from

0 °C to temperatures lower than −50 °C. There are many types of programmed freezer. We use stock freezers for cooling down to −60 °C. The stock freezer is equipped with a temperature control heater operated by a personal computer.

2.2 Materials

1. Use fresh plants under control (warm) growth conditions, under cold-acclimation conditions or in the dormant state.
2. In the case of dormant trees, stocked samples can also be used. For stock samples in dormant trees, remove 1–5-year-old twigs with length of about 30–50 cm from standing trees during winter, put about 20 twigs in a polystyrene bag together with a sufficient amount of snow, close the bag and store in a cold room kept at −10 °C for 6 months at maximum until use (*see* **Note 1**).

3 Methods

3.1 Preparation of Samples Before Freezing

In order to firmly fix frozen samples during fracturing and cryo-SEM observation, put fresh samples in a hole of the sample holder before freezing (*see* **Note 2**). During sample preparation, maintain the desired temperatures depending on samples. Do all the processes quickly to maintain the freshness of samples.

1. Write appropriate numbers by oil-based ink (*see* **Note 3**) indicating sample conditions, including name of the plant species, name of tissues, sampling conditions and freezing conditions, on the bottom or side walls of the sample holders.
2. Remove tissues from fresh or stocked plants.
3. Excise removed tissues with a sharp knife (*see* **Note 4**) to appropriate sizes that can be placed in a hole of the specimen holder with slight protrusion from the hole (*see* **Note 5**).
4. Put a small amount of distilled water in each of the holes in sample holders using a pipette (*see* **Note 6**).
5. Using forceps, put samples in a hole of the sample holder in contact with water in the hole. Arrange the samples so that small areas protrude from the top surface of the sample holder. Since the protruding parts are cut (fractured) with a cold knife that moves in parallel with the top surface of sample holder, arrange the samples to make the fractured plane correspond to the desired plane to be observed.
6. Prepare all of the samples (hereafter, samples in holders will be simply called "samples") just before freezing.

3.2 Cryo-Fixation of Reference Samples

For comparison of structures with those after controlled-freezing, structures before freezing need to be observed for reference samples. Although such reference structures can be provided by various kinds of cryo-fixation using rapid freezing, cryo-fixation by plunging samples into cooled Freon 22 is recommended for the study of freezing behavior of plant tissue cells (*see* **Note 7**). The process of cryo-fixation by plunge freezing with cooled Freon 22 is described below.

1. Fill a small Dewar vessel (1-L) with LN_2.
2. Insert a copper-made well into LN_2 in the Dewar vessel by grasping the well firmly with crucible tongs. Avoid getting any LN_2 in the well.
3. Inject Freon 22 slowly into the cooled well in LN_2.
4. When about half of the well is filled with Freon 22 (about 5 mL), stop injection.
5. Freon 22 in the well is soon frozen.
6. Insert metal sticks into the frozen Freon 22 in order to make a mixture of solid and liquid Freon 22 (−160 °C).
7. Quickly plunge a fresh sample (provided by the process described under Subheading 3.1) by grasping the sample holder with forceps and putting it into the liquid part of Freon 22 for at least 5 s.
8. Quickly transfer the sample to a basket filled in with LN_2 and release it from the forceps.
9. Store samples in a LN_2 stock container until use (*see* **Note 8**).

3.3 Controlled Freezing of Samples

In order to understand freezing responses of plant tissue cells, it is necessary to freeze samples at controlled cooling rates (generally by slow cooling to mimic temperature reduction in the field) to the desired temperature, and finally the frozen samples are cryo-fixed to keep conditions at the final freezing condition. Here, the general processes for controlled slow freezing are described.

1. Put several fresh samples (provided as described under Subheading 3.1) in a petri dish and cover with a lid at 4 °C or other desired non-freezing temperature (*see* **Note 9**).
2. Previously cool an empty petri dish in a programmed freezer kept at the starting temperature of freezing for the samples and put a small amount of ice chips (originating from frost) obtained by scratching from side walls of the programmed freezer with cooled forceps in the cooled petri dish.
3. Transfer the petri dish with the samples to a programmed freezer kept at −3 °C or other desired freezing temperature depending on the purpose (*see* **Note 10**).
4. Wait for more than 30 min to obtain a temperature equilibrium of the samples.
5. After a temperature equilibrium at −3 °C or desired freezing temperature has been reached, open the lid of the petri dish and put ice chips on the samples using cooled forceps as above.

6. Confirm freezing of the samples (*see* **Note 11**).
7. Keep the samples at −3 °C or desired freezing temperature for a further 30 min or more to reach temperature equilibrium after freezing.
8. Start cooling at a controlled rate depending on the purpose (*see* **Note 12**) to the desired final freezing temperature (*see* **Note 13**).
9. After the desired final freezing temperature has been reached, cryo-fix the samples immediately or after the desired time at the final temperature by plunge-freezing with cooled Freon 22 (by the process described under Subheading 3.2) (*see* **Note 14**).
10. Store the samples in a LN_2 stock container until use.

3.4 Cryo-Scanning Electron Microscopy

Frozen samples including both reference and controlled-freezing samples are observed by a cryo-SEM. Although each cryo-SEM has a different construction, common processes for cryo-scanning electron microscopy are described here. For details of the methods for specimen preparation and observation, see the procedures described in the manufacturer's instruction manual for each cryo-SEM.

1. With the SEM in operation (at fully high vacuum), cool a cold stage and a cold trap in the SEM column as well as a cold stage and a cold knife in the preparation chamber.
2. Wait for about 1 h to achieve full cooling (lower than −160 °C) of the above-mentioned cryo-SEM apparatuses.
3. Set the temperature of a cold stage in the pre-evacuation chamber to −105 °C. Keep maximum cooling in other apparatuses.
4. Transfer stocked frozen samples in the LN_2 stock container into a small Dewar vessel filled with LN_2.
5. Cool a specimen carrier by dipping into LN_2 in a LN_2 Dewar until bubbling has stopped (about 30 s).
6. Transfer the frozen samples to the LN_2 Dewar together with the cooled specimen carrier in LN_2.
7. Put a sample in a hole of the specimen carrier using forceps and firmly fix the samples in the hole of the carrier by a screw using a screwdriver under LN_2.
8. Cover the samples with the cold trap for decontamination, if there is one, under LN_2.
9. Quickly transfer the covered specimen carrier with the sample to the pre-evacuation chamber.
10. Quickly evacuate the pre-evacuation chamber.
11. After completion of full evacuation in the pre-evacuation chamber, open the gate to the preparation chamber, transfer

the specimen carrier with the sample, and fix the specimen carrier on a cold stage of the preparation chamber kept at −105 °C (*see* **Note 15**).

12. After the temperature has reached equilibrium at −105 °C, confirm removal of frost that may have covered samples by the naked eye or by using binoculars (*see* **Note 16**).
13. Fracture protruding areas of the samples by moving a cold knife in parallel with the top surface of the sample holder (*see* **Note 17**).
14. After fracturing the sample, cover the fractured plane with a cold knife kept at −160 °C as a cold trap for decontamination.
15. Keep fractured samples for a few minutes at −105 °C for making slight etching (*see* **Note 18**).
16. To stop further etching, cool the cold stage to a temperature lower than −120 °C (*see* **Note 19**).
17. Remove the cold knife covering the samples.
18. Coat the fractured face by metal evaporation. Depending on the apparatus, refer to manufacturer's instructions for appropriate coating (*see* **Note 20**).
19. Transfer the carrier with samples from the cold stage of the preparation chamber to a cold stage in the SEM column kept at −160 °C, fix the carrier, and cover the periphery of the samples by a cold trap.
20. Observe fracture planes by using secondary emissions by the SEM (*see* **Note 21**).
21. Take photographs of desired areas following the procedure described in the manufacturer's instruction manual.

3.5 Methods for Analyzing Effects of Freezing

Several examples of specific methods to determine effects of freezing on plant tissue cells are described. In most cases, prepare samples from at least three separate freezing experiments with more than three samples in each separate experiment and observe more than 100 cells in total from more than six different samples in order to determine the tendency of freezing-induced changes.

3.5.1 Detection of Extracellular Freezing

The usual method for observing structural changes induced by extracellular freezing under equilibrium freezing conditions is described using wheat leaves under control (warm) growth conditions (Fig. 2).

1. Prepare samples at room temperature by the process described under Subheading 3.1 and cryo-fix the samples at room temperature as reference samples by the process described in Subheading 3.2.
2. For equilibrium freezing of the samples, provide the samples prepared at room temperature by the process described in

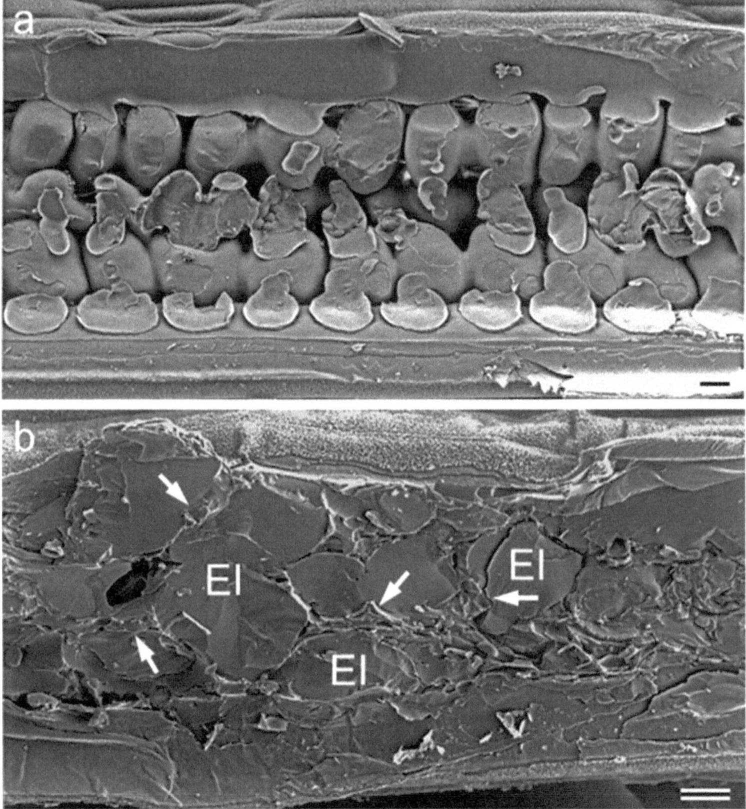

Fig. 2 Cryo-SEM photographs showing parts of freeze-fractured wheat leaves (**a**) in a reference sample cryo-fixed at room temperature and (**b**) in a sample slowly frozen from −3 to −10 °C at a cooling rate of 0.2 °C/min. Equilibrium slow freezing to −10 °C caused distinct shrinkage of cells (*arrows*) with occupation of extracellular spaces by large extracellular ice crystals (EI). Bars = 10 μm

Subheading 3.1 and slowly freeze the samples by the process described in Subheading 3.3, in which the samples are kept at −3 °C for 30 min, inoculated with ice, kept at −3 °C for 30 min, cooled at a rate of 0.2 °C/min to −10 °C, and cryo-fixed immediately after reaching −10 °C.

3. Take photographs of cryo-fixed reference and cryo-fixed slowly frozen samples using a cryo-SEM as described in Subheading 3.4.
4. Compare the structures of a reference sample (Fig. 2a) and a sample slowly frozen to −10 °C (Fig. 2b) (*see* **Note 22**).

3.5.2 Detection of Temperature Limit of Deep Supercooling

The method for examining the temperature limit of deep supercooling is described using xylem parenchyma cells in mulberry trees harvested in summer (Fig. 3).

1. Prepare samples at room temperature by the process described in Subheading 3.1 and cryo-fix samples from room temperature as described in Subheading 3.2.

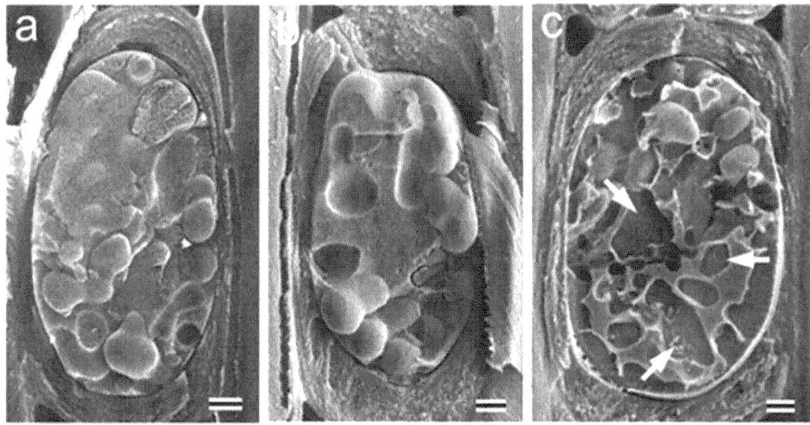

Fig. 3 Cryo-SEM photographs showing temperature limit of deep supercooling in xylem parenchyma cells of birch harvested in summer. (**a**) Reference samples cryo-fixed at room temperature. (**b**) Cryo-fixed samples after very slow cooling to −15 °C at a rate of 5 °C/day. (**c**) Cryo-fixed samples after very slow cooling to −20 °C at a rate of 5 °C/day. While very small intracellular ice crystals (which are difficult to detect or are seen as numerous small holes due to sublimation of ice) are produced in reference samples (**a**) and samples slowly frozen to −15 °C (**b**), large intracellular ice crystals (*arrows*) are produced in cells slowly frozen to −20 °C (**c**). The sizes of intracellular ice crystals differ depending on whether they are produced by rapid freezing due to cryo-fixation of liquid water (**a** and **b**) or produced during very slow cooling (5 °C/day) (**c**). Distinct differences in sizes of intracellular ice crystals indicate temperature limit of supercooling between −15 and −20 °C. Bars = 2 μm

2. To prepare frozen samples, put samples in a petri dish in a programmed freezer kept at −5 °C overnight and then lower the temperature very slowly in a stepwise manner by 5 °C/day to −50 °C (*see* **Note 23**).
3. At each −5 °C decline in temperature, cryo-fix the samples by the process described in Subheading 3.2.
4. Take photographs of cryo-fixed reference and treated samples cooled from −5 to −40 °C by the cryo-SEM (*see* Subheading 3.4).
5. Compare sizes of intracellular ice crystals (*see* **Note 24**) in the reference sample (Fig. 3a) and samples very slowly cooled to different freezing temperatures (Fig. 3b, c).

3.5.3 Detection of Freezable Water Under the Condition of Slow Freezing by a Recrystallization Experiment (Fig. 4)

In order to examine whether freezable water has remained in plant cells under the condition of slow freezing to the desired temperature, the method for a recrystallization experiment is described using tissue cells of dormant buds in katsura tree [29, 32].

1. Prepare samples at 4 °C as described in Subheading 3.1 and cryo-fix reference samples at 4 °C (Subheading 3.2).
2. To prepare slowly frozen samples, prepare samples at 4 °C as described in Subheading 3.1 and slowly freeze the samples from −3 °C at a cooling rate of 5 °C/day to −30 °C (Subheading 3.3) and then cryo-fix the samples at −30 °C as in Subheading 3.2.

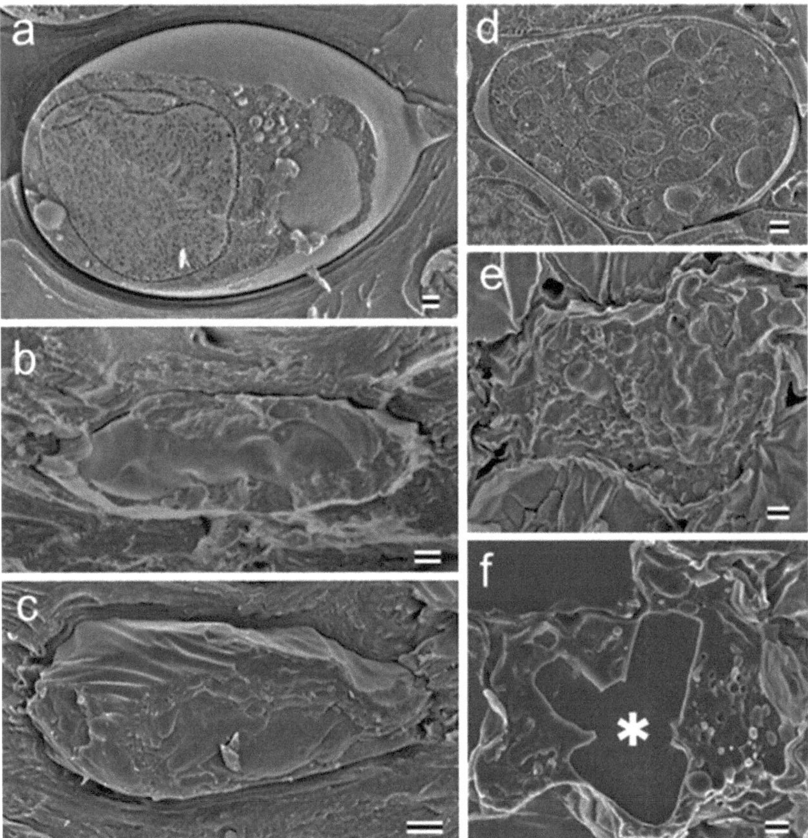

Fig. 4 Cryo-SEM photographs showing a recrystallization experiment in tissue cells of dormant buds of katsura tree in order to determine the presence or absence of freezable water after slow freezing to desired temperatures. (**a–c**) Cells in inner scales. (**d–f**) Cells in flower primordia. (**a** and **d**) Reference samples cryo-fixed at 4 °C. (**b** and **e**) Cryo-fixed samples after very slow freezing from −3 to −30 °C at a cooling rate of 5 °C/day. (**c** and **f**) Cryo-fixed samples after a recrystallization experiment in which samples in (**b** and **e**) were rewarmed at −20 °C. Formation of large intracellular ice crystals (*asterisk*) in flower primordia cells after the recrystallization experiment indicates the presence of freezable water even after slow cooling to −30 °C (**f**), whereas the absence of such intracellular change in cells in inner scales indicates complete dehydration of freezable water by slow freezing to −30 °C (**c**). Freezable water remaining at −30 °C after very slow cooling produced very small intracellular ice crystals by cryo-fixation and rewarming of such cells produced large intracellular ice crystals as a result of recrystallization. Bars = 1 μm

3. For the recrystallization experiment, transfer the above samples which have been slowly frozen to −30 °C and cryo-fixed to a petri dish in a freezer kept at −20 °C, keep overnight to allow recrystallization, and again cryo-fix (Subheading 3.2).
4. Take photographs of cryo-fixed reference samples, cryo-fixed samples slowly frozen to −30 °C and recrystallization experiment samples using a cryo-SEM (Subheading 3.4).
5. Compare cell shapes and especially the sizes of intracellular ice crystals (*see* **Note 24**) among reference samples (Fig. 4a, d), samples slowly frozen to −30 °C (Fig. 4b, e) and recrystallization experiment samples (Fig. 4c, f).

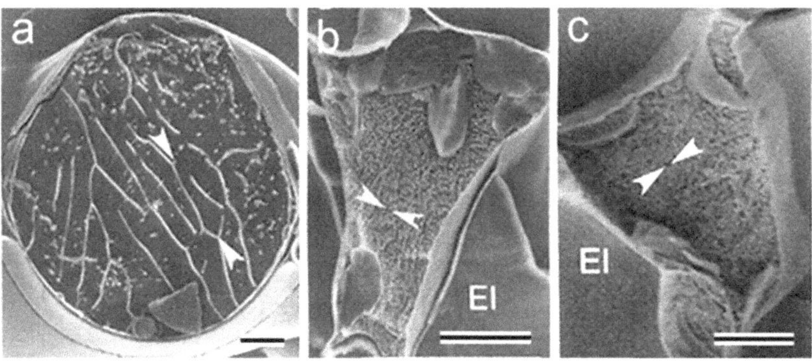

Fig. 5 Cryo-SEM photographs showing equilibrium freezing of *Arabidopsis* leaves. (**a**) Reference sample cryo-fixed at room temperature. (**b**) Cryo-fixed sample after freezing at −2 °C for 1 h. (**c**) Cryo-fixed sample after freezing at −2 °C for 3 days. Compared with the reference samples (**a**), samples frozen at −2 °C exhibited shrinkage by dehydration and far smaller intracellular ice crystals as a result of cryo-fixation (**b** and **c**). The sizes of intracellular ice crystals (shown between *arrowheads* in the maximum diameter) are similar in cryo-fixed samples after freezing at −2 °C for 1 h (**b**) and 3 days (**c**). Since the size of intracellular ice crystals depends on the degree of cellular hydration, the result indicates that equilibrium dehydration occurred with freezing at −2 °C for 1 h. EI = extracellular ice. (Reproduced from Nagao et al. [16] with permission from Springer) Bars = 10 μm

3.5.4 Confirmation of Equilibrium Freezing

The method to determine whether the cooling rates or freezing conditions used produced equilibrium dehydration is described using non-acclimated leaves of *Arabidopsis thaliana* (Fig. 5) [16].

1. Prepare reference samples at room temperature (Subheading 3.1) and cryo-fix the reference samples at room temperature (Subheading 3.2).
2. To prepare one type of controlled-freezing samples, prepare samples at room temperature (Subheading 3.1), keep the samples at −2 °C for 30 min, inoculate with ice, keep at −2 °C for 1 h, and immediately cryo-fix as described in Subheading 3.2.
3. For a second type of controlled-freezing samples, prepare samples at room temperature and then keep them at −2 °C for 30 min as above. Inoculate the samples with ice and keep at −2 °C for 3 days, then cryo-fix (Subheading 3.2; *see* **Note 25**).
4. Take photographs of cryo-fixed reference samples and cryo-fixed samples at −2 °C that have been kept for 1 h and 3 days.
5. Compare sizes of intracellular ice crystals in reference samples (Fig. 5a) and samples kept at −2 °C for 1 h (Fig. 5b) and for 3 days (Fig. 5c).

3.5.5 Evaluation of the Cell Wall as a Barrier Against Penetration of Extracellular Ice Crystals (in the Case of Chilling-Sensitive Plant Tissue Cells)

The method to estimate the effect of cell walls as a barrier against extracellular ice crystals is described using chilling-sensitive leaves of *Saintpaulia grotei* Engl. (Fig. 6) [21].

1. Prepare reference samples from room temperature (Subheading 3.1) and cryo-fix the samples at room temperature (Subheading 3.2).

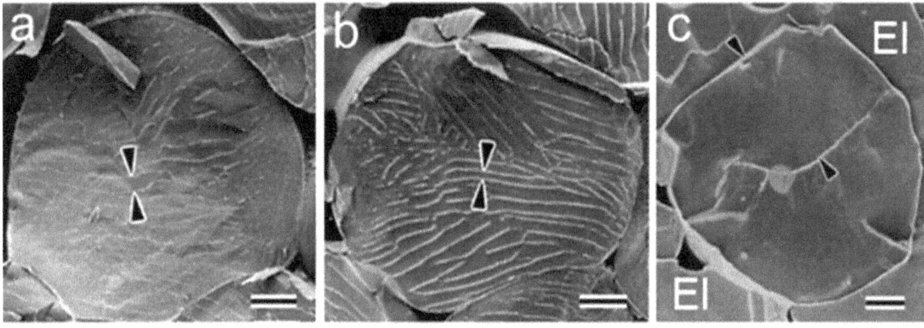

Fig. 6 Cryo-SEM photographs showing the effect of the cell wall against the presence of neighboring extracellular ice crystals in chilling-sensitive *Saintpaulia grotei* Engl. leaves. (**a**) Reference samples cryo-fixed at room temperature. (**b**) Cryo-fixed samples with supercooling at −2 °C for 1 h. (**c**) Cryo-fixed samples frozen at −2 °C for 1 h. While reference cells (**a**) and supercooled cells (**b**) exhibited small intracellular ice crystals (shown between *arrows* in the minimum diameter), frozen samples produced large intracellular ice crystals due to penetration of extracellular ice crystals through the cell walls. EI = extracellular ice. Bar = 20 μm. (Reproduced from Yamada et al. [21] with permission from Springer)

2. Supercooled samples are prepared as described in Subheading 3.1, put in a programmed freezer kept at -2 °C for 1 h under the condition of supercooling (*see* **Note 26**) and then cryo-fixed at −2 °C (Subheading 3.2).
3. Frozen samples are prepared as described in Subheading 3.1, put into a petri dish in a programmed freezer kept at −2 °C for 30 min, frozen by ice inoculation, kept for a further 30 min at −2 °C, and cryo-fix at −2 °C (Subheading 3.2).
4. Take photographs of cryo-fixed reference samples, supercooled samples, and samples frozen at −2 °C using a cryo-SEM (Subheading 3.4).
5. Compare intracellular ice crystal sizes in the reference sample (Fig. 6a), sample supercooled at −2 °C (Fig. 6b), and sample frozen at −2 °C.

3.5.6 Evaluation of the Cell Wall as a Barrier Against Penetration of Extracellular Ice Crystals (in the Case of Chilling-Resistant Plant Tissue Cells)

The method to estimate the effect of cell walls as a barrier against extracellular ice crystals is described using cells with freeze-damaged protoplasts in leaves of an orchid (Fig. 7) [21].

1. Prepare samples at room temperature as described in Subheading 3.1.
2. For breaking protoplasts together with plasma membranes, freeze samples by direct immersion in LN_2 for about 1 min and then thaw at room temperature for 30 min and cryo-fix (Subheading 3.2; *see* **Note 27**).
3. To produce frozen samples, put frozen-thawed samples in a petri dish (*see* **Note 28**), transfer to a programmed freezer kept at −2 °C for 30 min, freeze by ice inoculation, keep for a further 30 min at −2 °C, and cryo-fix (Subheading 3.2; *see* **Note 29**).

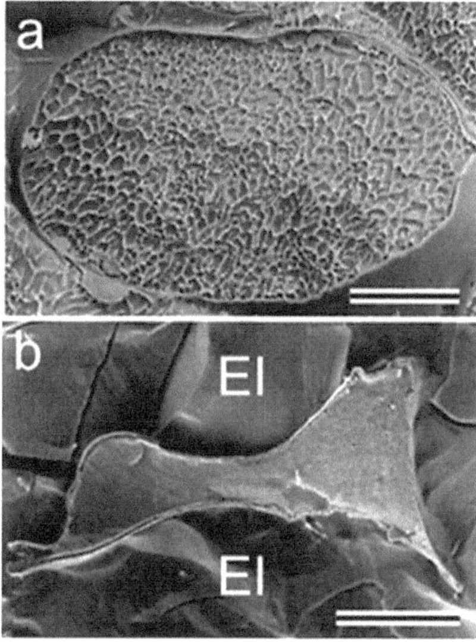

Fig. 7 Cryo-SEM photographs showing the effect of the cell wall against the presence of neighboring extracellular ice crystals in chilling-resistant orchid leaves. Before examination, the samples were freeze-thawed to destroy protoplasts. (**a**) Protoplast-broken cells cryo-fixed at room temperature. (**b**) Cryo-fixed protoplast-broken cells after freezing at −2 °C for 30 min. Freezing at −2 °C caused distinct cell shrinkage with small intracellular ice crystals by dehydration through cell walls, indicating that cell walls even after protoplast destruction act as a complete barrier against penetration of extracellular ice crystals. EI = extracellular ice. Bar = 20 μm. (Reproduced from Yamada et al. [21] with permission from Springer)

4. Take photographs of cryo-fixed freeze-thawed samples for reference and cryo-fixed freeze-thawed samples after further freezing at −2 °C using a cryo-SEM (Subheading 3.4).
5. Compare cell shapes before (Fig. 7a) and after slow freezing at −2 °C (Fig. 7b).

4 Notes

1. Add snow every month to prevent drying of twigs during the storage period. Long-term preservation of twigs exceeding 6 months may cause physiological changes.
2. Since loose fixation of frozen samples may result in drifting of samples under SEM observation, especially at higher magnification, close contact of samples in the sample holder is required. When samples are in the holder, physically firm fixation of the holders to the carrier holes by a screw is possible. If frozen plant tissues are directly fixed to the specimen carrier using a screw under LN_2, strong physical stress by fixation of the samples may

cause crushing of the samples. On the other hand, loose fixation of samples to the specimen carrier may produce drifting of samples during observation due to elimination of small ice crystals (originating from frost on surfaces of the samples) by sublimation during processes of cryo-SEM observation.

3. Do not use a magic marker with water-soluble ink, which may disappear after thawing.
4. Physical damage caused by cutting of samples may result in structural changes. Observation should be done in areas further away from the cutting.
5. If changes in survival of frozen samples are concomitantly measured in parallel with cryo-SEM observation, use the same sizes of samples in cryo-SEM observation and survival assays for comparison between them. Furthermore, if freezing responses of samples are measured by different methods, such as differential thermal analysis, also use the same sizes of samples for comparison.
6. Water is used to firmly fix samples to the holes of holders after freezing. If necessary, use an appropriate buffered solution instead of distilled water. For samples that need to be kept frozen for long periods (a few to several days), the samples should be completely embedded in water in order to prevent drying due to sublimation. For samples that need to avoid contact with water, use a small amount of glycerol for fixation of the samples under freezing to the sample holder instead of water.
7. Cryo-fixation by plunge-freezing with cooled Freon 22 (9,000 °C/s) produces very small intracellular ice crystals in many plant tissue cells, though comparatively large intracellular ice crystals are often detected in highly hydrated samples by cryo-SEM observation. However, for detection of freezing responses, it is necessary to use the same cryo-fixation method for reference and controlled-freezing samples. For cryo-fixation of samples previously frozen by controlled freezing, only plunge-freezing can be used. When reference samples without effects of ice crystals need to be obtained, other techniques including propane jet freezing [39], spray freezing [40], impact freezing [41], and high-pressure freezing [42] are useful. In these techniques, vitrification of sample water is possible, though this is restricted to very small volumes. *See* ref. 33 for manuals of these cryo-fixation techniques.
8. With a constant supply of LN_2 in the stock container, frozen samples can be stored for more than 1 year.
9. Keep the temperature at 0–4 °C for preparation of cold-acclimated samples or samples from dormant plants. For samples from growing plants, use room temperature for preparation.
10. Change the temperatures for the start of freezing depending on samples. For freeze-sensitive plant tissue cells, starting tem-

perature for freezing of −1 or −2 °C is necessary because a lower temperature may cause intracellular freezing. In plant tissues with moderate freezing tolerance both before and after cold acclimation, a starting temperature for freezing of −3 °C is used. In very cold hardy plant tissue cells, starting cooling at −5 °C is possible.

11. Freezing of samples can be confirmed by the naked eye or by physical fixation of samples to the specimen holder by touching with cooled forceps.
12. For most samples, a cooling rate of 0.2 °C/min is appropriate to achieve equilibrium freezing. In some samples, equilibrium freezing is obtained at far lower cooling rates. We sometime use a cooling rate of 5 °C/day as the lowest cooling rate. All cooling rates can be programmed in a programmable freezer.
13. By changing the final freezing temperature, the process of freezing in parallel with reduction of freezing temperature can be observed.
14. For cryo-fixation of samples after controlled freezing, put a small Dewar with cooled Freon 22 in a programmed freezer just before cryo-fixation, quickly grasp the controlled-frozen samples with cooled forceps, plunge into partially melted Freon 22 and transfer to an LN_2 Dewar. When it is necessary to exactly maintain the final freezing temperature, we use a programmed freezer placed in a cold room in which the temperature is also controlled to be the same as that in the programmed freezer.
15. A sample temperature of −105 °C under high vacuum in a cryo-SEM allows for slight sublimation from frozen samples in order to remove frost that has unintentionally covered samples during transfer of the samples from outside the cryo-SEM.
16. In a cryo-SEM, indicated temperatures for samples are generally measured in a specimen carrier. Keep samples for several minutes after temperature equilibrium at −105 °C with visual confirmation of removal of frost from samples.
17. Fracture by one cut is favorable for reducing areas where the cold knife had contact with the fracture plane. Contact areas of knife should be avoided for observation. For cryo-SEM observation, making a macroscopically flat fracture plane is favorable for prevention of charging during cryo-SEM observation. If fractured pieces remain near fractured samples, remove such debris by reversing the specimen carrier using the rod of the specimen carrier, because the presence of such debris may become a source of contamination and charging.
18. For cryo-SEM observation of frozen samples, a slight degree of etching is recommended to make structures clear. Freeze-fracture without etching is not recommended for specimen preparation of a cryo-SEM due to high risk for contamination

of fractured faces with frost. Vapor pressure of ice differs in relation to temperature [43]. When the temperature of samples has a higher vapor pressure than the vapor pressure of the SEM (i.e., vacuum in the SEM), etching by sublimation of ice occurs. On the other hand, when the temperature of samples has a lower vapor pressure than vacuum, no etching occurs.

19. In a vacuum of a general cryo-SEM, etching does not occur in samples kept at temperatures lower than −120 °C.
20. In all types of metal coaters, even coating of fracture faces is necessary. Coating with three-dimensional rotation of samples is recommended (except for sputter coating).
21. Use appropriate accelerating voltages. Observation of frozen samples by a cryo-SEM should be completed within 2 h in order to avoid contamination, increased charging and temperature rise in samples. Images by other emissions such as reflection electrons are also used.
22. By changing the final freezing temperature under equilibrium slow freezing, the process of extracellular freezing can be observed.
23. Without ice inoculation, most samples kept at −5 °C overnight are frozen. Even though a few samples may remain unfrozen at −5 °C, transfer of samples to −10 °C will ensure that all samples are frozen. In deep supercooling cells, such differences of starting temperature of freezing do not make a difference in the freezing behavior of cells.
24. Measure minimum or maximum diameter of ice (between eutectics) that appeared in fracture faces depending on the purpose. When minimum diameter of ice is measured, the difference among treatments becomes smaller (*see* Fig. 6), whereas when maximum diameter is measured, the difference becomes larger (*see* Fig. 5).
25. Although freezing at −2 °C is shown here, it is also possible to cool samples to lower temperatures.
26. Under this condition at −2 °C for 1 h, no samples are frozen without ice inoculation.
27. Freezing by LN_2 causes intracellular freezing that results in distinct destruction of protoplasts together with plasma membranes. Thus, the effect of only cell walls as a barrier of penetration of extracellular ice crystals is analyzed.
28. Remove water from outside wall of the sample holders in freeze-thawed samples. The presence of water and resultant formation of ice crystals makes it difficult to fix the sample holders in the holes of specimen carriers.
29. Although freezing at −2 °C is shown here, it is also possible to cool samples to lower temperatures.

Acknowledgments

One of authors (S.F.) sincerely appreciates the strong support by JEOL Co. Ltd. for improvement and development of a cryo-SEM since my start of studies using a cryo-SEM in 1974. The authors also appreciate the excellent works by Mr. K. Shinbori, Institute of Low Temperature Science, Hokkaido University, for making many apparatuses of a cryo-SEM.

References

1. Echlin P (1971) The examination of biological material at low temperatures. In: Johari O, Corvin I (eds) Scanning electron microscopy. IITRI, Chicago, p 225–232
2. Nei T, Yotsumoto H, Hasegawa Y, Nagasawa Y (1973) Development of new cryo-unit attached to scanning electron microscope. J Electron Microsc 22:169–182
3. Sargent JA (1988) Low temperature scanning electron microscopy: advantages and application. Scanning Microsc 2:835–849
4. Fujikawa S, Suzuki T, Ishikawa T, Sakurai S, Hasegawa Y (1988) Continuous observation of frozen biological materials with cryo-scanning electron microscope and freeze-replica by a new cryo-system. J Electron Microsc 37: 315–322
5. McCully ME, Canny MJ, Huang CX (2009) Cryo-scanning electron microscopy (CSEM) in the advancement of functional plant biology: morphological and anatomical applications. Funct Plant Biol 36:97–124
6. Utsumi Y, Sano Y, Ohtani J, Fujikawa S (1996) Seasonal changes in the distribution of water in the outer growth rings of *Fraxinus mandshurica* var. *Japonica*: a study by cryo-scanning electron microscopy. IAWA J 17:113–124
7. Utsumi Y, Sano Y, Fujikawa S, Funada R, Ohtani J (1998) Visualization of cavitated vessels in winter and refilled vessels in spring in diffuse-porous trees by cryo-scanning electron microscopy. Plant Physiol 117:1463–1471
8. Sano Y, Fujikawa S, Fukazawa K (1995) Detection and features of wetwood in *Quercus mongolica* var. *grosseserrata*. Trees Struct Funct 9:261–268
9. Johnson DM, Meinzer FC, Woodruff DR, McCulloh KA (2009) Leaf xylem embolism, detected acoustically and by cryo-SEM, corresponds to decreases in leaf hydraulic conductance in four evergreen species. Plant Cell Environ 32: 828–836
10. Cochard H, Bodet C, Ame'glio T, Cruiziat P (2000) Cryo-scanning electron microscopy observations of vessel content during transpiration in walnut petioles. Facts or artifacts? Plant Physiol 124:1191–1202
11. Nei T, Fujikawa S (1977) Freeze-drying process of biological specimens observed with a scanning electron microscope. J Microsc 111: 137–142
12. Fujikawa S, Miura K (1986) Plasma membrane ultrastructural changes caused by mechanical stress in the formation of extracellular ice as a primary cause of slow freezing injury in fruit-bodies of basidiomycetes (*Lyophyllum ulmarium* (Fr.) Kuhner). Cryobiology 23:371–382
13. Fujikawa S (1990) Cryo-scanning electron microscope and freeze-replica study on the occurrence of slow freezing injury. J Electron Microsc 39:80–85
14. Pearce RS (1988) Extracellular ice and cell shape in frost-stressed cereal leaves: a low temperature scanning-electron-microscopy study. Planta 175:313–324
15. Pearce RS, Ashworth EN (1992) Cell shape and localization of ice in leaves of overwintering wheat during frost stress in the field. Planta 188:324–331
16. Nagao M, Arakawa K, Takezawa D, Fujikawa S (2008) Long- and short-term freezing induce different types of injury in *Arabidopsis thaliana* leaf cells. Planta 227:477–489
17. Ball MC, Canny MJ, Cheng X, Huang CX, Heady RD (2004) Structural changes in acclimated and unacclimated leaves during freezing and thawing. Funct Plant Biol 31:29–40
18. Roden JS, Canny MJ, Huang CX, Ball MC (2009) Frost tolerance and ice formation in *Pinus radiata* needles: ice management by the endodermis and transfusion tissues. Funct Plant Biol 36:180–189
19. Endoh K, Fujikawa S, Arakawa K (2012) Freezing behavior of cells in evergreen needle

leaves of fir (*Abies sachalinensis*). Cryobiol Cryotechnol 58:125–134

20. Ashworth EN, Pearce RS (2002) Extracellular freezing in leaves of freezing-sensitive species. Planta 214:798–805
21. Yamada T, Kuroda K, Jitsuyama Y, Takezawa D, Arakawa K, Fujikawa S (2002) Roles of the plasma membrane and the cell wall in the responses of plant cells to freezing. Planta 215: 770–778
22. Ashworth EN, Echlin P, Pearce RS, Hayes TL (1988) Ice formation and tissue response in apple twigs. Plant Cell Environ 11:703–710
23. Fujikawa S, Kuroda K, Ohtani J (1996) Seasonal changes in the low-temperature behaviour of xylem ray parenchyma cells in red osier dogwood (*Cornus sericea* L.) with respect to extracellular freezing and supercooling. Micron 27:181–191
24. Kuroda K, Ohtani J, Fujikawa S (1997) Supercooling of xylem ray parenchyma cells in tropical and subtropical hardwood species. Trees Struct Funct 12:97–106
25. Fujikawa S, Kuroda K, Ohtani J (1997) Seasonal changes in dehydration tolerance of xylem ray parenchyma cells of *Stylax obassia* twigs that survive freezing temperatures by deep supercooling. Protoplasma 197:34–44
26. Kuroda K, Ohtani J, Kubota M, Fujikawa S (1999) Seasonal changes in the freezing behavior of xylem ray parenchyma cells in four boreal hardwood species. Cryobiology 38:81–88
27. Fujikawa S, Kuroda K, Jitsuyama Y, Sano Y, Ohtani J (1999) Freezing behavior of xylem ray parenchyma cells in softwood species with differences in the organization of cell walls. Protoplasma 206:31–40
28. Fujikawa S, Kuroda K (2000) Cryo-scanning electron microscopic study on freezing behavior of xylem ray parenchyma cells in hardwood species. Micron 31:669–686
29. Kuroda K, Kasuga J, Arakawa K, Fujikawa S (2003) Xylem ray parenchyma cells in boreal hardwood species respond to subfreezing temperatures by deep supercooling that is accompanied by incomplete desiccation. Plant Physiol 131:736–744
30. Kasuga J, Arakawa K, Fujikawa S (2007) High accumulation of soluble sugars in deep supercooling Japanese white birch xylem parenchyma cells. New Phytol 174:569–579
31. Kasuga J, Endoh K, Yoshiba M, Taido I, Arakawa K, Uemura M, Fujikawa S (2013) Roles of cell walls and intracellular contents in supercooling capability of xylem parenchyma cells of boreal trees. Physiol Plant 148:25–35
32. Endoh K, Kasuga J, Arakawa K, Fujikawa S (2009) Cryo-scanning electron microscopic study on freezing behaviors of tissue cells in dormant buds of larch (*Larix kaempferi*). Cryobiology 59:214–222
33. Fujikawa S (1991) Freeze-fracture techniques. In: Harris JR (ed) Electron microscopy in biology: a practical approach. IRL Press, Oxford, pp 173–201
34. Robards AW, Crosby P (1979) A comprehensive freezing, fracturing and coating system for low temperature scanning electron microscopy. Scanning Electron Microsc 2:325–343
35. Pawley J, Norton JT (1978) A chamber attached to the SEM for fracturing and coating frozen biological samples. J Microsc 112: 169–182
36. Bastacky J, Hook GR, Finch GL, Goerke J, Hayes TL (1987) Low-temperature scanning electron microscopy of frozen hydrated mouse lung. Scanning 9:57–70
37. Fujikawa S, Suzuki T, Sakurai S (1990) Use of micromanipulator for continuous observation of frozen samples by cryo-scanning electron microscopy and freeze replicas. Scanning 12: 99–106
38. Fujikawa S, Suzuki T, Ishikawa T, Sakurai S (1988) Continuous observation of frozen biological materials with cryo-scanning electron microscope and freeze-replicas by a new cryo-system. J Electron Microsc 37:315–322
39. Muller M, Meister N, Moor H (1980) Freezing in a propane jet and its application in freeze-fracturing. Mikroskopie 36:129–140
40. Bachmann L, Schmitt WW (1971) Improved cryofixation applicable to freeze etching. Proc Natl Acad Sci U S A 68:2149–2152
41. Heuser JE, Reese TE, Dennis MJ, Jan Y, Jan L, Evans L (1979) Synaptic vesicle exocytosis captured by quick freezing and correlated with quantal transmitter release. J Cell Biol 81: 275–300
42. Moor H, Bellin G, Sandri C, Akert K (1980) The influence of high pressure freezing on mammalian nerve tissue. Cell Tissue Res 209: 201–216
43. Umrath W (1983) Calculation of the freeze-drying time for electron-microscopical preparations. Mikroskopie 40:9–34

Chapter 11

Three-Dimensional Reconstruction of Frozen and Thawed Plant Tissues from Microscopic Images

David P. Livingston III and Tan D. Tuong

Abstract

Histological analysis of frozen and thawed plants has been conducted for many years but the observation of individual sections provides only a 2-dimensional representation of a 3-dimensional phenomenon. Most techniques for viewing internal plant structure in three dimensions are either low in resolution or the instrument cannot penetrate deep enough into the tissue to visualize the whole plant. Techniques with higher resolution are expensive and equipment often requires time-consuming training. We present a relatively simple and less-expensive technique using pixel-based (JPEG) images of histological sections of an *Arabidopsis thaliana* plant and commercially available software to generate 3D reconstructions of internal structures. The technique has proven to work just as effectively for images from medical histology.

Key words 3D reconstruction, *Arabidopsis thaliana*, Adobe After Effects, Freezing tolerance, Histology, Safranin, Microscopy, Color keying

1 Introduction

Since the development of X-rays, imaging internal structures of biological systems has provided information to help explain a variety of external symptoms of the organism. However, many of the techniques used for human and animal analysis such as magnetic resonance imaging (MRI) and positron emission tomography/computed tomography (PET/CT) are limited in resolution and cannot provide enough detail to be useful in plant systems. While syncrotron radiation imaging (SRI) produces images at considerably better resolution [1], it is rarely used for plants because of its expense and the limited number of facilities available for routine use.

Confocal laser scanning microscopy (CLSC) is used for imaging compounds within tissues that fluoresce and at higher resolution

Electronic supplementary material The online version of this article (doi:10.1007/978-1-4939-0844-8_11) contains supplementary material. This video is also available to watch on http://www.springerimages.com/videos/978-1-4939-0843-1. Please search for the video by the article title.

Dirk K. Hincha and Ellen Zuther (eds.), *Plant Cold Acclimation: Methods and Protocols*, Methods in Molecular Biology, vol. 1166, DOI 10.1007/978-1-4939-0844-8_11, © Springer Science+Business Media New York 2014

than MRI or CT (*see* ref. 2 for a review). However, the use of CLSC is limited by the ability to penetrate tissues beyond about 0.4 mm [3]. Software used in conjunction with MRI, PET/CT, SRI, and CLSC all allow optical sectioning of the organism and provide excellent 3D reconstruction in a nondestructive manner. Livingston et al. [3] review the positive and negative aspects of these and other techniques for viewing and reconstructing biological systems.

The limitations of resolution, tissue penetration, and expense make these techniques impractical for routine 3D reconstruction of plants. A 3D reconstruction technique initially described elsewhere [3] and described here in detail, will not replace the aforementioned techniques (MRI, CAT, CLSC). It is meant to be used on samples too small for MRI and CAT and too big or too thick for CLSC such as crown tissue of grasses. Advantages of this technique are that it is inexpensive, requiring only a digital camera, a microscope, and the Adobe System software, After Effects. Another advantage is that tissues of any size and images of any magnification can be used in the reconstruction. Disadvantages are that tissues must be sampled destructively to collect images and a certain degree of manual manipulation is required to align images that have been captured. Also, since images are pixel-based (not vector), boundaries of individual images will be visible and will create somewhat of a ladder effect in the final reconstruction. For most purposes this will not interfere with visualization of the final reconstruction.

This chapter is not meant to be a tutorial on histological techniques nor in the use of Adobe After Effects. A basic knowledge of histology is required to apply this technique. After Effects (AE) is a somewhat complicated program used for post production in the film industry. Despite its complexity, anyone with a familiarity of Adobe Photoshop can learn what is needed to apply the 3D reconstruction technique described here. The reader is referred to http://www.linda.com for comprehensive online tutorials on learning to use AE.

2 Materials

1. Chemicals and equipment for fixing, dehydrating, and embedding plant tissue in paraffin [4]. Equipment for sectioning, staining, and covering tissues on microscope slides.
2. Microscope or light table on which to photograph sequential images.
3. Tripod or camera mount for light table photography.
4. Sony DSC707 or another consumer-grade camera.
5. Adapter to allow the camera to mount to the microscope (available from Martin Microscope, Easley, SC, USA).
6. Mac or PC computer with at least 4 Gb of memory.
7. Adobe After Effects CS6.

8. Downloaded AE Plugin which allows AE to generate JPEG2000 images. Go to http://www.fnordware.com/j2k/ download the file "j2k.plugin" and place it into the Applications (programs) folder AfterEffects>Plug-ins>Format. *See* **Note 1** for why this download is important.
9. Downloaded AE Script called "AlignLayers" at http://www.james-cheetham.com/Downloads/Tools/AESCRIPTS/. Select AlignLayers from the list on this site. Once the file is unzipped, move it to your Applications (programs) folder AE CS6>scripts>ScriptUI Panels. Now open AE and go to file>scripts>run script file and find the .jsx file called AlignLayers. Click on it and it will open a panel in AE (Fig. 1). To make the panel open automatically every time you open AE, go to "window" and the script, AlignLayer.jsx will be at the bottom of the window. Make sure a check mark is next to it. This script will allow you to evenly distribute your images in Z space. You will use only z-alignment (i.e. not X or Y) for this reconstruction.

3 Methods

3.1 Preparation of Sections

1. The color of the stain used to highlight objects of interest in sections should stand out from the rest of the tissue either in hue or in color saturation or both. Consistency of stain throughout the entire sequence is important since all routines that depend on color recognition will use the same hue/saturation setting for each image.
2. Consistently undistorted sequential sections have a direct impact on the quality of the reconstruction. Great care should be taken in transferring ribbons/sections to slides and drying sections on the slide.
3. Sections can be as thin as possible. For high magnification reconstructions (200× or higher) 10 μm or thinner will give better images due to a reduced focal length. However, if the tissue is thick (one or more cm), consideration must be given to the number of sections necessary to section the whole piece of tissue. We routinely section at 20 μm for a winter cereal crown which gives from 200 to 300 total sections. While theoretically there is no restriction in the number of images that can be processed, attempting to use more than 500 images that are 3–5 MB each will reduce the performance of the program significantly.

3.2 Photography

1. Any sequential set of digital images can be used in this 3D reconstruction technique. For example, a sequence of images can be photographed on a light table and assembled and processed in AE. A microscope is necessary only if tissues need to be magnified to view them adequately. The Sony DSC707

(released by Sony in 2000) is still one of the most highly recommended consumer-grade cameras for microscopic photography. It can be purchased used or new online for less than $500 US. Martin Microscope (GA) manufactures an adapter for the camera to allow its use on any microscope.

2. Experimenting with white balance settings is important to see which setting gives the most contrast between tissue colors and the background color of the image. The auto focus feature on the Sony camera works remarkably well and allows images to be captured very quickly. A remote shutter-release is recommended to prevent vibration when taking the picture. A disadvantage of using the Sony camera is that no real-time imaging on a computer is available. Images must be captured to a memory card and then transferred to the computer.

3.3 Importing Images and Converting Them to JPEG2000

1. Open AE and double click anywhere in the Project panel on the upper left of the screen (Fig. 1). This will open the File Manager and allow you to select the images to import. Find the folder with the images and click on a single image; all images in that folder will be imported. Make sure "JPEG Sequence" has a check mark next to it in the menu in the bottom–middle of the file manager screen (Fig. 2); this will ensure images are imported as a sequence (*see* **Note 2** for difference between importing images as "Footage" or "Sequence"). Click "Open."
2. A single line showing the file names of all the images will be listed in the Project panel. Click/hold and drag the sequence to the composition icon at the bottom of the project panel (the composition icon looks like a piece of film. *See* Fig. 1). This will bring all the images to the Composition panel (Fig. 1) at the bottom of the screen and show the first image in the preview panel at the center of the screen. To auto-scroll through all the images, press the space-bar. Press the space-bar again to stop scrolling.
3. At the AE main menu (Fig. 1) select File>Save As>Save As and save the project in the same folder with the JPEG source images with an appropriate name (*see* **Notes 3** and **4**). Be aware that saving the project is not the same as saving images to a folder. You will save images by using the "render queue."
4. Rendering: To save (render) the images in JPEG2000 format, highlight the sequence in the Composition panel (Fig. 1), go to AE main menu and select Composition>Add to Render Queue. The Render-Queue panel will open within the Composition panel at the bottom of the screen (Fig. 3). The composition containing the sequence of JPEG source images will become a tab at the top of the panel. In the Render panel click "Best Settings" (in blue) to the right of "Render Settings."
5. In the window that opens, go to the box to the right of "Resolution" and make sure you are rendering at full resolution. Leave all other settings the same and click OK.

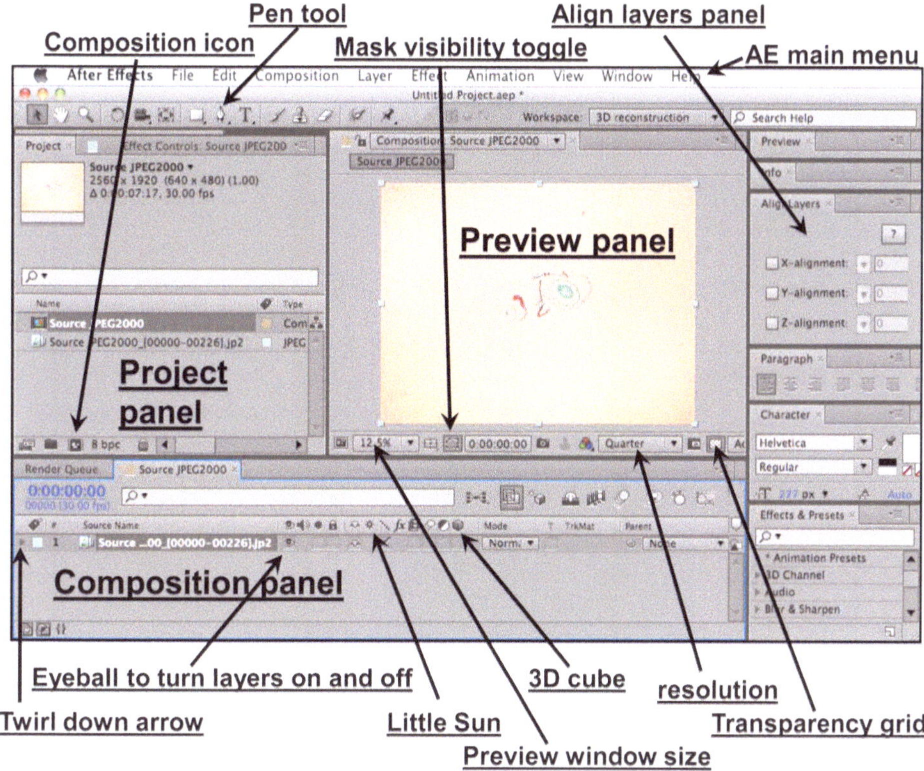

Fig. 1 The opening screen after starting Adobe After Effects. The different panels used in this demonstration are indicated as are various icons mentioned in the text

6. Back at the Render Queue panel click "Lossless" (in blue) next to "Output Module." In the window that opens, go to the Output Module Settings>Main Options window, click the box to the right of "Format" and select JPEG2000. Click the box to the right of "Channels" and select "RGB + Alpha." Leave all other options as they are and click OK.
7. At the Render Queue, click the blue file name to the right of "Output To." In the window that opens (Fig. 4) create a new subfolder (use the button at the lower left of the screen) in the same folder as the JPEG source files and AE Project. It is recommended that you name the folder and each image something similar to "source JPEG2000"; AE will automatically number each image sequentially.
8. Click "Save" and on the far right side of the Render Queue panel, click the "Render" button (Fig. 3). It will take a few minutes for the images to render into the new folder. A blue line across the top of the Composition panel will indicate the progress.

3.4 Aligning Images with Each Other

1. After images have been saved, double click anywhere in the project panel and click on one file in the JPEG2000 folder to bring all the images back into AE as a sequence of JPEG2000 images. Drag the new sequence to the Composition icon (Fig. 1).

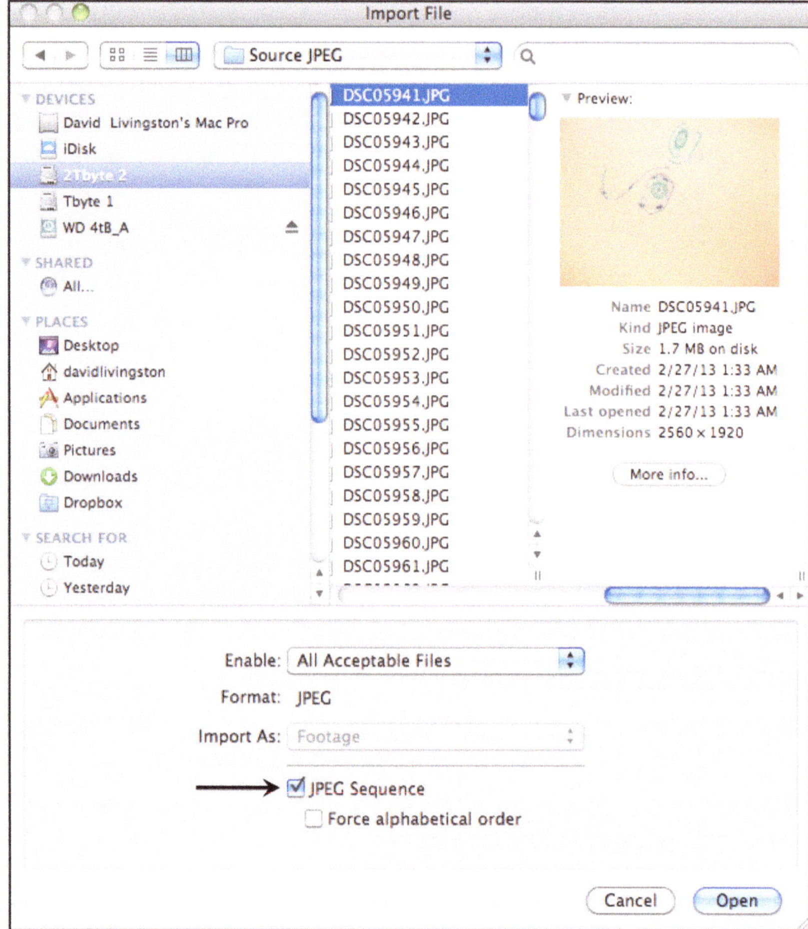

Fig. 2 The Import File window that opens (Mac) when the Project panel is double-clicked. Notice that only one file in the folder is selected and the JPEG Sequence option (*arrow*) is selected

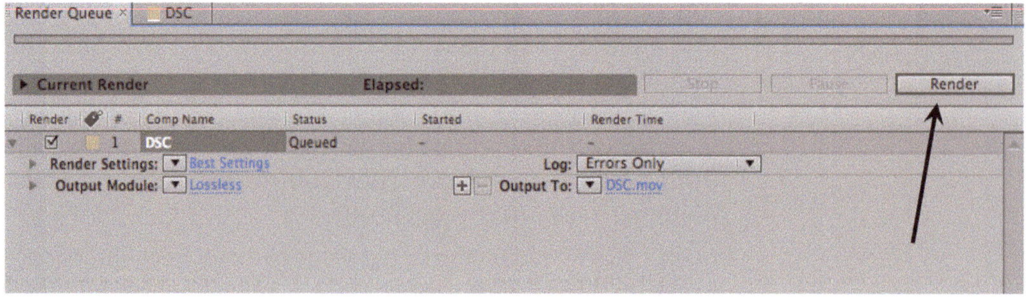

Fig. 3 The Render Queue panel that opens when Composition>Add to Render Queue is selected from the AE main menu. Note the text in *blue*. All three of the regions in *blue* text must be opened and the parameters adjusted as needed. Notice the Render button at the *far right* (*arrow*) (Color figure online)

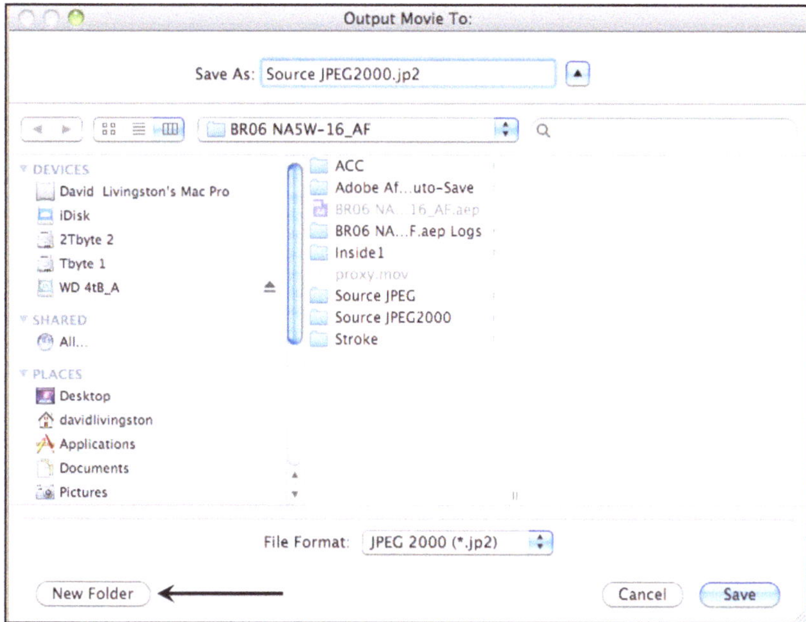

Fig. 4 The window (Mac) that opens when the "output to" is clicked in the Render Queue panel. Notice the New Folder button in the *lower left corner* (*arrow*)

2. With the new composition in the Composition panel highlighted, go to the AE main menu Effect>Distort>Warp Stabilizer. The Effects panel will open (Fig. 5) within the Project panel and the Project panel will revert to a tab.
3. Twirl down "advanced" in the Warp Stabilizer window by clicking the small gray triangular arrow on the left (Fig. 5) and be sure "Detailed Analysis" is checked. In the box to the right of "Method" select "Position, Scale, Rotation." In the box to the right of "Framing" select "Stabilize Only." Leave all other settings as they are.
4. The stabilization will take from 15 to 30 min depending on the number of images. Review the stabilization by placing the time needle at the start of the composition and press the spacebar. *See* **Note 5** and Subheading 3.13 if the alignment is not satisfactory.

3.5 Clearing the Background

1. Highlight the sequence of aligned images in the Composition panel and go to the AE Main Menu, Effects>Keying>Color Key. Click the eyedropper to the right of the "Key Color" line and move it to the color of the background you wish to remove, ideally somewhere near the tissue (*see* **Note 6**). Click the left button when you want to select a particular color. In the Effects panel, hover over the blue "0" to the right of "Color Tolerance" until the 2-way arrow appears and slide it

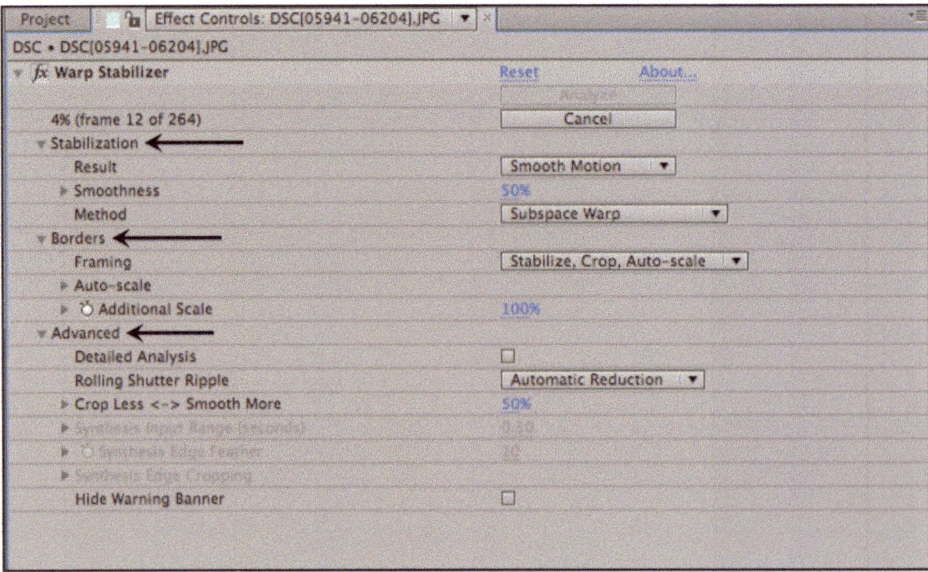

Fig. 5 The Effects panel with Warp Stabilizer opened and the three main parameters twirled down (*arrows*). The parameters in the *boxes* to the *right* of "Method" and "Framing" must be changed (*see* Subheading 3.4)

to adjust the amount of color removed. Toggle the Color Keying effect on and off by clicking "*fx*" to the left of "Color Key" in the Effects panel. Do not worry about remaining background at the corners of the image. A mask will be applied to the images to remove vignetting in the corners as well as most of the noise in the background.

2. Color Keying will be applied automatically to all images in the sequence in the same way. The color will not be removed in the same way for any image or images within the sequence that was stained either darker or lighter. This underscores the need to insure that all images were stained identically (*see* Subheading 3.1).

3.6 Cleaning Images Using a Mask

1. Even the best color-keying routine will not be able to remove all noise from individual images due to dust and other particles adhering to slides. To clean images a "garbage matte" will be drawn around the region of interest which will hide aberrant colors from unwanted particles on the image.
2. At the top of the AE window click on the fountain pen icon (the "pen tool" Fig. 1); also make sure the box to the right of the pen tool icon, labeled "RotoBezier" is checked.
3. Now the cursor (pen) in the preview window will be a small pen tip. Click the new cursor to one side of the tissue in the image in the preview panel, then move a little further and click again. Two small boxes will be drawn with a line connecting them. Continue this process until the tissue is completely outlined with about five or six boxes connected with lines. When you finally close the circle, all objects outside the line will disappear.

4. Now go to the Composition panel and click the gray triangular arrow to the far left (Fig. 1) to twirl down the layer. Twirl down "masks" and look for a box that says "Add." Click the arrow to the right and select "none." This will allow you to visualize everything in the image as the mask is adjusted. At the end of the masking process re-select "add" to again remove (hide) everything outside the mask.
5. Now make adjustments to the mask so it will remain just outside the tissue for each image and hide as much background noise as possible.
6. In the composition panel twirl down (click the gray triangular arrow to the left) "Masks" and then twirl down "Mask 1." Click the stop watch to the left of the line labeled "Mask Path." This will put a keyframe (small diamond) on your time line in your composition. Now click-hold the time needle and move it forward in your sequence until you see tissue beginning to migrate outside the mask. In the Preview panel command-click the mask tool while it is inside the mask, then move the arrow (cursor) over one of the boxes, click-hold and move the box so the tissue is once again inside the mask (*see* **Note** 7).
7. Work your way around the mask to ensure all the tissue is surrounded by the mask. Move the time needle ahead about 10 or 20 images and repeat this process. You should only have to do this about six or eight times over the course of the entire sequence unless the tissue is uneven or distorted. You should not have to reposition the mask for each image since masking with keyframes will reposition the mask gradually between each keyframe.
8. At the end of the sequence, move the time needle back to the beginning and press the spacebar to scroll through the sequence. Reposition the mask at any image where it was improperly placed. Keep in mind that a keyframe will automatically be added to your timeline each time you reposition the mask.
9. To see what the final image series will look like, change the mask type to "Add" in the box to the right of "Mask 1" under the Mask menu on the Composition panel. To hide the mask as you scroll through the sequence click the Mask Visibility icon (Fig. 1) at the bottom of the Preview panel.
10. When you are satisfied that the mask will remove as much noise as is practical then render the aligned/cleared/cleaned images into a new folder as described in Subheading 3.3. Then save the project.

3.7 Outlining Tissue Using the Stroke Function

1. To simulate shadows on the surface of the tissue in the final reconstruction and create a more distinct 3D object it will be necessary to place a thin, dark outline around all structures in each image (Fig. 6).

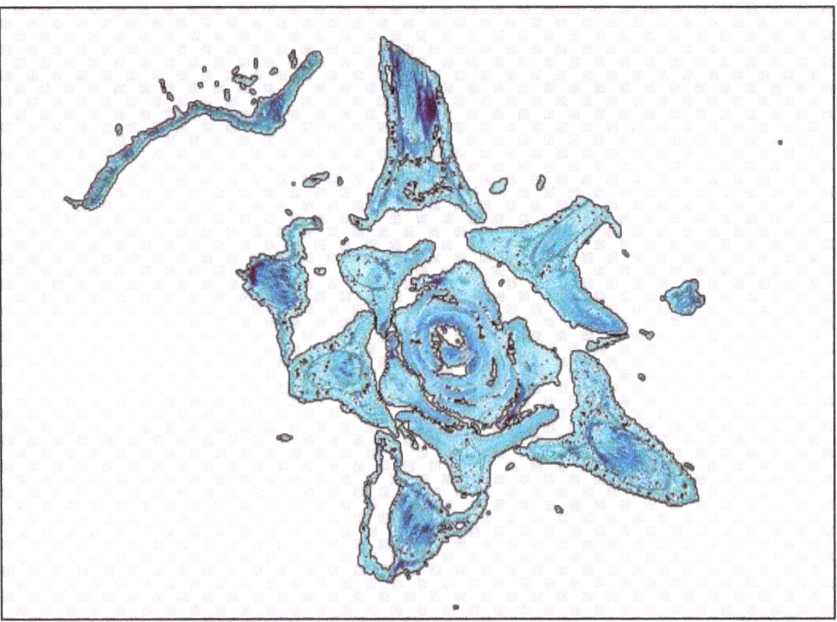

Fig. 6 A single section of *Arabidopsis* stained with Safranin and Fast Green. The background of the image has been cleared (Subheading 3.5) and is shown on top of the transparency grid in the Preview panel (Fig. 1). This image shows the *narrow black* outline around all the tissues called a "stroke" (Subheading 3.7)

2. Import the aligned and cleaned images as a sequence (double click the Project panel) and then drag the sequence to your composition icon. With the composition highlighted, go to the AE main menu, Layer>Layer Styles>Stroke.
3. In the Preview panel notice that a thin red line has appeared around all objects in the image. You will need to make this line somewhat thinner and darker. So, in the Composition panel, twirl down the sequence and then twirl down "Layer Styles" at the bottom. Then twirl down "Stroke." Click on the red box and click/hold the white circle in the upper right corner of the color panel. Drag the circle to the lower left corner (to the black corner) and release the mouse. Click OK. Now change the size of the stroke from 3 to 1 and reduce the opacity to 80 %. Leave position on "outside." To see the black stroke (Fig. 6), click the Transparency Grid icon (Fig. 1) at the bottom of the Preview panel.
4. Render the sequence (as described in Subheading 3.3) into a new folder with an appropriate name.

3.8 Distribute Aligned, Cleared Images in Z-Space

Now the aligned and cleaned images with the added stroke must be distributed in Z space to allow 3D manipulation.

1. Double click in the Project panel and highlight the folder containing the aligned and cleaned images (with a stroke). This time do not select an individual image. By highlighting the folder, all images will be imported as footage within the folder (*see* **Note 2** for a description of the difference between footage and sequence). To see the individual images once the folder is imported into the Project panel, click the folder.
2. To speed up the next few steps, press the "Caps-Lock" key. This will disable AE from refreshing the preview screen each time a change is made.
3. Select all the images in the folder (click the top image and then shift-click the last one) and drag them to the Composition icon. A window will open in the middle of the screen called "New Composition from Selection." In the "Still Duration" box type 20,000 for a 2 min duration. Leave other settings as is and click OK.
4. In the AE Main Menu go to Layer>New>Null Object. Then select all the images in the composition window, except the Null Layer. To quickly select all layers click one image and then type "Command (control on a PC) A" to de-select the Null Layer, Command-Click on it.
5. Click the coil icon just under the heading "Parent" in the composition panel. Drag the coil to the layer labeled "Null 1." Now all the selected images will be under the control of the Null Layer.
6. Click the 3D Cube box (Fig. 1) for the null layer and one of the selected image layers. The box will automatically be checked on all selected layers.
7. Next, with all layers, except the null selected, check the box next to "Z-alignment" in the AlignLayers Panel to the right of the preview panel (Fig. 1). In the rectangular box to the far right change the "0" to about 6 and press enter. There are no units to this number so you will need to experiment with it to get the proper overall height (or depth) of your 3D object.

3.9 Animation to Manipulate the Object in 3D Space

1. In the Composition panel twirl down the Null Layer and twirl down "Transform." On the line labeled "X-Rotation" change the number in blue to the far right to about 120. This will orient your 3D object at a slight angle.
2. Then click the stop-watch next to the line labeled Z-rotation to set a keyframe at the start of your composition. Move the time needle to the end of the composition and change "0×" to about "10×." This will allow ten complete rotations of your 3D object within the 2 min timeframe of your composition.

3. Now turn off the caps-lock button and your 3D object will appear in the Preview panel after a short wait. You will likely need to reposition the image to center it, or to reduce the size.
4. Twirl down "Transform" in the Null Layer and adjust the numbers in blue in the line labeled "Position." To change the size of the object adjust the blue numbers in the line labeled "Scale."

   You should experiment with the different angles and speed of rotation to create the most effective animation to visualize your 3D object.
5. Press the space bar to watch your object rotate. It will be slow since AE has to redraw each layer for every shift in the view. Once the animation has played through, a green line will appear at the top of the composition indicating that this portion of your animation has been rendered into RAM memory and will now play much faster. Reposition your time needle at the beginning of the green line and press the spacebar again to observe the animation in real time. Reducing the resolution (Fig. 1) of the Preview Panel will speed up the playback.
6. To hide the gray lines in the Preview Panel go to the AE Main Menu. Under "View," uncheck "Show Layer Controls." Save the project.

3.10 Adding Layers to View Internal Structures

1. To view internal structures repeat steps in the Subheadings 3.5, 3.7, and 3.8 using a different color to perform clearing described in Subheading 3.5. It will be necessary to clear all unwanted tissue and leave behind only the structures that are stained a color that is distinct from surrounding tissues. This works best with stains that are specific for specialized cells, such as Safranin for xylem vessels or immuno-histological stains that are specific for certain proteins.
2. Begin by importing the aligned and cleaned images from Subheading 3.6 as a sequence. Drag the sequence to the Composition icon and highlight the sequence in the Composition Panel. Go to Effect>Keying>Color Key and use the dropper to select the color that when removed will leave behind the structures you wish to reconstruct. You may need to experiment with precisely how much of the surrounding color to remove. A stroke can be applied to the sequence (*see* Subheading 3.7) but may not be necessary. Render the images into a new folder.
3. Repeat Subheading 3.8 for the images that show structures inside the tissue (*see* **Note 8**).
4. Now superimpose the images showing the whole plant with the images showing only the inside. The inside will be made visible by gradually reducing the opacity of the whole plant images during the animation.

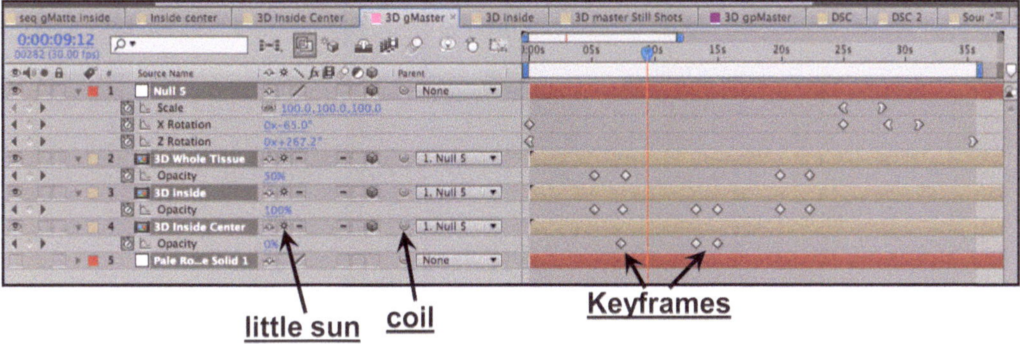

Fig. 7 The composition panel with three footage compositions nested (pre-comped—*see* Subheading 3.10) and under the control of a Null layer. Note the coil described in Subheading 3.8 and that the "little sun" icon (Collapse Transformations) is turned on for the three compositions. Keyframes have been added (*small diamonds* on the *right* side of the panel) to signal when changes to the various parameters (positional parameters for the Null layer and opacity to the 3D layers) are to begin

5. Click the tab for the 3D composition you made in Subheading 3.9. Under the "Transform" menu of the Null Layer click the blue "Reset." Click the stopwatch on all the lines where keyframes were set to remove them.
6. In the Project panel select both 3D compositions and drag them to the Composition icon (this process is called "pre-composing" or just pre-comping). Now you will have two layers in the Composition panel that have all your 3D layers embedded in them. Rename the layer "3D master" or something similar, by left-clicking the composition in the Project panel and selecting "Rename."
7. Add a Null layer to the new composition and then after selecting the two compositions (in the Composition panel) pull the coil to the null layer so both layers will be subject to manipulations from the null (*see* Subheading 3.9). Click the box with the icon that looks like a little sun (Fig. 1—"Collapse Transformations") on each layer (Fig. 7, arrow).
8. Twirl down the null layer as described in Subheading 3.9 and experiment with different rotations.
9. To reduce the opacity of the surrounding tissue and visualize internal structures (Fig. 8b), twirl down the whole tissue layer, then twirl down "Transform." Make sure the Time Needle is at the beginning of your time line. In the line labeled "Opacity," click the stop watch to the left to set a key frame on the Time Line. Move the time needle ahead to where you want to begin viewing internal structures. Click the gray diamond to the far left to set a keyframe. This will let the program know where to begin the reduction in opacity. Then move the time needle forward a few seconds and change the opacity from 100 to

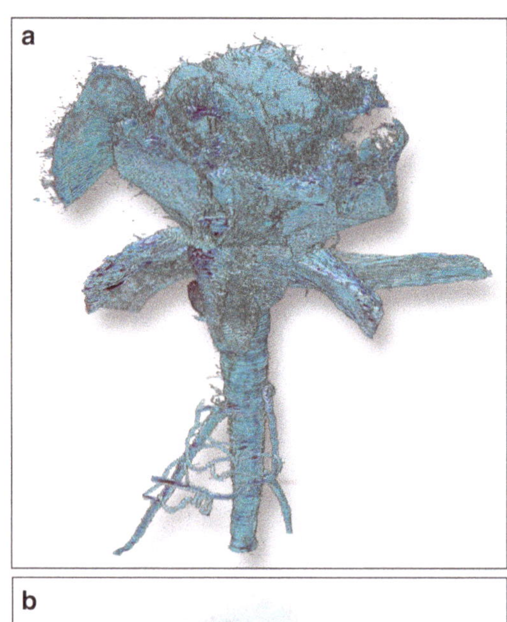

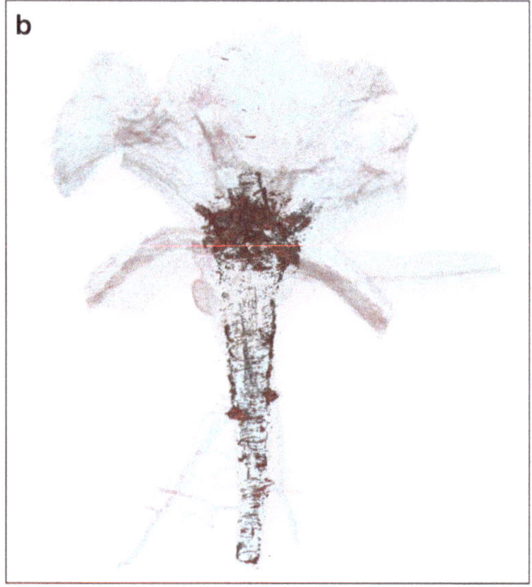

Fig. 8 3D reconstruction of *Arabidopsis* (cv Columbia-0) 7 days after freezing at −14 °C. (**a**) Reconstruction of the whole plant. Note the tattered appearance of the dead leaves. (**b**) The same reconstruction as in (**a**) except that tissue surrounding the putative freeze-response (in *red*) has been digitally cleared. Note the *circular* shape of the *red* staining region in the *center* of the crown (*arrow*). Its appearance as a donut-shape is more obvious in the video in Appendix A

about 5 or 10 % depending on how transparent you want the external tissue to be (Fig. 7).

10. When the composition is played, the first few seconds will show the whole tissue rotating and when the time needle reaches the opacity keyframe, the external tissue will gradually become transparent showing the internal structure in 3D (Fig. 8a, b). Keyframes can be repositioned anywhere in the

composition by click-hold and dragging them or they can be removed by clicking and deleting.

3.11 Rendering the 3D Animation as a Movie

1. You can preview your animation by rendering it as a movie at a much lower resolution than you will in your final video. Sometimes this is faster than waiting for AE to load the animation into RAM memory.
2. Follow the directions for rendering (Subheading 3.3) except in the Render Settings reduce the resolution to Quarter or Half. Click OK. In the Output Module make sure Format is set to "QuickTime" and Channels are set to RGB + Alpha. Click OK. In the Output To window have your video rendered to the same folder as all your other files related to the project.
3. When the video is finished rendering, you can view it outside of AE in QuickTime or some other video viewing program that will play .mov files.
4. Use this reduced resolution version of your animation to decide the best rotations and opacity changes. Make appropriate changes to your composition and then re-render it at full resolution.
5. A disadvantage of this 3D reconstruction technique is that the final video is not interactive. Once the animation is rendered to a video the 3D object cannot be manipulated. It is therefore important to select as many angles in the animation process as may be important to clearly understand the tissue in 3D.

3.12 Adding a Background and Drop-Shadow

1. To add a background color to your video, import the full resolution .mov file rendered in Subheading 3.11. Drag it to the Composition icon. In the AE main menu go to Layer>New>Solid. A solid colored layer will be placed on top of your .mov layer in the Composition panel hiding your video. Click and drag the solid layer below the .mov layer to allow you to see the video on top of the background (Fig. 8). Change the color of the background by highlighting the solid layer in your Composition panel and in the AE main menu go to Layer>Solid Settings. Click the color box to change the color.
2. To add a drop shadow, highlight the Video layer and go to the AE main menu Layer>Layer Styles>Drop Shadow. In the Composition panel, a Layer Styles menu will be added to the composition. Twirl down "Drop Shadow" and experiment with the settings. We have found good settings to be Opacity = 50, Distance = 50 and size = 50. Leave the other settings as they are.
3. Text can also be added to your video and the text can be animated in a variety of ways. Descriptions of how to do this are complicated and are best described in video tutorials. See www.linda.com.
4. When you are ready, render the 3D video at full resolution.

3.13 Manual Image Alignment

During manual alignment there is no reason to worry that you will drift in one direction or another as you manipulate each image. It is not known why but if a center point is used in each image as a kind of visual fiducial, the resulting 3D volume will be anatomically accurate.

1. Double click in the Project panel and import the JPEG2000 images from Subheading 3.3 as a sequence. While not required, it will be easier to manually align images that have been cleared of background. Follow the procedure above (Subheading 3.5) for digitally clearing the background. Render the cleared images into a new folder using composition>add to render>queue as described above in Subheading 3.3.
2. Double click in your project panel and select the folder (i.e., *not* an individual image this time) where the cleared images were rendered and click "open." All images will be imported within the folder as footage (i.e. not as a sequence).
3. Click on the folder in the project panel and all the individual images will be listed. Select the topmost image and shift>click the last one. Click and drag the entire set of images to the composition icon (Fig. 1) to create a footage composition in the Composition panel (Fig. 1). This will allow all images to be viewed simultaneously on top of each other. To view only a single image select all the images in the composition (command or control A) and click on the small eyeball (Fig. 1) in one image. This will hide all images from view. With the Composition panel highlighted press the "t" key to open the transparency parameters of all the images. Change the transparency number from 100 % to about 60 %, then press "t" again to close the transparency parameter window.
4. To begin manual alignment, click anywhere in the Composition panel to deselect all the layers. Then scroll to the bottom of the list of images and click the eyeball of the bottom two images to turn on only those images. You will be able to see the bottommost image through the second one and can now adjust the position and rotation of the second image so that it falls directly on top of the bottom image (which remains stationary). Be sure the eyeball for both images is showing but *only* highlight the top image in the Composition panel. Whichever image is highlighted will be the image that will be manipulated (*see* **Note 9**).
5. Press the "w" key to turn on the rotation parameter within the Preview pane (*see* **Note 10**). Click-hold the left mouse button in the Preview panel to rotate the image. If you hold down the "command" (control on a PC) button and mouse click/hold you will be able to move the top image positionally. By alternating rotation and positional manipulation you will be able to place the top image directly over the bottom one.

6. To see the progress of your adjustments, click the eyeball on the top image on and off. When you are satisfied with your progress, save the project (command>s) and click the eyeball of the bottom image to turn it off, then click the eyeball of the next image above to turn it on. Be sure and highlight the new image, then begin your adjustment with the mouse. After some practice you should be able to make manual adjustments to about 60 images in an hour.
7. When you are finished aligning, select all images, click the letter "t" and bring the transparency of all images back to 100 %.
8. Because all images are currently on top of each other, they must be distributed as a sequence before rendering.
9. Select all the images (the image you click first will determine the order in which the next process will occur) and then in the AE main menu go to Animation>Keyframe Assistant>Sequence layers. A small window will open asking if you want any overlap between images. Make sure the overlap box is not checked and click OK.
10. In the Composition panel at the bottom all your images will be distributed automatically in a "stair-step" pattern. If you click the space-bar, the sequence will automatically play through one image at a time.
11. Once you are satisfied with the alignment, render the images to a different folder (*see* Subheading 3.3 "Rendering") with a name such as "aligned."
12. After images have been aligned and rendered go to Subheading 3.6, "cleaning images using a mask" and proceed with the remaining steps.

3.14 Images Used in the Demonstration (See Note 11)

Arabidopsis thaliana (cv Columbia-0) was frozen at −14 °C for 3 h and was allowed to recover from freezing for 7 days. Plants were fixed, dehydrated, and embedded in paraffin, as described elsewhere [4]. The paraffin-embedded plants were cut in cross section at 20 μm. Sections were photographed at 40× with the camera described above and imported into Adobe AE on a Mac Pro computer. Except for the bottom 1 or 2 mm of the root, it took 330 sections to photograph the entire plant used here.

Freezing *Arabidopsis* plants at −14 °C under our conditions [5] resulted in a survival of 40 % of the population. The plant shown in Fig. 2 is one that would eventually fully recover and regrow because meristematic tissues in the center of the rosette survived freezing (Fig. 9). This is the region of the plant from which new growth appears, even though leaves that have been frozen are dead. Dead leaves in this case resulted in a somewhat wrinkled view of the top part of the plant in the 3D reconstruction (Fig. 8a). The crown region of an *Arabidopsis* plant is analogous to

Fig. 9 A surviving *Arabidopsis* plant after being frozen at −14 °C for 3 h and allowed to recover for 7 days. Note the dead leaves surrounding the live tissue regrowing from the *center*

the apical meristems of winter cereals that are also the hardiest tissues in the plant [6, 7].

Individual 2D cross sectional images show sporadic Safranin-stained tissue in seemingly random regions of the crown (not shown). When a sequential series of those images were aligned and viewed in 3D, a ring structure resembling a donut became visible in the center of the crown region of the plant (Fig. 8a, the "donut" structure is easier to see in the video). This structure was not apparent when individual images were viewed in 2D. It was only when viewed in 3D that the donut-like structure became apparent. The ring structure is similar to that found in oats recovering from freezing [8] but dissimilar in that the ring in a 3D reconstruction of oat forms a somewhat spherical shape. The *Arabidopsis* 3D reconstruction also shows continuity throughout most of the taproot for vessel plugging (*see* ref. 5 for 2D longitudinal view) which was also similar to that observed in oat.

Safranin commonly stains lignin/phenolic compounds [9] a deep red color as shown here. It is not unusual for specific cells within plants to produce phenolic compounds in response to abiotic or biotic stress [10]. Because of the absence of any similar Safranin staining regions in unfrozen plants it seems reasonable to assume that the Safranin shown here stained tissue/cells that were responding in some way to freezing stress. Confirmation that these regions are regions of phenol producing cells as a freezing response is in progress. Alternatively, since the Safranin staining regions do not become visible until about 3 days after freezing, the plant could be responding to secondary infection by bacteria or fungi [11, 12] as described by Beckman [10].

Whatever the composition of the red-staining material, this example illustrates how 3D reconstruction can be used to demonstrate continuity in 3D space of substances that are distributed in a seemingly random manner when viewed in 2D. The technique is well suited for in situ hybridization analysis to detect mRNA in an anatomical context and show its continuity in 3D space. In addition, it has been used successfully to 3D image cancerous and healthy tissues in veterinary pathology studies [3].

4 Notes

1. The reason for converting to JPEG2000 is that J2K images are better quality in a smaller file and will be easier to manipulate in AE than either original JPEG or TIFF images.
2. Images can be imported as either a sequence or as footage. Sequences can be thought of as a horizontal line of images. One change will be applied to the entire series of images. To Color Key all your images simultaneously they must be imported as a sequence. Footage can be thought of as a vertical column of images. Changes can be made to individual images without affecting any other. Manual Alignment requires importing your images as footage so that each image can be separately manipulated.
3. It is important to decide on a permanent organizational structure before starting. For example, the project file as well as all folders containing images generated by the project should be in the same folder. If folders and/or files are rearranged or deleted, reopening a project associated with the deleted or rearranged files/folders will cause a conflict; i.e., AE will not know where to find the files. If that happens go to the AE main menu File>replace footage and select the folder where you moved the images.
4. Save the project regularly. Under After Effects>Preferences> Auto-save specify a time for AE to save automatically. This may cause some annoyance but it is well worth it, if a major problem causes AE to freeze or crash. This will be a rare occurrence but it does happen.
5. Warp stabilizer will produce an alignment automatically but a small amount of resolution will be compromised in each image. Warp Stabilization works well with most crown tissue sections from winter cereals but the *Arabidopsis* plant used in this demonstration had to be aligned manually. If Warp Stabilization does not produce an adequate alignment, images can be aligned manually. This is a somewhat time-consuming process since each image has to be manipulated individually. With practice, about 60 images can be manually aligned in an hour. *See* Subheading 3.13 for instructions on manual alignment.

6. When Color keying, as many color keys as needed can be added to the Effects panel to remove all of an unwanted color or several different colors in the same sequence.
7. To change the preview window from a black background to a transparent background click on the transparency grid at the bottom of the preview window (Fig. 1).
8. Use command (control)>Z to back up, redo.
9. It is not unusual to loose a panel due to a slip of the mouse or click of a button. To get back to your original panel structure click the "workspace" box at the top right side of the AE window (Fig. 1) and select "all panels."
10. The size of the Preview panel can be increased or decreased using the > and < keys. To move the Preview panel spacebar-click in the preview panel to reveal a hand. Use the mouse to reposition the window.
11. The demonstration video can be found at extras.springer.com. The video shows a three-dimensional reconstruction of an *Arabidopsis* plant that had been frozen at −14 °C for 3 h and allowed to recover from freezing for 7 days. This video was made with JPEG images from 330 sequential cross sections of a plant photographed at 40× magnification on a light microscope and processed with Adobe After Effects as described in this chapter. An internal freeze-response was made visible by staining sections with Safranin and Fast Green. Note the doughnut shaped ring in the center of the crown towards the end of the video which is similar to that described in oats recovering from freezing [8].

References

1. Peyrin F (2009) Investigation of bone with synchrotron radiation imaging: from micro to nano. Osteoporos Int 20:1057–1063
2. Ferrando M, Spiess WEL (2000) Review: confocal scanning laser microscopy. A powerful tool in food science. Food Sci Tech Int 6: 267–284
3. Livingston DP III, Tuong TD, Gadi SRV, Haigler CH, Gelman RS, Cullen JM (2010) 3D reconstructions with pixel-based images are made possible by digitally clearing plant and animal tissue. J Microsc 240:122–129
4. Livingston DP III, Tuong TD, Haigler CH, Avci U, Tallury SP (2009) Rapid microwave tissue processing of winter cereals for histology allows identification of separate zones of freezing injury in the crown. Crop Sci 49:1837–1842
5. Livingston DP III, Herman EM, Premakumar R, Tallury SP (2007) Using *Arabidopsis thaliana* as a model to study subzero acclimation in small grains. Cryobiology 54:154–163
6. Tanino KK, McKersie BD (1985) Injury within the crown of winter wheat seedlings after freezing and icing stress. Can J Bot 63:432–435
7. Olien CR, Marchetti BL (1976) Recovery of hardened barley from winter injuries. Crop Sci 16:201–204
8. Livingston DP III, Henson CA, Tuong TD, Wise ML, Tallury SP, Stanley H, Duke SH (2013) Histological analysis and 3D reconstruction of winter cereal crowns recovering from freezing: a unique response in oat (*Avena sativa* L.). PLoS One 8:e53468. doi:10.1371/journal.pone. 0053468
9. Johansen DA (1940) Stains. In: Plant microtechnique. McGraw-Hill, New York, p 49–64
10. Beckman CH (2000) Phenolic-storing cells: keys to programmed cell death and periderm

formation in wilt disease resistance in general defense responses in plants? Physiol Mol Plant Pathol 57:101–110

11. Marshall D (1988) A relationship between ice-nucleation-active bacteria, freeze damage, and genotype in oats. Phytopathology 78: 952–957

12. Olien CR, Smith MN (1981) Extension of localized freeze injury in barley by acute post-thaw bacterial disease. Cryobiology 18:404–409

Chapter 12

Proteomic Approaches to Identify Cold-Regulated Soluble Proteins

Stefanie Döll, Rico Lippmann, and Hans-Peter Mock

Abstract

DIGE (differential in-gel electrophoresis) is a modified version of the widely used 2-D gel electrophoresis (2-DE) for separation of complex protein samples. Two extracts to be compared are differentially labeled using fluorescent cyanine dyes and then separated together by 2-DE. An internal standard labeled using a third dye is included. This approach avoids the pitfalls of gel distortions frequently observed in the standard procedure, which hamper the subsequent gel image analysis. Inclusion of an internal standard improves the quantitative evaluation of the protein patterns. Using the advantages of the DIGE approach, impact of minor temperature differences during cold stress treatment could be quantitatively monitored. We will describe the application of DIGE to monitor the impact of cold stress on the proteome pattern of Arabidopsis. In addition to the separation of proteins, we will also outline how plant growth is performed. Finally, we will also give protocols how proteins of interest can be identified by MALDI-TOF- as well as ESI-MS/MS.

Key words Proteomics, DIGE (differential in-gel electrophoresis), Cold stress, 2-D Gel electrophoresis, Plants, LC-MS, MALDI-TOF/TOF

1 Introduction

The most frequent method to investigate differential protein abundance in large-scale proteomics experiments on crude protein mixtures is two-dimensional gel electrophoresis (2-DE). Differential in-gel electrophoresis (DIGE) is a variant of 2-DE with several advantages. For DIGE, three fluorescent cyanine dyes (Cy2, Cy3, Cy5) are used to label the ε-amino group of lysine residues in proteins. Two of the dyes (Cy3, Cy5) are used to label protein samples to be compared, the third (Cy2) to label an internal standard. All three are loaded on the same 2-D gel allowing minimization of gel-to-gel variability. For detection, each of the dyes is excited separately with a different wavelength. The reader is referred to the earlier protocol describing the sample processing for conventional 2-DE [1] for comparison. In order to benefit from the capability of the

Dirk K. Hincha and Ellen Zuther (eds.), *Plant Cold Acclimation: Methods and Protocols*, Methods in Molecular Biology, vol. 1166, DOI 10.1007/978-1-4939-0844-8_12,

DIGE technique to monitor small quantitative changes, a reproducible setup for plant growth is a mandatory prerequisite. We will outline our experimental design in detail as an example, being aware that modifications of individual parameters might be necessary in the context of other research. The important issue is to control the specific settings to obtain reproducible results across biological replicates. We then describe the identification of candidate proteins resulting from image analysis. Instead of picking spots from the DIGE gels, we prepare additional gels then stained by Coomassie, allowing higher amounts of protein to be loaded. We then compare the image patterns to select equivalent spots. We have observed easier spot identification with this strategy compared to the direct picking of protein spots from the DIGE gels. For identification we describe the procedures applied for our instruments. Protocols will be easily transferred to instruments from other vendors. An outline of the workflow is shown in Fig. 1.

2 Materials

2.1 Plant Cultivation, Treatment, and Harvest

1. Equipment for controlled plant growth (standardized substrate, pots or multi-well trays, plant incubators with light- and temperature control).

2.2 Protein Extraction, Quantification, and Labeling

1. Precipitation solution (PS): 10 % (w/v) trichloroacetic acid (TCA) in acetone and 0.07 % (w/v) 2-mercaptoethanol (2-ME) according to ref. 2. Take 1 g TCA, fill up to 10 mL with acetone and add 6.3 μL 2-ME (= amount for one sample). Prepare directly before use.
2. Washing solution (WS): 0.07 % (w/v) 2-ME in acetone. Add 12.6 μL 2-ME to 20 mL acetone (= amount for one sample). Prepare directly before use.
3. Lysis buffer (LyB): 8 M urea, 30 mM Tris–HCl, 2 % (w/v) CHAPS; pH 8.5. Mix 9.6 g urea, 0.4 g CHAPS, 3.63 g Tris and dissolve in 12 mL ultrapure water. Set pH by adding 1 M HCl and fill up to 20 mL. You can aid dissolution by stirring in a 25 °C water bath or by shaking over night. Filter solution if undissolved particles still remain.
4. 2D Quant Kit (GE Healthcare Life Sciences, Little Chalfont, UK).
5. Spectrophotometer (at 480 nm).
6. CyDye DIGE Fluor Cy2 minimal dye, 5 nmol, CyDye DIGE Fluor Cy3 minimal dye, 5 nmol, CyDye DIGE Fluor Cy5 minimal dye, 5 nmol (from GE Healthcare or equivalent from other companies like NH DyeAGNOSTICS GmbH, Halle, Germany).

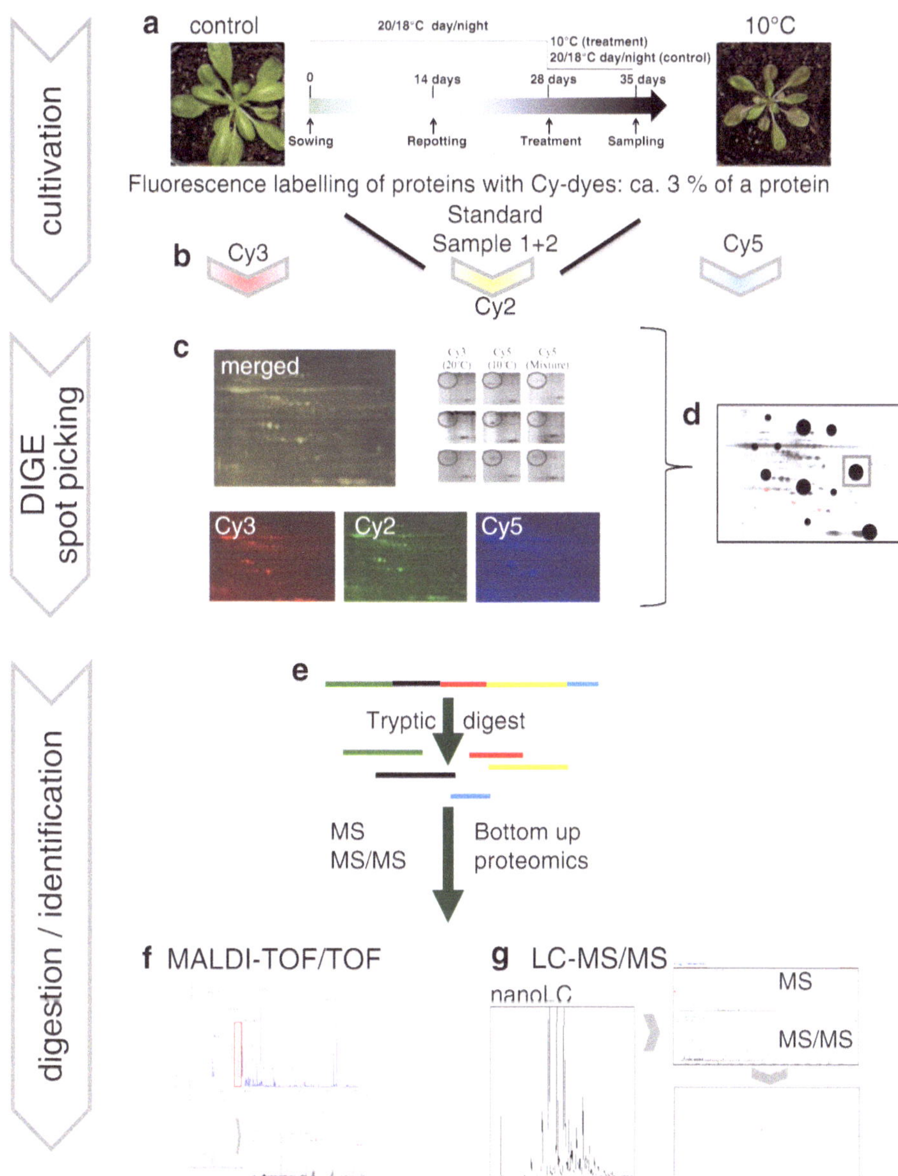

Fig. 1 Simplified workflow. **(a)** Precultivation and treatment of plants. **(b)** Differential labeling of protein extracts of treatment, control, and internal standard. **(c)** 2D-gel electrophoresis of DIGE-labeled samples and merging of scanned images. **(d)** Spot picking of differentially expressed proteins. **(e)** In-gel digestion of the excised spot with protease to cleave the protein into specific peptide fragments. **(f)** Representative MALDI peptide mass fingerprint spectrum (PMF) and the subsequent sequencing of a selected mass in the MS/MS. **(g)** Example base peak chromatogram of separated peptides and selected MS and MS/MS spectra followed by the sequencing of fragment spectra and the database search

7. Dimethylformamide (DMF), high quality anhydrous (specifications <0.005 % H_2O, >99.8 % pure).
8. 10 mM lysine.
9. 0.45 μm filter unit for centrifugation.

2.3 2-D Gel Electrophoresis

2.3.1 Isoelectric Focusing

1. IPG buffer (GE Healthcare/Amersham Biosciences).
2. Bromophenol blue stock solution 1 % (w/v): 0.5 g bromophenol blue, fill up to 50 mL with 50 % (v/v) isopropanol. Store at room temperature.
3. 1 M dithiothreitol (DTT) stock solution: 154 mg DTT per mL ultrapure water. Store at −20 °C.
4. Rehydration solution (8 M urea, 2 % (w/v) CHAPS, 0.005 % (w/v) bromophenol blue, 20 mM DTT, 0.5 % (v/v) IPG). Prepare rehydration stock solution with 8 M urea, 2 % (w/v) CHAPS, 0.005 % (w/v) bromophenol blue. Mix 9.6 g urea, 0.4 g CHAPS, 0.1 mL bromophenol blue stock solution and fill up to 20 mL with ultrapure water. Store in aliquots of 1 mL at −20 °C. Directly before use mix 250 μL rehydration stock solution with 1.25 μL IPG buffer and 5 μL DTT stock solution.
5. Isoelectric Focusing System (Ettan IPGphor 3 IEF System; GE Healthcare).
6. Immobiline DryStrip pH 4–7, 13 cm (GE Healthcare).
7. Immobiline DryStrip Cover Fluid (GE Healthcare).
8. 1.5 M Tris–HCl, pH 8.8. 181.5 g Tris, 600 mL ultrapure water, adjust pH to 8.8 with 1 M HCl. Fill up to 1,000 mL with ultrapure water. Filter through a bottle top filter, 0.45 μm. Store at room temperature.
9. 10 % (w/v) SDS. 100 g SDS, fill up to 1,000 mL with ultrapure water, dissolve by heating to not more than 68 °C and store at RT.
10. Equilibration buffer (50 mM Tris–HCl, 6 M urea, 30 % (v/v) glycerol, 2 % (w/v) SDS, 20 mM DTT, 0.01 % (w/v) bromophenol blue. Prepare stock solution with 6.7 mL of 1.5 M Tris–HCl pH 8.8, 72 g urea, 73.5 g of 99.5 % glycerol, 40 mL of 10 % (w/v) SDS, 0.5 mL of 1 % (w/v) bromophenol blue. Store in aliquots of 9.8 mL at −20 °C. Before use, add 200 μL of 1 M DTT stock solution to one aliquot.

2.3.2 SDS-PAGE

1. Separation gel buffer according to Laemmli [3], pH 8.8 (240 mL 1 M HCl, 183 g Tris, 40 mL 10 % SDS), fill up to 500 mL with ultrapure water. Control pH (8.8), but do not correct it. Store at 4 °C.
2. Stacking gel buffer according to Laemmli [3], pH 6.8 (5.98 g Tris, 4 mL 10 % SDS, 20 mL ultrapure water). Set pH to 6.8

by adding 1 M HCl (approximately 47–48 mL). Fill up to 100 mL with ultrapure water. Store at 4 °C.

3. Vertical electrophoresis system, 18 cm (e.g., SE600, Hoefer Inc. Holliston, MA, USA).
4. Ready-to-use acrylamide/bisacrylamide mixture: 30 % acrylamide/bisacrylamide mixing ratio 37.5:1 *(careful, acrylamide is highly neurotoxic and can penetrate the skin!)*.
5. 10 % (w/v) ammonium persulfate solution. Dissolve 100 mg ammonium persulfate in 1 mL water. Store at 4 °C not longer than 1 week.
6. Tetramethylethylenediamine (TEMED).
7. n-Butanol, saturated with water. Shake 150 mL n-butanol and 100 mL ultrapure water. Store at room temperature.
8. 10× electrode buffer according to Laemmli [3], pH 8.3. 30.28 g Tris, 144 g glycine, 100 mL 10 % SDS. Fill up to 1,000 mL with ultrapure water. Control pH (8.3–8.6), but do not correct it. Store at 4 °C. Prepare 1× electrode buffer working solution.
9. Agarose for mounting of strips. Mix 0.5 g agarose NA (agarose with low electroendosmosis (EEO), highly purified) with 200 μL 1 % bromophenol blue solution and 100 mL 1× electrode buffer. Heat until agarose is dissolved and aliquot in 1.2 mL portions.

2.4 DIGE Gel Imaging and Data Analysis

1. Scanner for imaging of CyDyes with three standard lasers: blue laser (473 nm), green laser (532 nm), and red laser (635 nm) and emission filters for 520 nm (BP40), 580 nm (BP30), and 670 nm (BP30) (e.g., Typhoon® 9400 imager GE Healthcare). Alternatively, a gel documentation system with the appropriate LEDs and filters can be used. They nearly reach the quality of the scanners.
2. 2D gel image analysis software such as DeCyder (GE Healthcare) or Progenesis SameSpots (Nonlinear Dynamics, Newcastle upon Tyne, UK).

2.5 Identification of Proteins by Mass Spectrometry

2.5.1 Coomassie Staining and Spot Picking

1. Coomassie staining solution after Kang et al. [4]. Five percent (w/v) aluminum sulfate hydrate ($(Al_2(SO_4)_3 \cdot (H_2O)_x$, $x = 14–18$), 0.02 % (w/v) Coomassie Brilliant Blue G 250 (CBB-G250), 10 % (v/v) ethanol, 2 % (v/v) *ortho*-phosporic acid. Dissolve 50 g aluminum sulfate (14–18) hydrate in 400 mL water. Mix 200 mg CBB-G250 with 10 mL ethanol, transfer this solution into the aluminum sulfate solution, mix and fill up to approx. 800 mL with water. After aluminum sulfate is dissolved completely, add 23.5 mL of 85 % *ortho-phosporic* acid *(careful, highly corrosive!)* and fill up to 1,000 mL. Particles are visible in this solution. This is the colloidal form of Coomassie.

Do not filter! Keep at room temperature and protected from light. Shake well before use.

2. Destaining solution. 10 % (v/v) ethanol, 2 % (v/v) *ortho-phosporic* acid. Add 100 mL ethanol and 23.5 mL 85 % *ortho-phosporic* acid *(careful, highly corrosive!)* to ultrapure water and fill up to 1,000 mL.
3. Spot picker (hand spot picker (e.g., from Biostep, Germany) or automated spot picker (e.g., Ettan Spot Picker; GE Healthcare)).
4. Gel spot washing solution. 10 mM ammonium bicarbonate, 50 % (v/v) acetonitrile *(careful, highly toxic upon inhalation! Only use inside of a ventilated fumehood)*. Dissolve 80 mg ammonium bicarbonate in 50 mL ultrapure water, add 50 mL acetonitrile. Store at 4 °C and replace regularly.

2.5.2 Tryptic Digest

1. 50 mM acetic acid. Prepare a 1 M acetic acid stock solution. Add 5.27 mL 100 % acetic acid to ultrapure water and fill up to 100 mL. Dilute 5 mL of 1 M acetic acid stock solution to 100 mL with water. Store at room temperature.
2. Trypsin stock solution (200 ng/μL). Dissolve 20 μg trypsin (Sequencing Grade) in 100 μL 50 mM acetic acid. Aliquot and store at −20 °C.
3. 5 mM ammonium bicarbonate with 5 % (v/v) acetonitrile. Dissolve 40 mg ammonium bicarbonate in 50 mL ultrapure water, add 5 mL acetonitrile and fill up to 100 mL with ultrapure water. Store at room temperature and replace regularly.
4. Trypsin working solution (10 ng/μL). Dilute trypsin stock solution to a final concentration of 10 ng/μL (1:20). Example for ten spots: add 4 μL trypsin stock solution to 76 μL 5 mM ammonium bicarbonate in 5 % (v/v) acetonitrile. Prepare fresh.
5. 1 % (w/v) trifluoroacetate (TFA). Prepare a 10 % (w/v) TFA stock solution. Store at room temperature. Prepare fresh 1 mL of 1 % (w/v) TFA working solution.

2.5.3 Protein Identification with MALDI-TOF Peptide Fingerprint

1. MALDI-TOF/TOF (Matrix-assisted laser desorption/ionization-Time-of-Flight) mass spectrometer (e.g., Ultraflextreme, Bruker, Germany).
2. MALDI target (e.g., AnchorChip™ standard targets (Ø = 800 μm).
3. 90 % (v/v) acetonitrile 0.1 % (w/v) TFA (TA90). Add 9 mL acetonitrile and 100 μL 10 % TFA to 900 μL ultrapure water. Store at room temperature.
4. 10 mM ammonium phosphate in 0.1 % (w/v) TFA. Dissolve 57.5 mg $(NH)_4H_2PO_4$ in 50 mL 0.1 % (w/v) TFA. Store at room temperature.

Table 1
Preparation of a MALDI peptide calibration standard from stock solutions

Peptide	Vol. stock [μL]	Conc [pmol/μL]	MW
Leucine-Enkephalin	3	3	556
Bradykinin	2	2	757.4
Angiotensin II	2	2	1,046.54
Angiotensin I	1	1	1,296.69
Substance P	1	1	1,347.74
Bombesin	2	2	1,619.82
ACTH 1-17	3	3	2,093.09
ACTH 18-3	3	3	2,465.2
Somatostatin	3	3	3,147.47
ACTH 1-39	8	8	4,539.266

5. Solvent for HCCA Matrix. Mix 900 μL TA90 with 100 μL 10 mM ammonium phosphate in 0.1 % (w/v) TFA. Prepare fresh.
6. Matrix stock solution (0.7 mg/mL HCCA in TA90/phosphate mix). Mix 7 mg HCCA (alpha-cyano-4-hydroxycinnamic acid) with 1 mL solvent for HCCA (TA90/phosphate mix).
7. MALDI peptide calibration standard (e.g., Peptide Calibration Standard II, Bruker, Germany) or prepare your own peptide standard. Prepare stock solutions of the following peptides with 100 pmol/μL each in 0.1 % (w/v) TFA in water:acetonitrile (2:1; e.g., 10 μL 1% TFA, 60 μL water, 30 μL acetonitrile). Then mix the stock solutions as detailed in Table 1. Add 72 μL 0.1 % (w/v) TFA.
8. Dilute this mixture 1:4 with TA30 (30 % acetonitrile, 0.1 % TFA; e.g., 3 mL acetonitrile, 100 μL 10 % TFA, 6.9 mL ultrapure water). Store aliquots of 20 μL at −20 °C.

2.5.4 Protein Identification by LC-MS/MS

Use only solvents of MS-Grade quality.

1. Solvent A: 0.1 % formic acid in ultrapure water.
2. Solvent B: 0.1 % formic acid in acetonitrile.
3. ESI-Q-TOF (Electron Spray Ionization–Quadrupole–Time-of-Flight) mass spectrometer (e.g., QTOF Premier, Waters Corporation, Milford, MA, USA) with a nanoscale LC (e.g., nanoAcquity UPLC system, Waters).

4. nanoACQUITY BEH C18 column (1.7 μm particle size, 100 μm × 100 mm) with a trap column C18 nanoACQUITY Trap (100 Å, 5 μm, 180 μm × 20 mm), Waters.
5. Calibration standard, e.g., [Glu1]—Fibrinopeptide B in 0.1 % (v/v) formic acid and 50 % (v/v) acetonitrile.
6. MS certified LC-vials (e.g., Clear Glass 12 × 32 mm Snap Neck Total Recovery Vial, Waters).
7. Protein identification software, e.g., Protein Lynx Global Server (PLGS) 2.3 software (Waters).

3 Methods

3.1 Plant Cultivation, Treatment, and Harvest

1. Germinate seeds in a climate chamber under standard conditions. For *Arabidopsis*, for example, these are short-day conditions (9/15 h light/dark cycle, 20/18 °C day/night, 140 μmol/m^2/s light intensity) on a low-nutrient substrate with daily irrigation. Repot after 2 weeks in single pots or multi-well trays into nutrient-rich substrate (*see* **Note 1**) and continue growth at the initial conditions. Plan for a sufficient number of biological replicates (*see* **Note 2**).
2. Four weeks after sowing (*see* **Note 3**), place half of the plants at 10 °C (*see* **Note 4**) in a 12/12 h light/dark cycle at 140 μmol/m^2/s light intensity (*see* **Note 5**), and the other half, as the control, into another incubator with the same conditions, except for the temperature which remains at 20/18 °C day/night.
3. After 7 days, harvest the plants by cutting of the aerial parts, freeze immediately in liquid nitrogen, and store at −80 °C.

3.2 Protein Extraction, Quantification, and Labeling

3.2.1 Protein Extraction (According to Refs. 2, 5)

1. Grind material to a fine powder under liquid nitrogen using mortar and pestle (*see* **Note 6**).
2. Mix 1 part of powder (1 g) with 10 parts (10 mL) of precipitation solution (PS) in a centrifugation tube.
3. Transfer suspension into 2 mL reaction tubes with a cutoff tip (aliquots of 1.8 mL). Cool tubes in liquid nitrogen for 15 s. Keep at −20 °C for 45 min (minimum) or up to 2 h. Vortex after 5, 10, and 15 min.
4. Centrifuge at 25,000 × *g* and 4 °C for 15 min (*see* **Note 7**). Remove supernatant with a fine syringe needle coupled to a Büchner flask and vacuum pump.
5. Resuspend pellet in 1.5 mL washing solution (*see* **Note 8**).
6. Cool tubes in liquid nitrogen for 30 s. Keep at −20 °C for 30 min.
7. Centrifuge at 25,000 × *g* and 4 °C for 15 min (*see* **Note 7**). Remove supernatant as described above. Resuspend pellet in 1.5 mL washing solution (*see* **Note 8**).

8. Cool tubes in liquid nitrogen for 30 s. Keep at −20 °C for 30 min or overnight.
9. Centrifuge at 25,000 × *g* and 4 °C for 15 min (*see* **Note 7**). Remove supernatant as described above.
10. Dry the pellet in a vacuum centrifuge at maximum 35 °C for 20–30 min (*see* **Note 9**) and proceed or store at −20 °C.
11. Determine weight of pellet and resuspend in lysis buffer (50 μL/mg; *see* **Note 10**).
12. Centrifuge at 25,000 × *g* and room temperature for 15 min (*see* **Note 7**). Keep supernatant and discard the pellet.
13. Clarify supernatant by centrifugation through a 0.45 μm filter unit for 10 min at 12,000 × *g* and room temperature. Use directly for determination of protein concentration or store at −20 °C.

3.2.2 Protein Quantification (After Ref. 6)

1. Prepare the needed amount of working color solution provided by the 2D-Quant Kit. To 100 parts color reagent A add 1 part color reagent B. One milliliter per measurement is needed. Samples are measured in duplicates. Add 12 mL for the standard.
2. Prepare standard curves using BSA (2 mg/mL) as a duplicate. In 2 × 6 tubes pipette 0, 5, 10, 15, 20, or 25 μL BSA (*see* **Note 11**).
3. Prepare two tubes with two different volumina per sample (between 1 and 50 μL). Dilute samples with an expected high concentration beforehand (*see* **Note 12**).
4. Add 500 μL of precipitant, vortex and leave for 2–3 min at room temperature.
5. Add 500 μL of co-precipitant and vortex.
6. Centrifuge for 5 min at 10,000 × *g* (minimum).
7. Remove supernatant completely (*see* **Note 13**).
8. Add 100 μL copper solution and 400 μL ultrapure water. Vortex for at least 10 s to dissolve the pellet completely (*see* **Note 14**).
9. Add 1 mL working color solution to each sample. Instantaneous mixing is achieved by introducing the solution as rapidly as possible (*see* **Note 15**).
10. Twenty minutes after pipetting the first sample, the absorption of the samples is measured at 480 nm in disposable 10 mm cuvettes with a spectrophotometer.

3.2.3 Protein-Cyanine Dye Labeling (After Ref. 7)

1. Prepare stock solution of the three CyDye DIGE Fluor minimal dyes. Add 5 μL fresh DMF to 5 nmol of dye solution (1 mM). Centrifuge at 12,000 × *g* for 5 min at room temperature. Store in the dark at −20 °C not longer than 3 months (*see* **Note 16**).

2. Prepare dye working solutions. Add 2 μL dye stock solution to 3 μL DMF. Store at −20 °C not longer than 1 week.
3. Produce an internal standard by mixing the same amounts of protein from all samples.
4. Adjust pH to 8.5 with NaOH (50 mM or higher) (*see* **Note 17**).
5. Add a volume of protein sample equivalent to 50 μg protein to a microcentrifuge tube.
6. Add 1 μL of working dye solution to each sample. Cy2 to the internal standard, Cy3 to control samples, and Cy5 to treatment samples. Mix by pipetting and vortexing.
7. Centrifuge briefly and leave on ice in the dark for 30 min.
8. Terminate the reaction by adding 1 μL of 10 mM lysine and incubation on ice for 10 min.
9. Proceed or store up to 3 months at −70 °C in the dark.

3.3 Two-D Gel Electrophoresis

This protocol follows ref. 1 which also contains instructions for other tissues than leaves.

3.3.1 First-Dimension Separation by Isoelectric Focusing According to Ref. 5

1. Mix 50 μg of one control sample, one treatment sample, and one internal standard (= 150 μg protein in total). Bring the volume to 250 μL by adding rehydration solution and load the mix on one Immobiline DryStrip by pipetting it into the sample application well of a 13 cm IPGphor strip holder. Remove the cover foil from the strip. Place the strip with the gel side down into the strip holder (*see* **Note 18**). Cover with Cover Fluid (*see* **Note 19**). Close cover of strip holder. Place holder into chamber and cover with lid (*see* **Note 20**).
2. Set the program to Step 1: 1 h, 250 V. Step 2: 1 h, 500 V. Step 3: 1 h, gradient from 500 to 4,000 V. Step 4: 5 h, 4,000 V, for a total of about 24 kVh. Start the program (*see* **Note 21**).
3. When finished, equilibrate strips in equilibration buffer for 15 min (*see* **Note 22**). Store strips in stoppered glass reaction tubes at −20 °C (*see* **Note 23**) or proceed with second-dimension separation.

3.3.2 Second-Dimension Separation by SDS-PAGE (After Ref. 8)

1. Prepare gel casting stand, use 1 mm spacers. Prepare the separation gel (11.25 % acrylamide). Mix 3 mL separation gel buffer, 9 mL 30 % acrylamide/bisacrylamide, 12 mL ultrapure water, 240 μL 10 % ammonium persulfate, and 12 μL TEMED (*see* **Note 24**). Fill the separation gel between the glass plates (approx. 2 cm under the upper rim). Overlay with at least 400 μL n-butanol (*see* **Note 25**). Let the gel polymerize for 15–30 min, siphon off the butanol and rinse three times with water (*see* **Note 26**).
2. Prepare the stacking gel (6 % acrylamide). Mix 1 mL stacking gel buffer, 0.8 mL 30 % acrylamide/bisacrylamide, 2.2 mL

ultrapure water, 25 μL 10 % ammonium persulfate and 8 μL TEMED. Fill it on top of the separation gel, height 1 cm. Overlay with n-butanol, let the gel polymerize, remove the butanol, and rinse as above.

3. Reheat agarose including bromophenol blue at 95 °C for mounting strips and pipette 0.5–1 mL on top of the stacking gel. Rinse equilibrated strip shortly in 1× electrode buffer and push it carefully down to the stacking gel. The pointed tip of the strip should be facing to the left.
4. Place gels into the electrophoresis chamber and fill up with 1× electrode buffer (*see* **Note 27**).
5. Perform electrophoresis. 30 min at 75 V for passing through the stacking gel, approximately 4 h 30 min at 150 V for passing through the separation gel. Keep at room temperature. Keep voltage constant. Stop separation shortly before the line formed by the bromophenol blue reaches the end of the gel. Disassemble the chamber. Keep the Cy-labeled gels between the glass plates and scan as soon as possible (*see* **Note 28**).

3.4 DIGE Gel Imaging and Data Analysis

Refer to the manual of your scanner. Procedure is described for a Typhoon Scanner (GE Healthcare).

1. Carefully clean the scanner, switch it on, and let it warm up for 30 min.
2. Place the gel on the scanner without air bubbles. When using low-fluorescent glass plates, the gel can remain between the plates for scanning. Close lid.
3. Set up the fluorescence scan parameters to (1) Blue laser (473 nm), emission filter 520 nm (BP40), (2) Green laser (532 nm) emission filter 580 nm (BP30), and (3) Red laser (635 nm), emission filter 670 nm (BP 30). PMT voltage 550, sensitivity normal.
4. Make a pre-scan with 1,000 μm pixel size to identify a suitable PMT voltage. For this, open the picture in the Image Quant software, select the most intense spots and quantify them with the Volume Review Tool. The maximum pixel value should not exceed 100,000 as this indicates signal saturation has been reached and this will prevent quantitative analysis. A target maximum pixel value of 50,000–80,000 is usually suitable. When adjusting the voltage, relatively small increments of 20–50 V are recommended. If only one or two spots show saturation then only slight downward adjustments of the PMT voltage setting are required. Once the voltage has been optimized for one gel in an experiment, these settings can be used for all similar gels within the same experiment. The maximum pixel value should be within the specified range for all gels, to enable accurate quantitation of spot volumes.
5. Scan the gels at a minimum resolution of 100 μm.

6. Use appropriate software for 2D gel image analysis to merge/warp the three images from each fluorescence channel and to quantify the spots after background subtraction and normalization with the loaded standard over all gels.

3.5 Coomassie Staining and Spot Picking

1. Prepare for each of your treatment/control sample pairs a mixed sample for a nonlabeled duplicate gel (or use the internal standard in three replicates) (*see* **Note 29**).
2. Perform isoelectric focusing and SDS-PAGE exactly as described above for the labeled samples. However, you can load 100–500 μg (*see* **Note 30**) of the mixed unlabeled sample on one Immobiline DryStrip.
3. For Coomassie staining [4], disassemble the gel electrophoresis chamber and rinse the gel twice for 10 min in distilled water (on a rocking shaker).
4. Place the gel into Coomassie Staining Solution. Cover and shake for 1–3 h.
5. Rinse twice (briefly) with water.
6. Place in destaining solution and shake for 30 min.
7. Rinse twice with water.
8. Pick the spots of interest (either by hand or with an automated spot picker) and transfer each into a 0.5 mL reaction tube (*see* **Note 31**).
9. Wash the gel plugs with 400 μL Gel Spot Washing Solution (WS2) for 30 min with vigorous shaking (e.g., vortex with adapter for reaction tubes).
10. Centrifuge briefly and remove the liquid by pipetting or with a fine needle coupled to a Büchner flask.
11. Dry gel plugs in a vacuum centrifuge at maximum of 35 °C (ca. 15 min) and store at −20 °C.

3.6 Tryptic Digest

1. To each gel plug (1–2 mm^3) pipette 7.5 μL trypsin solution and incubate at 37 °C (incubator) for 5 h.
2. Stop the reaction by adding 1 μL 1 % TFA.
3. Incubate at 4 °C overnight.

3.7 Protein Identification with MALDI-TOF

3.7.1 Preparation of Samples and Instrument

1. Pipette 0.5 μL of the Calibration Standard onto the central positions of the MALDI target plate and 0.5 μL of each digested sample in the sample positions surrounding the standard. Let the spots dry, but not completely.
2. Cover semi-dried spot with 1 μL of matrix working solution. Let it dry completely.
3. Place the target into the source of your MALDI instrument.

4. Run a method as described below with the following parameters. These are the values and methods for a Bruker Ultraflextreme MALDI-TOF/TOF. Adjust to your own instrument if necessary as described in its manual.
5. General settings:
   - High voltage. Ion source 1: 25 kV, ion source 2: 22.45 kV, lens: 7.5 kV, reflector 1: 26.5 kV, reflector 2: 13.4 kV.
   - Pulsed ion extraction: 80 ns.
   - Polarity: positive.
   - Matrix suppression: mode deflection—suppress up to 650 Da.
   - Mass range: low, 700–4,020.
   - Detector gain (reflector): 10 × 10.5.
   - Sample Rate: 2 GS/s.
   - Electronic gain: enhanced, 100 mV.
   - Realtime smooth: Off.
   - Spectrum size: 115,200 pts, delay 83,680 pts.
   - Laser frequency: 1,000 Hz.
   - Laser attenuator: offset 67 %, range 20 % (have to be set before each run).

3.7.2 Creating and Starting a New Automatic Run

1. Prepare your instrument for usage. Place the target into the source and teach the target via "Sample Carrier—Advanced—Teach". Adjust first position by setting spot A1 and click "GO". Centre the spot in the view window and finish with reach button. Repeat this for second (A24) and third position (P24). Save teaching file. Check the Laser intensity with 3,000 shots on at least two samples and adjust the laser power.
2. Use the autoXecute Run Wizard to create a new run.
3. Set the appropriate target geometry (e.g., MTP Anchor Chip 800-384) and continue with "Next".
4. "Laser power tuning": If you do not want to use the automatic power tuning, switch to "Next". The laser power can be adjusted within the autoXecute method before the run.
5. "Spot selection": Mark the spots for automatic measurement and choose "calibrate with predefined template" and continue.
6. "Run parameter": All parameters will be applied to the previously selected samples. The MS and MS/MS group consists of appropriate "AutoXecute", "flexAnalysis", and "Biotools" methods.
7. Within "AutoXecute" set the laser power intensities for calibration standard and sample MS method separately. Go to "edit method" and the "laser" tab (Initial Laser Power) and

set them according to the test measurements. To test the laser intensity, apply a test shot on at least two samples and one calibration spot. For the MS/MS method set approximately 5 % more laser intensity than in the MS method. In data destination specify the location where the acquired data shall be saved (with data directory as root folder and sample name as specific directory). For "Biotools" select the customer database you use for Arabidopsis. Continue with "Next".

8. Save the method and go to the control panel.
9. Start the autoXecute run.
10. After measurement the files are stored in your directory and are available for further analysis.
11. Start the "Biotools" program and go to "search"—"Mascot Batch Search"—"Task Editor" and open your tas-file.
12. Click on your samples within the Scout MTP window and the results for the selected spot are opened.

3.7.3 Alternative: Doing a Manual Measurement for a Single Spot

1. Start with measuring one calibration standard. Select it on the target, start the laser (3,000 shots), and record the spectrum. Calibrate the instrument by selecting your calibration standard from the Calibration Mass Control List, select mode "Cube Enhanced". Click through the list and select the monoisotopic peaks in the spectrum. The highest peak should not be cut off at the tip. If the error is acceptable (not greater than 15 ppm), apply the calibration, if not record another one.
2. Now measure all the samples around the first standard, then measure the second standard and the surrounding spots and so on. Record for every spot several spectra.
3. Do an external recalibration of your measurements. In the data analysis program, open one calibration standard and the surrounding samples. Use the standard for a recalibration (as above) of the surrounding samples and identify and remove background peaks.
4. Create a mass list of each sample using your analysis software.
5. Do a database search of your mass lists. You can use a program like Bio Tools (Bruker Daltonics, Germany) or use public databases like Uniprot (http://www.uniprot.org).

3.8 Protein Identification with LC-MS

If the MALDI measurement results in a nonsignificant hit or several protein IDs are overlapping within one spot, protein identification with LC-MS is an alternative and can also be used in addition for de-novo-sequencing.

1. Centrifuge the digested sample at 20,000 × g for 15 min.
2. Transfer the supernatant to a MS certificated LC vial.

3. Prepare the lock mass and calibration solution of [Glu1]-Fibrinopeptide B as described in Subheading 2.5.4.
4. Use 2–4 μL of the sample for an injection onto a Nano LC-MS system.
5. Preconcentrate and desalt the digested samples on the precolumn for 5 min at 5 μL/min with 100 % Solvent A.
6. Separate and elute with a gradient of 3–40 % Solvent B over 30 or 60 min at a constant flow rate of 600 nL/min.
7. Acquire mass spectra data for peptide identification on a Q-ToF mass spectrometer. For lock mass correction of the precursor and the product ions use 150 pmol/μL Glu Fibrinopeptide B in 0.1 % formic acid in acetonitrile/water (25:75, v/v).
8. Use the following parameters for a data-dependent acquisition:

Source	
Source temperature [°C]	80
Cone Gas Flow [L/h]	30
Nano Tip Voltage [kV]	Approx. 2.9
Polarity	ESI positive
TOF Mode	V-Mode

Calibrant [Glu1]-Fibrinopeptide B	
Scan frequency [s]	20
Scan time [s]	1
Reference cone voltage(V)	35
Collision energy [eV]	21
Temperature correction	Disabled

MS and MS/MS parameters	
Scan time [s]	0.95
Interscan time [s]	0.05
Start mass	50
End mass	1,700
Resolution	9,000
Trigger threshold	750
Signal threshold	20
Collision low energy [eV]	4

3.9 Data Analysis

1. Use a protein identification software for processing and identification of proteins from the obtained spectra.
2. Set the processing parameters as followed: mass accuracy with lock mass correction at MS: 785.8426 ± 0.2 Da and MS/MS: 684.3469 ± 0.2 Da; noise reduction with automatic threshold and deisotoping set to medium.
3. Use resulting sequences for database search against, e.g., UniProt Database containing all entries from Viridaeplantae including contaminants keratin and trypsin.
4. Set the search parameters for CID data as followed: peptide tolerance 10 ppm, fragment tolerance 0.05 Da, minimum peptides to match 1, primary digest trypsin, one missed cleavage site, oxidation (Met) and carbamidomethyl (Cys) as variable modifications.
5. Import the raw files to a virtual microtiter plate.
6. Add the processing parameter method and the workflow method to start the processing.
7. If no search hit is significant, perform BLAST homology search with the same database using the de novo Query tool (*see* **Note 32**).
8. Add the second workflow to the samples meant for de novo sequencing.
9. Do not change parameter for mass spectrum, fragment tolerance, digest reagent, etc.
10. Set the calibration error to maximum 10 ppm, maximum five hits and set validation result (*see* **Note 33**).
11. Blast the resulting peptide sequences within the program against your library (e.g., UniProt Viridiplantae).

4 Notes

It is recommended to read the cited manuals as they contain a plethora of explanations and additional hints.

1. The choice of substrate influences total amounts and relative changes of secondary metabolites reacting to cold stress and will therefore also have an influence on protein composition. As the influence of nutrient deficiency may affect one metabolite positively and another one negatively, it might also be feasible to use low-nutrient substrate only, but be consistent during your experiments.
2. Fifty plants of *Arabidopsis* make a good pool for one biological replicate. Make at least three biological replicates per conditions, e.g., three trays with 50 plants.

3. The most important factor for a clear and consistent reaction towards cold stress is the plant age. Young plants should be preferred. The ideal precultivation time for *Arabidopsis* is 4 weeks (or earlier) with a maximum difference in secondary metabolite changes. Already at 6 weeks precultivation time, physiological differences are drastically reduced. Also, the effects of side conditions like the usage of different incubators or substrates become stronger.
4. 10 °C is sufficient to elicit cold stress reactions in *Arabidopsis* and easy to realize in most plant incubators. However, some experimentators prefer to use 4 °C conditions.
5. Light quality and quantity are the second most important factors to consider for your experiment. Light stress is the major threat for plant survival at above-freezing cold stress. While most enzyme reactions are slowed down in the cold by van't Hoff's rule, the light reactions of the photosystems are not, leading to ROS production. It is not advisable to exclude the light factor by conducting cold stress experiments in the dark, because under natural conditions cold and light stress always intertwine. ROS production might be needed to trigger cold signaling chains and you will miss out on a large part of the naturally occurring physiological reactions. Short-day conditions also diminish physiological reactions, whereas long-day and high-light conditions (350 μmol/m^2/s and more) will increase them. Make sure that all plants get the same amount of light. *Technical note*: Operating lamps in cold temperatures diminishes their light output. Measure the light intensity at 10 and 20 °C conditions with a light meter and adjust to the same value.
6. Continue with protein extraction immediately after grinding.
7. Centrifugation can be extended if material is not pelleting.
8. Resuspending can be aided by vortexing, ultrasonic bath (5 min), ultrasonic homogenizer, or stirring with a fine glass rod.
9. Dry until odor of acetone is no longer perceptible (approximately 10 min).
10. Do not include primary amines, DTT, or ampholytes in the lysis buffer, as they might react with the NHS esters of the cyanine dyes. Resuspension is achieved by vortexing, then incubation in an ultrasonic bath for 5 min and stirring with a glass rod. Collect drops by centrifugation and incubate on a shaker at 37 °C for 1 h. Use positive displacement pipettes for the buffer if possible.
11. All steps are carried out at room temperature in 2 mL reaction tubes. Accurate pipetting is essential. Use positive displacement pipettes if possible. Average the two values for the standard curve.
12. For *Arabidopsis* seeds dilute samples 1:10 and use 5 and 10 μL. For *Arabidopsis* stem use 5 and 10 μL undiluted. For *Arabidopsis*

root use 10 and 20 μL undiluted. For barley leaf and root use 5 and 10 μL undiluted.

13. This step has to be carried out rapidly before the pellet dissolves. We use a vacuum pump to remove the supernatant.
14. After dissolving samples are stable for up to 1 h.
15. This step is critical! The time between pipetting and measurement (and also the ambient temperature) has a great influence on the binding of the copper. To minimize this effect we treat all samples individually. Before pipetting the next 1 mL working color solution we wait exactly 10 s. We also wait 10 s between measuring every sample at the spectrophotometer.
16. CyDyes are sensitive to light, ozone, and humidity. Keep dyes and labeled samples in the dark at all times by covering with aluminum foil. We also work in shaded labs when labeling. Ozone might become a serious problem in hot climates. If your area is affected by high ozone or your labeling looses sensitivity during certain seasons, you should consider measuring ozone [9].
17. Check pH of your samples with indicator paper. It should be exactly 8.5. Adjust with NaOH (50 mM or higher) if necessary.
18. At first, the pointed end of the strip goes into the pointed end of the holder. Distribute the solution evenly and remove air bubbles by lifting, lowering, and moving back and forward of the strip, and slight tilting of the strip holder. At last, the rectangular end of the strip is placed into the holder. Make sure that the gel touches the electrodes on both ends.
19. The strip is covered with Cover Fluid to prevent evaporation. Fill oil (1 mL) drop-wise into the holder starting from one end until the whole strip is well covered.
20. The IPGphor chamber has to be exactly horizontal. The pointed end of the holder lies on the anode, the rectangular one on the cathode. The contacts of holder and chamber have to be well matched and the lid should hold down each holder at least in two places.
21. Check if the bromophenol blue migrates towards the anode. If not, check the assembly and programming.
22. Stick exactly to the 15 min incubation time. It is a compromise between buffer exchange and protein leaching.
23. Strip holders and glass tubes need proper cleaning to not disturb later experiments. Directly after use put the holders into a warm detergent solution for 1 h. Then place them for 1 h in fresh 3 % IPGphor cleaning solution on a tilt shaker. Scrub them thoroughly with a tooth brush. Rinse them for 30 min in deionized water and again for 1 h in ultrapure water.

Dry completely before next use. Rinse glass tubes twice with a detergent solution and with ultrapure water.

24. For identifying the gels later place small written tags made of filter paper into the lower left corner of each gel before polymerization.
25. n-Butanol is the upper phase.
26. If you want to leave the gel overnight, mix 1 mL separation gel buffer with 7 mL water and fill on top of the gel. Store at room temperature in a moist chamber (wetted plastic bag).
27. The upper chamber of the electrophoresis system can also be filled with 2× electrophoresis buffer.
28. If it is not possible to scan the gels immediately, store them in 1× electrode buffer at 4 °C in the dark. However, already after 1 day the spots will show a significant diffusion.
29. For protein identification you can either post-stain the Cy-labeled analytical gels or make additional preparative gels. Analytical and preparative gels can be stained with Deep Purple total protein stain (GE Healthcare). These gels can also be scanned (green laser, 532 nm, 560 nm LP emission filter), post-stained gels with 457 nm laser and 610 nm emission filter. Image analysis programs (e.g., Decider 2D) can match the analytical and preparative gels and help creating a pick list. This can be exported to automated spot pickers. For picking by hand you can use the common Coomassie staining. In our lab we prefer to make a fresh preparative gel (and Coomassie staining) to get high quality samples for MALDI-TOF analysis.
30. Usually you load more than the 150 μg used for the analytical gel. However, when analyzing leaf samples, we reduce the amount loaded to 100 μg to minimize gel overload with the abundant RuBisCO protein. Otherwise it will overlap with many of the other spots.
31. Avoid getting dust and especially epidermal scales into your samples. Work in a dust-free environment. Rinse your lab ware with distilled water. Wear gloves and long-sleeved lab coats all the time. Otherwise keratin is constantly detected in your samples. Use glassware to prepare and store your solutions, because solvents like acetonitrile and TFA extract polymers out of plastic materials leading to background signals disturbing the measurements. Glassware should never be cleaned in the dishwasher. Instead, fill it with 25 % nitric acid and place in an ultrasonic bath for 30 min. Then rinse three times with ultrapure water and three times with the solvent you want to store in it.
32. Use this type of analysis only as a second step in the workflow applied to sequence data not matching any known protein.

33. To have a validated peptide, a series of at least three consecutive y-ions are necessary.

Acknowledgment

Support of the proteomic research performed by the authors obtained from IPK Gatersleben and from BMBF research grants (FKZ 0315047A & FKZ 0315699) is gratefully acknowledged.

References

1. Schlesier B, Mock H-P (2006) Protein isolation and second-dimension electrophoretic separation. In: Salinas J, Sanchez-Serrano J (eds) Arabidopsis protocols. Humana Press, Totowa, NJ, pp 381–391
2. Damerval C, Devienne D, Zivy M, Thiellement H (1986) Technical improvements in two-dimensional electrophoresis increase the level of genetic-variation detected in wheat-seedling proteins. Electrophoresis 7:52–54
3. Laemmli UK (1970) Cleavage of structural proteins during assembly of head of bacteriophage T4. Nature 227:680–685
4. Kang DH, Gho YS, Suh MK, Kang CH (2002) Highly sensitive and fast protein detection with coomassie brilliant blue in sodium dodecyl sulfate-polyacrylamide gel electrophoresis. B Kor Chem Soc 23:1511–1512
5. Amme S, Matros A, Schlesier B, Mock H-P (2006) Proteome analysis of cold stress response in *Arabidopsis thaliana* using DIGE-technology. J Exp Bot 57:1537–1546
6. GE Healthcare (2009) 2-D Quant Kit. Little Chalfont: GE Healthcare. https://www.gelifesciences.com/gehcls_images/GELS/Related%20Content/Files/1314729545976/litdoc28954714AE_20110830215136.pdf
7. GE Healthcare (2005) Ettan DIGE system user manual, 18-1173-17 Edition AB. Uppsala: GE Healthcare. https://www.gelifesciences.com/gehcls_images/GELS/Related%20Content/Files/1314774443672/litdoc18117317AB_20110831091933.pdf
8. GE Healthcare (2010) 2-D Electrophoresis. Principles and methods. Little Chalfont: GE Healthcare. https://www.gelifesciences.com/gehcls_images/GELS/Related%20Content/Files/1335426794335/litdoc80642960_20140202231631.pdf
9. Branham WS, Melvin CD, Han T, Desal VG, Moland CL, Scully AT, Fuscoe JC (2007) Elimination of laboratory ozone leads to a dramatic improvement in the reproducibility of microarray gene expression measurements. BMC Biotechnol 7:8

Chapter 13

Proteomic Approaches to Identify Cold-Regulated Plasma Membrane Proteins

Daisuke Takahashi, Takato Nakayama, Yushi Miki, Yukio Kawamura, and Matsuo Uemura

Abstract

Plasma membrane is the primary determinant of freezing tolerance in plants because of its central role in freeze–thaw cycle. Changes in the plasma membrane proteins have been one of the major research areas in plant cold acclimation. To obtain comprehensive profiles of the plasma membrane proteomes and their changes during the cold acclimation process, a plasma membrane purification method using a dextran–polyethylene glycol two polymer system and a mass spectrometry-based shotgun proteomics method using nano-LC-MS/MS for the plasma membrane proteins are described. The proteomic results obtained are further applied to label-free protein semiquantification.

Key words Cold acclimation, Plasma membrane, Nano-LC-MS/MS, Shotgun proteomics, Label-free semiquantification, In-solution digestion

1 Introduction

Many plants that grow in temperate and subarctic regions increase in freezing tolerance when exposed to a nonfreezing, low temperature (Levitt 1980), which is known as cold acclimation. Cold acclimation results in diverse alterations in plant cell physiology, morphology, and molecular biology [1, 2]. In many cases, freezing results in ice formation extracellularly (extracellular freezing) and plant cells must keep ice crystals from entering into the cytoplasm. Because the plasma membrane (PM) plays a central role in water transport between the inside and the outside of the cell and functions as a barrier for separation of the cytoplasm from the extracellular region, there is a consensus that stabilization of the PM is a prerequisite for survival under freezing stress [3–5]. Thus, it is reasonable to consider that the PM composition responds to low temperature and changes during cold acclimation in order to withstand the upcoming stresses incurred during a freeze–thaw cycle.

Dirk K. Hincha and Ellen Zuther (eds.), *Plant Cold Acclimation: Methods and Protocols*, Methods in Molecular Biology, vol. 1166, DOI 10.1007/978-1-4939-0844-8_13, © Springer Science+Business Media New York 2014

In fact, there are a number of reports that describe dynamic changes in PM components, both in protein and in lipid compositions [6–11]. In addition, several studies demonstrated that specific proteins in the PM were functionally involved in cold acclimation [12–17]. With a recent advance of protein separation and identification techniques (such as nano-liquid chromatography), increased availability of protein analysis equipment (such as mass spectrometers) and development of user-friendly software and excellent genome/protein databases (such as MASCOT, SEQUEST, X!! Tandem and PEAKS), it became possible to reveal proteomic responses to cold acclimation on a large scale in a relatively short period.

In this chapter, we introduce the procedures for cold acclimation, plasma membrane isolation and then protein mass analysis to identify cold-regulated proteins associated with the PM. The procedures described below are easily able to adapt to *Arabidopsis* plants [18, 19] but, in general, it will be applied for other plants as experimental subjects. We have been using the procedures slightly modified for monocotyledonous plants such as rye and oat [20, 21] and *Brachypodium* [22], woody plants such as poplar [23] and *Arabidopsis* suspension cultured cells [24]. Lastly, detailed protocols for protein identification of the PM and microdomains in the PM using nano-LC MS/MS were described elsewhere [25] (Fig. 1).

2 Materials

Prepare all solutions using ultrapure water (prepared by purifying deionized water by Millipore apparatus to attain a resistance of 18.2 MΩ cm at 24 °C) and analytical grade reagents. Prepare and store all reagents at room temperature (unless indicated otherwise). Carefully follow all waste disposal regulations determined by local authorities when disposing of waste materials.

2.1 Plant Growth Components

1. Plant seeds: *Arabidopsis* seeds can be obtained from Arabidopsis stock centers such as ABRC, NASC, and SASSC (http://www.arabidopsis.org/portals/mutants/stockcenters.jsp) or purchased from Lehle Seeds (Round Rock, Texas, USA). Several accessions have been used but Columbia ecotype (Col-0) is one of the most popular ecotypes for cold acclimation studies.
2. Plant bedding mix: two parts of vermiculite and one part of perlite.
3. Nutrient solution A (10× stock): 60 mM KNO_3, 40 mM $Ca(NO_3)\cdot 4H_2O$, 20 mM $NH_4H_2PO_4$, 10 mM $MgSO_4\cdot 7H_2O$.
4. Nutrient solution B (50× stock): 25 mM KCl, 12.5 mM H_3BO_3, 1 mM $MnSO_4\cdot 5H_2O$, 1 mM $ZnSO_4\cdot 7H_2O$, 0.25 mM $CuSO_4\cdot 5H_2O$, 0.25 mM H_2MoO_4.

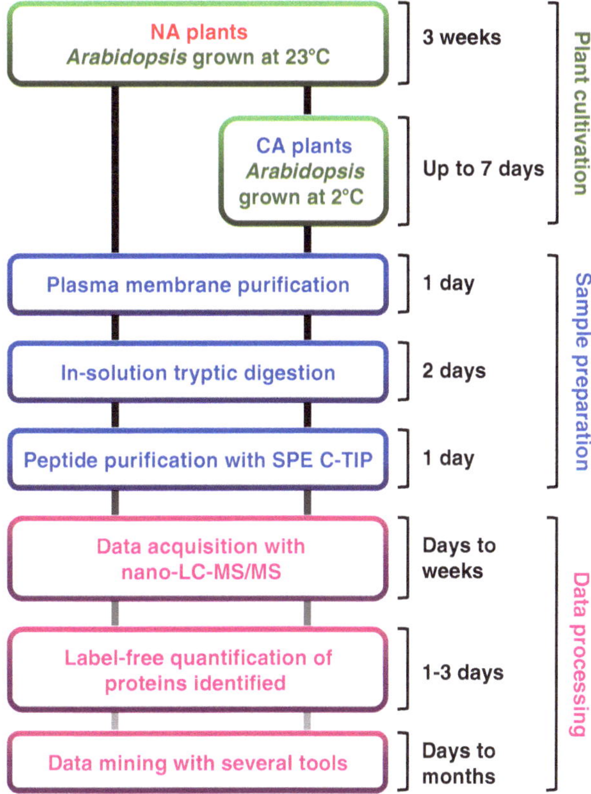

Fig. 1 A representative workflow for shotgun proteomics of *Arabidopsis* PM proteins. It consists of three parts; plant cultivation, sample preparation, and data processing. All plants are grown at 23 °C for 3 weeks (NA plants) and some of the NA plants are then transferred to a cold chamber for cold acclimation treatments (CA plants). Plasma membrane (PM) fractions are isolated using a two-phase partition system, and proteins in the PM fractions are digested with trypsin. Subsequently, peptides obtained are purified and concentrated with SPE-C-TIP. Peptides are then subjected into a nano-LC-MS/MS system and software for label-free identification and semiquantification of PM proteins is used. Data obtained are processed with several tools for mining of novel and/or known factors

5. Nutrient solution C (50× stock): 8 mM Na_2 EDTA, 8 mM $FeSO{\cdot}7H_2O$.
6. Working nutrient solution: mix the nutrient solution A (1 L), B (200 mL), and C (200 mL) and add tap water (8.6 L) to make 10 L working solution.

2.2 Plasma Membrane Purification Components

Several items, including 2 L of ultrapure water, a Polytron homogenizer, centrifuge rotors and ultracentrifuge rotors, should be precooled in a refrigerator.

1. Homogenizing medium: 0.5 M sorbitol, 50 mM Mops-KOH (pH 7.6), 5 mM EGTA, 5 mM EDTA, 1.5 % (w/v)

polyvinylpyrrolidone 40 (molecular weight 40,000), 0.5 % (w/v) BSA, 2 mM phenylmethanesulfonyl fluoride (PMSF), 4 mM salicylhydroxamic acid (SHAM), 2.5 mM 1,4-dithiothreitol. Store at 4 °C (*see* **Note 1**).

2. Polytron generator (PT10SK, Kinematica, Inc., Lucerne, Switzerland).
3. Gauze or cheesecloth.
4. Microsome (MS)-suspension medium: 10 mM KH_2PO_4/K_2HPO_4 (K-P) buffer (pH 7.8), 0.3 M sucrose. Store in a refrigerator (*see* **Note 2**).
5. Electric Teflon-glass homogenizer.
6. NaCl medium: 100 mM NaCl in MS-suspension medium. Store in a refrigerator.
7. Plasma membrane (PM)-suspension medium: 10 mM Mops-KOH (pH 7.3), 1 mM EGTA, 0.3 M sucrose. Store in a refrigerator.
8. Two-phase medium: weigh 1.4 g of polyethylene glycol 3,350 and 1.4 g dextran in a 40 mL centrifuge tube (5.6 % [w/w] polymers in final solution with microsomal suspensions). Add 9.4 mL MS-suspension medium and 7.3 mL NaCl medium (30 mM NaCl in final solution) to the centrifuge tube and mix well by shaking. Prepare three tubes per sample. Store in a refrigerator overnight to completely dissolve the polymers.
9. BioRad Protein Assay Kit (BioRad Laboratories, Hercules, CA): store in a refrigerator.

2.3 In-Solution Tryptic Digestion Components

All preparations must be carefully performed in a clean bench with gloves and a clean lab coat to avoid contamination from keratin, dust, and other exogenous proteinaceous materials.

1. MPEX PTS reagents kit (GL Science, Inc., Tokyo, Japan): make solution B according to the manufacturer's instruction manual. Solution B can be stored in a refrigerator. Prepare DTT solution, IAA solution, and trypsin solution according to the manufacturer's instruction manual freshly immediately before use.
2. Acetonitrile–TFA solution: 5 % (v/v) acetonitrile and 0.1 % (v/v) TFA in water. Mix well (*see* **Note 3**). *(Careful, both solvents are highly toxic upon inhalation! Only use in a ventilated fumehood.)*
3. Pierce BCA protein assay kit (Thermo Fisher Scientific, Waltham, MA, USA).
4. Vial and vial insert (National Scientific, http://www.nationalscientific.com/).

2.4 Peptide Purification Components

1. SPE C-TIP T-300 (Nikkyo Technos Co., Ltd., Tokyo, Japan).
2. 1.5 mL microtubes: make a hole of 3 mm in diameter in the cap with a soldering iron. Prepare two tubes per sample.
3. Solution A: 80 % (v/v) acetonitrile and 5 % (v/v) TFA in water. Mix well (*see* **Note 3**).
4. Solution B: 4 % (v/v) acetonitrile and 0.5 % (v/v) TFA in water. Mix well (*see* **Note 3**).
5. 0.1 % (v/v) TFA solution: quickly add 1 μL of TFA into 999 μL of water and mix well (*see* **Note 3**).

2.5 Instruments and Software for Nano-LC-MS/ MS

1. LC instrument: ADVANCE UHPLC system (MICHROM Bioresources, Auburn, CA).
2. MS instrument: LTQ Orbitrap XL mass spectrometer (Thermo Fisher Scientific, Waltham, MA).
3. Ion source: ADVANCE spray source (MICHROM Bioresources).
4. Trap column for peptide concentration: L-column Micro 0.3 × 5 mm (CERI, Japan).
5. Column for peptide separation: Magic C18 AQ nano column (0.1 × 150 mm; MICHROM Bioresources).
6. Data conversion software: Proteome Discoverer (ver. 1.1.0.263, Thermo Fisher Scientific).
7. Search engine for protein identification: MASCOT search engine (version 2.3.02, Matrix Science, London, UK).

3 Methods

3.1 Plant Growth and Cold Acclimation

1. Plant seeds in a moist Vermiculite–Perlite mix (2:1) in plastic pots and place the pots in a controlled-environment chamber at 23 °C under continuous light (100 μmol/m^2/s).
2. Add nutrient solution occasionally from the bottom of the pot (usually twice a week).
3. Grow for approximately 3 weeks to obtain non-acclimated plants (*see* **Note 4**).
4. Cold acclimate by transferring non-acclimated plants to a cold growth chamber at 2 °C under 12-h light condition (100 μmol/m^2/s) for up to 7 days.

3.2 Plasma Membrane Purification

Wear gloves and a clean lab coat throughout the experiments to avoid contamination by keratin, dust, and other exogenous proteinaceous materials. It is preferable to use low protein absorption microtubes at all stages. Perform all steps on crushed ice (unless indicated otherwise). Centrifuges should be prechilled at 4 °C.

1. Cut off the aerial parts of *Arabidopsis* seedlings and weigh them (10 g or more in fresh weight is desirable for the plasma membrane purification). Put the harvested plant material on a plastic container and wash with chilled distilled water. Wash twice and then drain on paper towels. Keep the harvested plants wrapped with paper towels on crushed ice.
2. Put plant samples into four volumes of chilled homogenizing medium and cut into small pieces with a pair of scissors.
3. Homogenize with a chilled Polytron generator until the samples are broken into tiny pieces (speed 5–6 for 60–90 s). Filter the homogenates through four layers of gauze and squeeze thoroughly. Put the filtrate into 40 mL centrifuge tubes.
4. Centrifuge at 5,000 × *g* for 15 min with a chilled rotor to remove debris and heavy membrane fractions. Transfer the supernatants into ultracentrifuge tubes by decantation. Discard precipitates.
5. Centrifuge at 231,000 × *g* for 50 min with a chilled ultracentrifuge rotor to precipitate microsome fractions. Discard supernatants by decantation.
6. Add appropriate volume of MS-suspension medium to each tube (usually 2–3 mL per tube) and homogenize the pellets with a Teflon-glass homogenizer. Collect the microsomal suspensions with a large-aperture Pasteur pipette into ultracentrifuge tubes. Balance ultracentrifuge tubes in pairs with MS-suspension medium.
7. Ultracentrifuge at 231,000 × *g* for 50 min as described in **step 4**. After centrifugation, discard the supernatant with an aspirator.
8. Put 5 mL of MS-suspension medium in a Teflon-glass homogenizer and mark the solution surface on the glass homogenizer as an indication of 5 mL volume. Discard the medium.
9. Add 2 mL of MS-suspension medium onto microsomal pellets in the ultracentrifuge tubes. Break up the precipitated pellets with a glass rod. Transfer into a Teflon-glass homogenizer using a large-aperture Pasteur pipette. Put 2 mL of MS-suspension medium into the same ultracentrifuge tubes and break up the remaining pellets by pipetting. Transfer into the Teflon-glass homogenizer already containing the first part of the resuspended pellet and add MS-suspension medium up to 5 mL. Homogenize well with an electric Teflon-glass homogenizer (moving up and down five times) on ice (*see* **Note 5**).
10. Put all of the homogenate in a centrifuge tube containing two-phase partition medium (tube A). Add 5 mL of MS-suspension

medium to the other two two-phase partition mixtures (tubes B and C). Chill on crushed ice for 10 min. During this time, mix well every 2 min.

11. Centrifuge tubes A and B at 440 × *g* for 5 min in a chilled rotor. Two phases should be observed to have settled in both tubes. Discard the upper phase of tube B with a Pasteur pipette and transfer the upper phase of tube A into tube B. Chill on crushed ice for 10 min. During this time, mix well every 2 min (*see* **Note 6**).
12. Centrifuge tubes B and C at 440 × *g* for 5 min in a chilled rotor. Discard the upper phase of tube C with a Pasteur pipette and transfer the upper phase of tube B into tube C. Balance tube C with another centrifuge tube filled with water. Chill on crushed ice for 10 min. During this time, mix well every 2 min (*see* **Note 6**).
13. Centrifuge at 440 × *g* for 5 min and split the resultant upper phase of tube C into two ultracentrifuge tubes. Fill up the tubes with PM-suspension medium and balance them. Ultracentrifuge at 231,000 × *g* for 50 min, as described in **step 4** (*see* **Note 6**).
14. Discard the supernatant with an aspirator. Add 1 mL of PM-suspension medium to each tube. Homogenize the pellets with a glass rod. Transfer into an electric Teflon-glass homogenizer and homogenize well (moving up and down five times). Collect the plasma membrane suspensions with a Pasteur pipette into ultracentrifuge tubes. Balance ultracentrifuge tubes in pairs with PM-suspension medium. Ultracentrifuge again at 231,000 × *g* for 35 min.
15. Discard the supernatant with an aspirator. Add minimal volume of PM-suspension medium to the plasma membrane pellets. Homogenize the pellets with a glass rod. Transfer into an electric Teflon-glass homogenizer and homogenize well (moving up and down five times) with cooling on ice. Transfer into a 1.5 mL microtube.
16. Measure protein content using the Bradford assay (BioRad Protein Assay Kit). Use 100 μg of protein for tryptic digestion and LC-MS/MS analysis. The remaining PM fractions should be divided into aliquots, frozen in liquid nitrogen immediately and stored at −80 °C.

3.3 In-Solution Tryptic Digestion

All of these procedures must be performed at a clean bench whenever possible and at room temperature unless otherwise specified.

1. Precipitate 100 μg of PM protein by ultracentrifugation (231,000 × *g*, 4 °C, 50 min).
2. Discard supernatant by decantation. Add solution B to the PM pellets. Homogenize the pellets with a glass rod. Transfer into

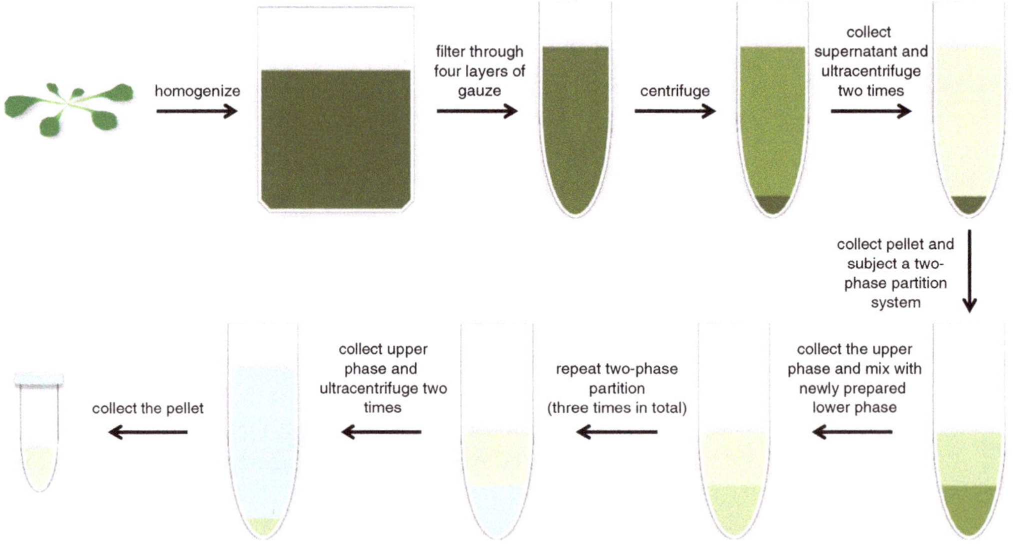

Fig. 2 A schematic overview of PM extraction from plants. Leaves are homogenized in a homogenizing medium and then passed through four layers of gauze to remove debris. Subsequently, the filtrates are centrifuged at 5,000 × *g* and then at 231,000 × *g* to obtain microsomal fractions. Microsomal fractions are suspended in MS-suspension medium and then recentrifuged twice for washing. The resultant microsomal fractions are subjected to a two-phase partition system that consists of polyethylene glycol 3,350 and dextran T500 in MS-suspension medium with NaCl. After repeating two phase partitioning three times to increase the purity of the PM in the upper phase, PM fractions are recovered, diluted with PM-suspension medium, and centrifuged (231,000 × *g*) twice to remove polymers

an electric Teflon-glass homogenizer and homogenize well (moving up and down five times) with cooling on ice. Transfer to 1.5 mL microtubes.

3. Solubilize samples and measure protein concentration with a Pierce BCA protein assay kit according to the instruction manual.
4. Transfer 5 μg of PM protein to another 1.5 mL microtube. Make up to 20 μL with solution A.
5. Perform reductive alkylation and tryptic digestion according to the instruction manual and then store at −30 °C (*see* **Note 7**).

3.4 Peptide Purification

All of these procedures must be performed at a clean bench whenever possible and at room temperature unless otherwise specified.

1. Insert a SPE C-TIP into the 3 mm hole in the top of a microtube (Fig. 2).
2. Add 30 μL of solution A to the upper side of the SPE C-TIP for preconditioning. Centrifuge at 1,000 × *g* for 30 s to get solution A through the tip column.

3. Add 30 μL of solution B to upper side of SPE C-TIP for preconditioning. Centrifuge at 1,000 × *g* for 30 s to get solution B through the tip column.
4. After confirming that the column is moist, add the entire trypsin digested peptide sample to the upper side of the SPE C-TIP for column absorption. Centrifuge at 1,000 × *g* for 30 s to get the sample solution through the tip column.
5. Add 30 μL of solution B to upper side of SPE C-TIP for cleaning. Centrifuge at 1,000 × *g* for 30 s to get solution B through the tip column.
6. Put a vial insert for each LC-MS/MS sample into another microtube with a hole in the cap. Transfer the SPE C-TIP into the microtube.
7. Add 30 μL of solution A to the upper side of the SPE C-TIP for elution. Centrifuge at 1,000 × *g* for 30 s to get solution A through the tip column. Discard the SPE C-TIP.
8. Dry the eluted samples using a centrifugal concentrator for 15 min. Add 15 μL of 0.1 % (v/v) TFA. Put the vial insert into the vial and close the lid. Store at −30 °C (*see* **Note 8**).

3.5 Nano-LC-MS/MS Analysis

An example of nano-LC-MS/MS and database search settings for *Arabidopsis* PM proteins are described below.

3.5.1 Settings of Nano-LC-MS/MS

1. Mobile phase for peptide elution from trap column: 0.1 % (v/v) formic acid in acetonitrile.
2. Mobile phase for peptide separation: linear gradient of acetonitrile from 5 % (v/v) to 45 % (v/v).
3. Flow rate and analysis time: 500 nL/min for 120 min.
4. Spray voltage for peptide ionization: 1.8 kV.
5. Mass spectrometer control settings: scan range, 400–1,800 *m/z*; resolution, 30,000; Collision induced dissociation, five most intense ions with a threshold above 500.

3.5.2 Settings for Data Conversion, Protein Identification, and Quantification

1. Parameters for conversion from raw files to mgf files (Proteome Discoverer software): precursor mass range, *m/z* 350–5,000; highest and lowest charge state, 0; lower and upper RT limit, 0; the minimum total intensity of a spectrum, 0; and the minimum number of peaks in a spectrum, 1.
2. Parameters for identification of proteins (Mascot search engine): database, Arabidopsis TAIR 10 protein database; allowance of missed cleavage, 1; fixed modification, carbamidomethylation (C); variable modification, oxidation (M); peptide mass tolerance, 5 ppm; MS/MS tolerance, 0.6 Da; peptide charges, +1, +2, +3.

4 Notes

1. Mops-KOH (pH 7.6), EGTA (pH 8.0), EDTA (pH 8.0) should be prepared as 0.5 M stock solutions and stored at 4 °C. EGTA and EDTA can be dissolved by adding KOH, but the pH of the solutions should be adjusted at 8.0. When BSA is dissolved, BSA powder should be pre-equilibrated at room temperature. PMSF and SHAM should be separately prepared as 1 and 1.6 M stock solutions in DMSO, respectively, and stored at 4 °C. DTT should be stored at −20 °C as a 1 M stock solution. PMSF, SHAM, DTT should be diluted only as needed just before use.
2. KH_2PO_4/K_2HPO_4 (K-P) buffer (pH 7.8) should be prepared as a 0.5 M stock solution and diluted to make the MS-suspension medium. First, 200 mL of 0.5 M K_2HPO_4 and 30 mL of 0.5 M KH_2PO_4 are prepared. The pH of the 0.5 M K_2HPO_4 is adjusted to 7.8 by adding 0.5 M KH_2PO_4.
3. TFA evaporates quickly. Thus, solutions containing TFA should be freshly prepared just immediately before use.
4. Non-acclimated plants should be harvested before bolting. It may be necessary to adjust how long plants are kept before harvesting.
5. In this step, homogenization should not be too long or too vigorous because harsh homogenization can severely disrupt membrane integrity.
6. Two-phase partitioning is the most important step for preparing highly purified PM. When the upper phase of the two-phase partition medium is removed, the Pasteur pipette should be moved from left to right near the boundary of the two phases to prevent taking lower phase.
7. Digested and purified peptides should be analyzed by nano-LC-MS/MS within 1 week.
8. At this stage, dehydrated, compressed, and completely bleached gels should be observed. If the gels do not change, repeat this step twice.

Acknowledgement

This work was supported in part by a Research Fellowship for Young Scientists (#247373 to D.T.) and Grants-in-Aid for Scientific Research (#22120003 and #24370018 to M.U. and Y.K.) from JSPS, Japan.

References

1. Guy CL (1990) Cold acclimation and freezing stress tolerance: role of protein metabolism. Annu Rev Plant Physiol Plant Mol Biol 41: 187–223
2. Thomashow MF (1999) Plant cold acclimation: freezing tolerance genes and regulatory mechanisms. Annu Rev Plant Physiol Plant Mol Biol 50:571–599
3. Steponkus PL (1984) Role of plasma membrane in cold acclimation and freezing injury in plants. Annu Rev Plant Physiol 35: 543–584
4. Webb MS, Uemura M, Steponkus PL (1994) A comparison of freezing injury in oat and rye: two cereals at the extremes of freezing tolerance. Plant Physiol 104:467–478
5. Uemura M, Tominaga Y, Nakagawara C, Shigematsu S, Minami A, Kawamura Y (2006) Responses of the plasma membrane to low temperatures. Physiol Plant 126:81–89
6. Yoshida S, Uemura M (1984) Protein and lipid compositions of isolated plasma membranes from orchard grass (*Dactylis glomerata* L.) and changes during cold acclimation. Plant Physiol 75:31–37
7. Uemura M, Yoshida S (1984) Involvement of plasma membrane alterations in cold acclimation of winter rye seedlings (*Secale cereale* L. cv Puma). Plant Physiol 75:818–826
8. Lynch DV, Steponkus PL (1987) Plasma membrane lipid alterations associated with cold acclimation of winter rye seedlings (*Secale cereale* L. cv Puma). Plant Physiol 83:761–767
9. Uemura M, Joseph RA, Steponkus PL (1995) Cold acclimation of *Arabidopsis thaliana*: effect on plasma membrane lipid composition and freeze-induced lesions. Plant Physiol 109:15–30
10. Kawamura Y, Uemura M (2003) Mass spectrometric approach for identifying putative plasma membrane proteins of *Arabidopsis* leaves associated with cold acclimation. Plant J 36:141–154
11. Minami A, Fujiwara M, Furuto A, Fukao Y, Yamashita T, Kamo M, Kawamura Y, Uemura M (2009) Alterations in detergent-resistant plasma membrane microdomains in *Arabidopsis thaliana* during cold acclimation. Plant Cell Physiol 50:341–359
12. Mazars C, Thion L, Thuleau P, Graziana A, Knight MR, Moreau M et al (1997) Organization of cytoskeleton controls the changes in cytosolic calcium of cold-shocked *Nicotiana plumbaginifolia* protoplasts. Cell Calcium 22:413–420
13. Orvar BL, Sangwan V, Omann F, Dhindsa RS (2000) Early steps in cold sensing by plant cells: the role of actin cytoskeleton and membrane fluidity. Plant J 23:785–794
14. Sangwan V, Foulds I, Singh J, Dhindsa RS (2001) Cold-activation of *Brassica napus BN115* promoter is mediated by structural changes in membranes and cytoskeleton, and requires Ca^{2+} influx. Plant J 27:1–12
15. Welti R, Li W, Li M, Sang Y, Biesiada H, Zhou HE et al (2002) Profiling membrane lipids in plant stress responses: role of phospholipase Dα in freezing-induced lipid changes in *Arabidopsis*. J Biol Chem 277:31994–32002
16. Li W, Li M, Zhang W, Welti R, Wang X (2004) The plasma membrane-bound phospholipase Dδ enhances freezing tolerance in *Arabidopsis thaliana*. Nat Biotechnol 22:427–433
17. Yamazaki T, Kawamura Y, Minami A, Uemura M (2008) Calcium-dependent freezing tolerance in *Arabidopsis* involves membrane resealing via synaptotagmin SYT1. Plant Cell 20: 3389–3404
18. Kondo M, Takahashi D, Minami A, Kawamura Y, Uemura M (2012) Function of *Arabidopsis* dynamin-related proteins during cold acclimation. Cryobiol Cryotechnol 58:105–111 (In Japanese with English summary)
19. Miki Y, Takahashi D, Kawamura Y, Uemura M (2014) Temporal analysis of plasma membrane proteome in *Arabidopsis* during cold acclimation and deacclimation. Cryobiol Cryotechnol (in press) (In Japanese with English summary)
20. Takahashi D, Kawamura Y, Yamashita T, Uemura M (2012) Detergent-resistant plasma membrane proteome in oat and rye: similarities and dissimilarities between two monocotyledonous plants. J Proteome Res 11:1654–1665
21. Takahashi D, Li B, Nakayama T, Kawamura Y, Uemura M (2013) Plant plasma membrane proteomics for improving cold tolerance. Front Plant Sci 4:90
22. Nakayama T, Takahashi D, Kawamura Y, Rahman A, Uemura M (2013) Compositional changes in plasma membrane proteins in *Brachypodium distachyon* during cold acclimation. Cryobiol Cryotechnol 59:61–65 (In Japanese with English summary)
23. Kasuga J, Takahashi D, Kawamura Y, Uemura M (2012) Proteomic analysis of seasonal cold-deacclimation process in poplar phloem and

xylem tissues. Abstract of plant and microbe adaptation to cold 2012 (O-18)

24. Li B, Takahashi D, Kawamura Y, Uemura M (2012) Comparison of plasma membrane proteomic changes of *Arabidopsis* suspension cells (T87 line) after cold and abscisic acid treatment in association with freezing tolerance development. Plant Cell Physiol 53:542–554

25. Takahashi D, Li B, Nakayama T, Kawamura Y, Uemura M (2013) Shotgun proteomics of plant plasma membrane and microdomain proteins using nano-LC-MS/MS. In: Novo JVJ, Komatsu S, Weckwerth W (eds) Plant proteomics: methods and protocols, 2nd edn. Springer Science + Business Media, LLC, New York, pp 481–498

Chapter 14

Profiling Methods to Identify Cold-Regulated Primary Metabolites Using Gas Chromatography Coupled to Mass Spectrometry

Frederik Dethloff, Alexander Erban, Isabel Orf, Jessica Alpers, Ines Fehrle, Olga Beine-Golovchuk, Stefanie Schmidt, Jens Schwachtje, and Joachim Kopka

Abstract

This book chapter describes the analytical procedures required for the profiling of a metabolite fraction enriched for primary metabolites. The profiling is based on routine gas chromatography coupled to mass spectrometry (GC-MS). The generic profiling method is adapted to plant material, specifically to the analysis of single leaves from plants that were exposed to temperature stress experiments. The described method is modular. The modules include a rapid sampling and metabolic inactivation protocol for samples in a wide size range, a sample extraction procedure, a chemical derivatization step that is required to make the metabolite fraction amenable to gas chromatographic analysis, a routine GC-MS method, and finally the procedures of data processing and data mining. A basic and extendable set of standardizations for metabolite recovery and retention index alignment of the resulting GC-MS chromatograms is included. The method has two applications: (1) the rapid screening for changes of relative metabolite pools sizes under temperature stress and (2) the verification of cold-regulated metabolites by exact quantification using a GC-MS protocol with extended internal and external standardization.

Key words Gas chromatography, Time-of-flight mass spectrometry, GC-MS, TOF-MS, Metabolomics, Metabolite profiling, Metabolism, Relative quantification, Absolute quantification

1 Introduction

Metabolite profiling methods are the basis of modern metabolomic approaches that aim for comprehensive analyses of biological systems [1, 2]. Targeted and nontargeted metabolic profiling methods that are in part automated and technically robust have been developed to investigate various parts of metabolism. A GC-MS-based method that covers a wide range of primary metabolism has made a strong impact. The analytical procedures and the means to identify metabolites within the generated complex GC-MS data

Dirk K. Hincha and Ellen Zuther (eds.), *Plant Cold Acclimation: Methods and Protocols*, Methods in Molecular Biology, vol. 1166, DOI 10.1007/978-1-4939-0844-8_14, © Springer Science+Business Media New York 2014

were easily transferable between labs, e.g., ref. 3. The method covers among others, sugars, amino acids, amines, organic acids, phosphorylated metabolites and includes small secondary metabolites. The molecular size of the covered metabolites ranges between small 2-carbon metabolites, such as glycolate, glyoxylate or glycine, and 18-carbon trisaccharides, such as raffinose. The coverage of the GC-MS profiling method that is described in the following has been thoroughly studied. The metabolite coverage and compound identifications are frequently updated and accessible online via the Golm Metabolome Database (http://gmd.mpimp-golm.mpg.de/). Libraries that provide mass spectra and retention indices of more than 1,000 metabolites and tools for mass spectral analysis have been made publicly available [4–7].

Metabolite profiling methods are fast and efficient postgenomic tools that screen for relative changes of metabolite pool sizes. Comparisons of the effects of environmental changes, such as stress by low temperature [8–11], or genetic modifications are typically made relative to control plants of a wild type genotype that is cultivated under optimal standard conditions. The speed of profiling allows extended experimental designs that involve typically more than 100 samples and may comprise more than 1,000 samples. As a consequence, many independent events of same genetic manipulation or breeding populations can be studied with high replication. Experimental designs that eliminate or suppress the influence of noncontrolled factors can be applied, and highly resolved time courses or dosage dependencies of metabolic responses explored. Whole experiments can easily be independently repeated. Thus the bottle neck of metabolic physiological studies is moved back to the sound performance of well-designed physiological experiments.

For most physiological questions information on the changes of relative pool sizes or on patterns of metabolic changes is sufficient to diagnose the effect of the experimental intervention. But some questions, not least those raised by the demands of systems modelling, require information on exact metabolite concentrations. Exact quantification requires additional experiments, such as costly and time consuming standardization of compound recovery. In addition compound-specific quantitative calibration samples are necessary sometimes in numbers that can easily exceed the number of samples to be quantified. For this reason exact quantification should never be performed before metabolite profiling has shown that the metabolite of interest is indeed among the most relevant within the screened metabolic fraction.

A second requirement after obtaining information on relevant candidate metabolites by profiling methods is the verification of metabolite identity [12]. This is a basic but indispensable requirement because metabolite profiles are complex and may contain hundreds of known metabolites but also a large fraction of still non-identified metabolites. Metabolic products also comprise a large number of chemical isomers, such as the different mono-, di-, and trisaccharides. Such isomers can be hard to distinguish and to selectively quantify. To identify compounds by GC-MS authenticated reference substances are needed [4–7] which are employed in standard addition experiments to test for match of mass spectrum and chromatographic retention. Such standard addition experiments can be efficiently used for two purposes as they also allow the determination of quantitative recovery and are a prerequisite of exact quantification that test so-called matrix effects caused by the physicochemical nature of the sample.

In the following protocol we describe the basic analytical modules that encompass both relative and absolute quantification by the GC-MS-based analysis of a metabolite fraction enriched for primary metabolites. The method is modular, serves the demand for standardized reporting [13, 14], and uses a generalizable structure that can also describe other MS-based metabolite profiling methods or other variants of the GC-MS method, e.g., GC-MS methods that use different extraction procedures, GC capillary columns, or GC settings [1, 2, 15–17]. The method starts with sampling and ends with the identification of relevant metabolites by mass spectral analysis and standard addition. Additional actions that are required to upgrade the basic profiling method from analysis of relative pool size changes to the quantification of absolute pool sizes are added. The method is generally applicable to plant material. The adaptations that are specifically required to investigate cold-regulated metabolites are indicated.

2 Materials

Use ultrapure or bi-distilled water at approximately 0.055 μS/cm. Purchase analytical grade reagents and all chemicals in best available purity. Buy small packages to avoid contaminations and loss of reagent reactivity. Buy authenticated reference substances for internal and external quantitative calibration in highest available purity and in amounts suitable for the accurate gravimetric determination of stock solutions. Diligently follow recommended procedures for the safe handling of chemicals and for waste disposal.

2.1 Extraction and Standardization

1. Methanol.
2. Chloroform.
3. $^{13}C_6$-Sorbitol (CAS 121067-66-1), nonadecanoic methyl ester (CAS 1731-94-8) and other authenticated reference substances for quantitative internal and external calibration as required.
4. 1.5 mL safe-lock, tapered bottom plastic micro vials.
5. 2.0 mL safe-lock, round bottom plastic micro vials.

6. Centrifuge for micro vials.
7. Oscillating ball mill with adaptors that hold 5 or more 2 mL micro vials.
8. 5 mm stainless steel balls.
9. Vacuum concentrator system with rotors for micro vials.
10. Silica gel.
11. Argon.
12. Calibrated regularly checked pipetting devices in adequate volume ranges and respective disposable pipette tips.
13. Calibrated and regularly checked balance with at least ±0.1 mg precision.
14. Heated shakers for micro vials.
15. Vortex mixer.

2.2 Chemical Derivatization

1. Methoxyamine hydrochloride (CAS 593-56-6).
2. Pyridine (CAS 110-86-1).
3. 4-(Dimethylamino)pyridine (CAS 1122-58-3).
4. *N*,*O*-Bis(trimethylsilyl)trifluoroacetamide (CAS 25561-30-2).
5. *n*-Alcanes: *n*-decane (CAS 124-18-5), *n*-dodecane (CAS 112-40-3), *n*-pentadecane (CAS 629-62-9), *n*-octadecane (CAS 593-45-3), *n*-nonadecane (CAS 629-92-5), *n*-docosane (CAS 629-97-0), *n*-octacosane (CAS 630-02-4), *n*-dotriacontane (CAS 544-85-4), *n*-hexatriacontane (CAS 630-06-8).
6. GC glass vials with crimp or screw caps and chemically inert septa.
7. Crimp cap sealer.

2.3 GC-MS

1. GC-MS system with electron impact- and/or chemical ionization. The MS system can have nominal mass resolution or better. Electron impact ionization and atomic mass unit resolution is preferred for use with conventional GC-MS mass spectral libraries. Other systems will require the establishment of custom spectral libraries based on authenticated reference compounds.
2. Split/splitless injector with electronic pressure control.
3. Low-bleeding septa.
4. Inert conical single taper liner with glass wool for split/less injection.
5. Low-bleeding GC capillary column suitable for hyphenation to mass spectrometry systems. The stationary phase needs to be stable in the presence of trimethylsilylation and methoxyamination reagents. A 5 %-phenyl–95 %-dimethylpolysiloxane

fused silica capillary column with 30 m length, 0.25 mm inner diameter, 0.25 μm film thickness, and an integrated 10 m precolumn is preferred for the separation of chemically derivatized primary metabolites and for use with retention index libraries that are exchangeable between metabolite profiling laboratories. The use of more polar or other stationary phases will require the establishment of custom retention index libraries based on authenticated reference compounds.

6. Helium 5.0 carrier gas.
7. *n*-Hexane.
8. Ethylacetate.

3 Methods

3.1 Sampling and Gravimetric Determination of the Sample Amount

Sampling methods for metabolic profiling must be performed in situ with minimal disturbance of the plant's environment. Specifically the temperature and illumination of leaves must not be changed prior to or during sampling. Metabolic inactivation must be immediate and needs to be maintained during subsequent sample processing. All other factors that influence metabolism must be controlled by the experimental design. The experimental design must include a randomization or arraying strategy to account for residual experimental factors that cannot be completely controlled. For relative quantification of metabolite pool sizes the amount of all samples must be in the same range with a defined maximal tolerance. For absolute quantification the exact amount of the sample must be determined.

1. Prepare 1.5 or 2.0 mL micro vials prior to sampling.
2. Number vials using a permanent marker.
3. Precool and keep vials in liquid nitrogen.
4. Determine the empty weight of each vial while frozen including the hoarfrost that may form during the process. Return vial to liquid nitrogen.
5. Take a vial from liquid nitrogen, cut a sample from a plant and seal sample into vial. Return the loaded vial to liquid nitrogen within 10 s from cutting or faster.
6. Determine the weight of the loaded vial while frozen including the hoarfrost that may form (*see* **Note 1**). Return loaded vial to liquid nitrogen. Samples can be stored at this step at −80 °C.
7. The sample amount of a single leaf for absolute quantification must be not lower than 2.5 mg (Fig. 1) and must not exceed 125.0 mg fresh weight for extraction in 2 mL micro vials. The sample amount for relative quantification should be kept constant ideally with a tolerance of ±5–10 % (*see* **Note 2**).

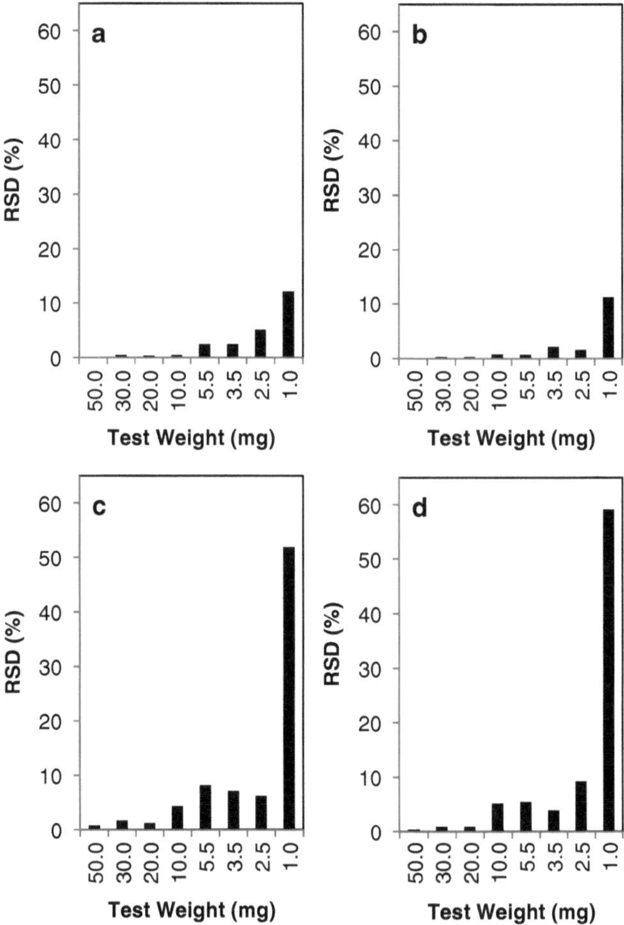

Fig. 1 Precision of gravimetric weight determination of samples that were shock frozen in liquid nitrogen. The relative standard deviation (RSD) of the direct weighing of test weights was determined at room temperature using a balance with a tolerance of ±0.1 mg. Testing weights were made of aluminum foil and covered the weight range of ~1.0–50.0 mg. Direct weighing (**a**) was compared to differential weighing using 2.0 mL micro vials at room temperature (**b**) and to weighing using 1.5 mL (**c**) or 2.0 mL (**d**) micro vials while the test weights and the micro vials were kept frozen throughout. Note that the precision of weighing while frozen was below 2 % RSD between 20.0 and 50.0 mg ($n=6$) test weight and better that 10 % RSD between 2.5 and 20.0 mg ($n=6$). The weight of 1.0 mg test weights cannot be determined precisely when frozen

8. Add a pre-cooled stainless steel ball to each vial. The curvature of the steel ball must not exceed the curvature of the bottom of the micro vial. Load micro vials into a precooled mounting adaptor of an oscillating ball mill. Homogenize samples to a fine powder by 1.0 min bursts at 15/s frequency (*see* **Note 3**). Keep samples below −60 °C throughout the process. If necessary return the loaded adaptors to liquid nitrogen between bursts. Store homogenized samples without removing the steel balls at −80 °C until further processing (*see* **Note 3**).

9. Keep, store, and continue to process in parallel through all subsequent steps a set of 5–10 empty micro vials from the batch used for each respective experiment. These vials are processed as non-sample controls and allow the determination of the specific chemical laboratory background of each experiment (*see* **Note 4**).
10. Prepare a set of quality control samples that is used to control the performance of the GC-MS profiling system. Specifically control for system drifts when comparing between independently repeated experiments that cannot be processed together. For this purpose prepare a large representative leaf sample from immediately frozen material that combines all sample types of the experimental design. Homogenize thoroughly in a mortar under liquid nitrogen, avoid hoarfrost during homogenization. Generate equal frozen aliquots of the average sample fresh weight of the experiment with a tolerance of ±5–10 %. Keep, store, and continue to process in parallel a set of 5–10 quality control samples through all subsequent steps of the analysis. Store surplus quality control samples at −80 °C until processing of the next and subsequent independently repeated experiments.

3.2 Metabolite Extraction and Standardization

Extraction methods determine and delimit the chemical nature and coverage of metabolite profiling methods. The extraction method that is described in the following generates a fraction that is enriched for polar primary metabolites while volatiles, highly lipophilic metabolites, and complex lipids are removed. Extraction efficiencies can vary between metabolites depending on the choice of solvent. Extraction efficiency can also depend on variations of the physical properties or chemical composition between the compared sample types. So far we did not observe variable matrix effects in temperature stress experiments. However, frequent control experiments that test the recovery of metabolites are advised. Moreover, knowledge of metabolite recovery is required to verify temperature regulated metabolites by exact quantification. Isotope-labelled internal standards can be added during extraction to determine the specific recovery of each quantified metabolite from each investigated sample. Alternatively, non-labelled authenticated reference compounds can be added at a constant concentration to representative samples. This process estimates specific constant factors of metabolite recoveries for each type of profiled sample. Both types of recovery experiments will test the overall recovery of the method including matrix effects that may occur at the subsequent steps of the method.

1. For the extraction of a set of approximately 150 samples, prepare 50 mL fresh 90 % methanol:water (v/v) extraction solvent and add methanol soluble internal standards. For routine profiling dissolve 0.02 mg/mL $^{13}C_6$-sorbitol in this solvent.
2. Prepare 35 mL fresh chloroform solvent with chloroform soluble internal standards. Dissolve 0.25 mg/mL nonadecanoic

acid methyl ester in chloroform to test for the presence or absence of lipophilic compounds.

3. Prepare 60 mL fresh bi-distilled water.
4. Dissolve other additional internal standards, e.g., stable isotope labelled or non-labelled xenobiotic internal standards, according to their solubility in the methanol solvent, the chloroform solvent, or the bi-distilled water solvent. Adjust the concentration of internal standards to concentrations that approximate the expected endogenous metabolite concentrations (*see* **Notes 5** and **6**).
5. Prepare dilution series for quantitative external calibration of metabolites. Prepare a stock mixture of authenticated reference substances at up to 10 mg/mL in 90 % methanol, in chloroform, or in bi-distilled water according to compound solubility. Choose the relative amount of each metabolite in a stock mixture to mimic the composition that is expect within the analyzed type of sample. Prepare dilution series to cover the highest and lowest expected amount. Perform prior test experiments to adjust the calibration series appropriately. Process calibration series in parallel through all subsequent steps of the method.

3.2.1 Methanol Extraction with Liquid Partitioning into Chloroform

Methanol extracts may contain an excess of complex lipids and chlorophyll. These compound classes are non-volatile and can cause frequent maintenance of the GC-MS system unless liquid partitioning into chloroform removes these compound classes.

1. Add 330 μL methanol solvent containing the internal standards to a micro vial with 100 mg ±5–10 % frozen sample powder (*see* **Note 7**). Do not remove the steel ball during extraction.
2. Mix thoroughly using a vortex-mixer and shake all samples simultaneously for 15 min at 70 °C. Vent micro vials after 1 min at 70 °C to release excess vapor pressure.
3. Cool to room temperature.
4. Add 230 μL chloroform solvent containing internal standards.
5. Mix thoroughly using a vortex-mixer and shake all samples simultaneously for 5 min at 37 °C.
6. Add 400 μL bi-distilled water containing internal standards, if added.
7. Mix thoroughly using a vortex-mixer and separate liquid and solid phases by centrifugation for 5 min at 14,000 rpm (*see* **Note 8**).
8. Transfer an 80 μL aliquot of the upper phase, which contains the polar metabolic complement of the sample, to a 1.5 mL safe-lock micro vial or directly to a GC vial and dry in a vacuum

concentrator. Prepare micro vials with two or more backup samples of each extract, e.g., of a second 80 μL and a 160 μL aliquot. For external calibration series take identical aliquots and prepare micro vials to run at least two calibration series in parallel to each experimental sample set and prepare multiple backups for use in subsequent experiments (*see* **Note 8**).

9. Seal dried samples with closed caps under inert gas, e.g., argon, in plastic bags with silica gel. Store at −20 °C or colder until further processing. Before reopening the plastic bags allow equilibration to room temperature and remove condensed water.

3.2.2 Methanol Extraction Without Liquid Partitioning into Chloroform

This method is used to include small lipophilic metabolites and may require frequent maintenance of the GC-MS system. This method can be applied to small, dilute samples or to samples that contain a smaller proportion of complex lipids and chlorophyll.

1. Prepare methanol extraction solvent with methanol soluble internal standards. Dissolve 0.005 mg/mL $^{13}C_6$-sorbitol in 100 % methanol.
2. Dissolve other stable isotope labelled or non-labelled xenobiotic internal or external standards in methanol at concentrations that approximate the expected endogenous metabolite concentrations (*see* **Note 5**).
3. Extract samples that vary in fresh weight with a constant 40:1 (v/w) solvent to fresh weight ratio. In detail, add 400 μL methanol solvent to micro vial with 10 mg frozen sample powder. Scale up extraction volume in proportion to sample fresh weight (*see* **Note 9**). Do not remove the steel ball during extraction.
4. Mix thoroughly using a vortex-mixer and shake all samples simultaneously for 15 min at 70 °C. Vent micro vials after 1 min at 70 °C to release excess vapor pressure.
5. Cool to room temperature and centrifuge for 5 min at 14,000 rpm.
6. Transfer the complete supernatant, i.e., a 350 μL aliquot of the liquid phase, to a 1.5 mL safe-lock micro vial or directly to a GC vial and dry in a vacuum concentrator.
7. Seal dried samples with closed caps under inert gas, e.g., argon, in plastic bags with silica gel. Store at −20 °C or colder until further processing. Before reopening the plastic bags allow equilibration to room temperature and remove condensed water.

3.2.3 Methanol: Chloroform Extraction Without Liquid Partitioning

This method is used to include a higher fraction of small lipophilic metabolites and may require more frequent maintenance of the GC-MS system. To include more lipophilic compounds compared to Subheading 3.2.2 chloroform is added. This method can be applied to small, dilute samples or to samples that contain an even smaller proportion of complex lipids and chlorophyll.

1. Perform all steps described in Subheading 3.2.1 but do not add 400 μL bi-distilled water. Centrifuge 5 min at 14,000 rpm to separate solids from liquid supernatant.

3.3 Retention Index Standardization for GC Analysis

Gas chromatography is subject to drifts of retention time [17]. Such drifts interfere with the retention time alignment of GC-MS chromatograms, especially if an experiment comprises a high number of samples. Retention index standards are used to align GC chromatograms and to improve compound identification by matching to retention index libraries of authenticated reference compounds.

1. Prepare retention index (RI) standard mixture of *n*-alkanes in pyridine. Combine *n*-decane (RI 1,000), *n*-dodecane (RI 1,200), *n*-pentadecane (RI 1,500), *n*-octadecane (RI 1,800), *n*-nonadecane (RI 1,900), *n*-docosane (RI 2,200), *n-octacosane* (RI 2,800), *n*-dotriacontane (RI 3,200), and *n*-hexatriacontane (RI 3,600) at a final concentration of 0.22 mg/mL each, except *n*-decane and *n*-hexatriacontane which are added at 0.44 mg/mL.
2. To calculate retention indices use the following definition, RI of an *n*-alkane equals the number of carbons multiplied by 100, see above, and apply the method for linear temperature programmed gas chromatography [18].

3.4 Chemical Derivatization for GC Analysis

Chemical derivatization reactions are required to modify the structure of non-volatile compounds to form volatile products that can be analyzed by GC. The choice of derivatization reactions determines and delimits the coverage and sensitivity of metabolite profiling methods, e.g., ref. 19. The chemical derivatization method that is described in the following is essential as initially described [1, 2, 15, 16]. The reactions have low specificity, high yields for almost complete conversion and generate volatile derivatization products, i.e., the analytes, of most stable primary metabolites (*see* **Note 10**).

1. Prepare fresh methoxyamine reagent daily. Dissolve first 5 mg/mL 4-(dimethylamino)pyridine in pyridine then add methoxyamine hydrochloride to a final concentration of 40 mg/mL (*see* **Note 11**).
2. Prepare trimethylsilylation reagent. Mix frcsh *N*,*O*-*bis*(trimethylsilyl)trifluoroacetamide and retention index standard mixture dissolved in pyridine (*see* Subheading 3.3) in a 7:1 (v/v) ratio. Avoid humidity and do not store opened bottles of *N*,*O*-bis(trimethylsilyl)trifluoroacetamide or preparations of trimethylsilylation reagent (*see* **Note 12**).

3. Mix dried extract thoroughly with 40 μL methoxyamine reagent using a vortex-mixer and shake all samples simultaneously for 90 min at 30 °C (*see* **Note 13**).
4. Add 80 μL trimethylsilylation reagent and mix thoroughly using a vortex-mixer. Incubate for 30 min at 37 °C (*see* **Note 13**).
5. Perform reactions either in GC vials or transfer 80 μL to a GC vial. Avoid humidity, close vial immediately, and keep vials at room temperature on a GC auto-injector system until injection (*see* **Note 14**). If a chemically derivatized sample is reanalyzed by GC-MS exchange the septum of the used GC vial after first analysis. Do not store used GC vials for extended periods or GC vials with punctured septa (*see* **Note 12**).

3.5 Injection for GC Analysis

The available gas chromatography settings for GC-MS systems are typically inbuilt features of the employed GC-MS system. In the following we will use an Agilent 6890N gas chromatograph with split/splitless injector and electronic pressure control up to 150 psi (Agilent, Böblingen, Germany) to exemplify the analytical procedures and to report the details that should be considered and reported when publishing GC-MS profiling data. Other GC systems are equally amenable to metabolite profiling analyses. The choice of injection technology for gas chromatography modifies the amount of the chemically derivatized sample that is transferred onto the capillary column. The temperature, pressure, and gas flow during injection may influence peak shape.

1. Mount a 10 μL syringe on the GC injection system.
2. Mount a new conical single taper split/splitless liner with glass wool into injector port of the GC system before analyzing a new experimental set of samples.
3. Before each sample injection, clean syringe by full volume draws of pure ethylacetate and *n*-hexane.
4. Perform at least five injections of *N*,*O*-Bis(trimethylsilyl)trifluoroacetamide after a change of syringe, liner or GC column (*see* Subheading 3.6).

3.5.1 Splitless Injection for GC Analysis

1. Inject 1 μL of chemically derivatized sample at 250 °C in splitless mode with helium carrier gas flow set to 0.6 mL/min.
2. Adjust purge time to 1 min with purge flow set to 20 mL/min flow.
3. Keep the flow rate constant and electronic pressure control enabled (*see* **Note 15**).

3.5.2 Split Injection for GC Analysis

1. Inject 1 μL chemically derivatized sample at 250 °C in split mode at a split flow ratio of at least 1:30 with helium carrier gas flow set to 0.6 mL/min.

2. Adjust purge time to 1 min at 20 mL/min flow.
3. Keep the flow rate constant and electronic pressure control enabled (*see* **Note 15**).

3.6 GC Analysis

The type and dimensions of capillary GC columns, the temperature ramping and the flow/pressure settings of the carrier gas determine the scope of the profiling method and the elution sequence of the analytes. Minor shifts of retention indices between GC-MS systems using identical column types can be mathematically compensated [17, 18].

1. Mount a 5 %-phenyl–95 %-dimethylpolysiloxane fused silica capillary column, with 30 m length, 0.25 mm inner diameter, 0.25 μm film thickness, and an integrated 10 m pre-column. Use a low-bleeding column suitable for mass spectrometry.
2. Operate system with helium carrier gas set to 0.6 mL/min constant flow.
3. Start the temperature program isothermal for 1 min at 70 °C, ramp to 350 °C at 9 °C/min, keep at 350 °C for 5 min. Cool and return to initial conditions as fast as instrument specifications allow.
4. Set the transfer line temperature to 250 °C.

3.7 Mass Calibration for Mass Spectrometry

The mass spectrometry settings for GC-MS systems are typically inbuilt features of the employed GC-MS system. In the following we will use a Pegasus III time-of-flight mass spectrometer (LECO Instrumente GmbH, Mönchengladbach, Germany) to exemplify the details of this module that should be considered and reported when publishing GC-MS profiling data. Other MS systems are equally amenable to metabolite profiling analyses. The mass calibration of GC-MS systems is typically part of inbuilt auto-tuning processes. Mass calibration and instrumental limitations define the mass range which can be chosen for analysis.

1. Trigger the auto-tuning process and perfluorotributylamine to calibrate the MS system before processing a set of samples comprising a metabolite profiling experiment. Set the mass range to $m/z = 70$–600.

3.8 Ionization for Mass Spectrometry

The ionization process determines the mass spectrum of compounds and thereby the availability and chemical nature of molecular ions and mass fragments that can be used for specific and selective analysis of compounds in complex mixtures. In the following we will use a Pegasus III time-of-flight mass spectrometer (LECO Instrumente GmbH, Mönchengladbach, Germany) to exemplify the details of this module that should be considered and reported when publishing GC-MS profiling data.

1. Use electron impact ionization, set the ion source temperature to 250 °C and the filament bias current to −70 eV. Optimize detector voltage depending on detector age to approximately 1,500–1,950 V (*see* **Note 16**).

3.9 Mass Spectrometric Analysis

The type of mass spectrometric analysis determines speed of mass spectral scanning and the accuracy and precision of mass recordings. The later may determine the availability of specific ions for the selective monitoring of compounds in complex mixtures.

1. Use a Pegasus III time-of-flight mass spectrometer (LECO Instrumente GmbH, Mönchengladbach, Germany) at nominal mass resolution with scanning rate set to 20 scans/s (*see* **Note 17**).

3.10 Chromatogram Data Processing and Generation of Comprehensive Peak Lists

The algorithms used to process GC-MS chromatogram data files and those applied to pick peak apices and respective peak responses, so-called mass features, can influence relative and absolute quantification.

1. Check the quality of raw chromatogram files without pre-processing using the manufacturer's file format and respective manufacturer's software. Avoid the following analytical artifacts through system maintenance: Column bleeding and chemical background caused by laboratory contaminations; chromatography artifacts, e.g., unusual peak shapes, peak tailing, over all retention drifts and shifts of peak position; mass spectral artifacts, e.g., absence of baseline responses, presence of positive or negative electronic spikes, drift of mass calibration; quantitative artifacts, e.g. peak overloading, drifts of overall recovery or changes of recovery of compound classes, and fast loss of detector sensitivity (*see* **Note 18**).
2. Eliminate single deviant chromatograms before subsequent data processing, reanalyze full experimental sets by split injection in cases of peak overload (*see* Subheading 3.5.2) or reanalyze full sets of backup samples (*see* Subheading 3.2.1).
3. Perform baseline correction above noise, apply smoothing algorithm set to five scans, and export a chromatogram file in an interchange format, e.g., CDF-format, using the manufacturer's software options (Fig. 2a).
4. Generate a comprehensive peak list of each chromatogram that contains all observed mass features above a signal to noise ratio of ≥2. Mass feature information contains monitored mass (m/z ratio), retention index (arbitrary RI units, *see* Subheading 3.3), and respective detected response (arbitrary

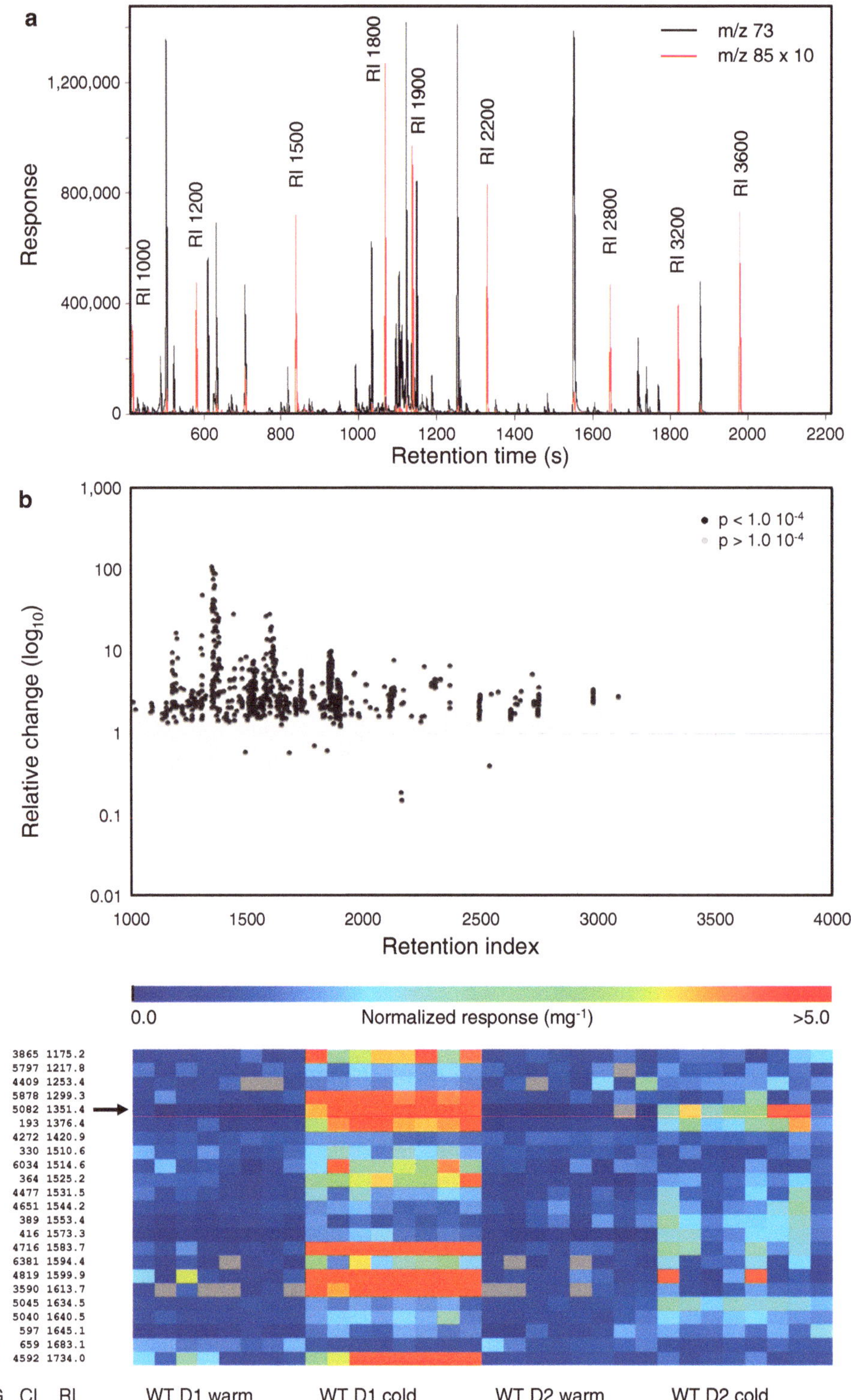

Fig. 2 Exemplary GC-MS chromatogram (**a**) of a methoxyaminated and trimethylsilylated primary metabolite fraction from *Arabidopsis thaliana* (Col-0) wild type rosettes which were grown under optimal growth temperatures. An overlay of two selected ion chromatograms is shown. The mass fragment, *m*/*z* 85, is characteristic of *n*-alkanes which are added internal retention index standards. The fragment, *m*/*z* 73, is characteristic of trimethylsilylated derivatives, so-called analytes, which are formed from primary metabolites by the chemical

intensity units). For this purpose use, for example, TagFinder software [20, 21] or other equivalent software (*see* **Note 19**). Convert manufacturer's peak response values, if necessary, into integers, round manufacturer's m/z values according to the instrument's mass precision or into integers representing nominal mass units and calculate retention indices with two decimal places from retention times recorded with three significant decimal places (*see* Subheading 3.3).

5. Archive the raw chromatogram files in manufacturer's data format. Archive the processed chromatogram files in the chosen interchange format. Archive the comprehensive peak list obtained from each chromatogram.

3.11 Mass Alignment

Comprehensive peak lists of multiple chromatograms are combined into tables that represent a matrix of detected responses of all mass features across all chromatograms of an experimental data set. The process requires a mass alignment step, more precisely alignment according to m/z ratio. This alignment ensures that the mass features that are merged across the different individual chromatograms have identical mass within the limits of the instrument's mass precision.

1. Round the m/z values from the preciseness provided by the GC-MS system to full nominal mass without decimals.
2. Align according to equal nominal mass (*see* **Note 20**).
3. Store nominal mass of the merged mass feature in the resulting tabular matrix of an experimental data set.

3.12 Chromatographic Alignment

Chromatographic alignment of comprehensive peak lists is required for the generation of tables that contain the detected responses of observed mass features across all chromatograms of an experimental data set. The chromatographic alignment ensures that the mass

Fig. 2 (continued) derivatization steps. (**b**) The relative fold-change plot indicates general trends of all aligned mass features (*grey*) and of significantly changed features at *t*-test $p < 1.0 \times 10^{-4}$ (*black*). A relative change of 10 indicates a tenfold increase of the normalized responses in the cold, 10 °C during the day and 8 °C at night, compared to the normalized response at optimal growth temperature, 20 °C during the day and 18 °C at night. A relative change of 0.1 indicates a tenfold decrease in the cold. Note that most mass features show an increase in the cold. (**c**) Part of the table of significantly changed mass features. Mass features are identified among other information specified in the text by time group (TG), cluster (CL), and average retention index (RI). The normalized responses, i.e., the responses of mass features divided by the response of the internal standard ($^{13}C_6$-sorbitol) divided by the sample fresh weight (mg), are displayed as a color-coded heat map. The warm to cold comparison was repeated at two developmental stages, D1 and D2, of the *Arabidopsis thaliana* rosette. The metabolites represented by the mass features can be identified top-down according to relevance, for example, according to significance, e.g., *t*-test *p* values, magnitude of the fold change, and repeatability, e.g., common significant changes at two developmental stages. The further analysis of cluster 5082 (*arrow*) in time group 54 at retention index 1,351.4 is demonstrated in the following (*see* Fig. 3)

features that are merged across the different individual chromatograms elute within an identical chromatographic window.

1. Use retention indices for chromatographic alignment (Fig. 2a). Calculate retention indices of all mass features within the peak lists of single chromatograms. Use the recorded retention times of *n*-alkane time standards that are added to each individual chromatogram (*see* Subheading 3.3).
2. Sort observed mass features with identical nominal mass from all peak lists constituting an experimental set according to retention indices. Group into common retention index intervals by defining the gap width between neighboring intervals and by defining the number of tolerated random observations within the gap; for details refer to previous publications [20, 21].
3. Store the alignment of merged mass features with information on minimal, average, and maximal observed retention index including retention index width of the interval in the resulting tabular matrix of an experimental data set (*see* **Note 21**).

3.13 Grouping of Aligned Mass Features

Mass features from mass spectrometric analyses carry redundant quantitative information, due to induced fragmentation of the molecular ion after ionization and due the presence of mass isotopomers that result from the presence of naturally occurring stable isotopes. Such redundancies can be reduced by grouping of mass features that represent the same analyte.

1. Group all mass features with overlapping retention index intervals, i.e., a time group of mass features.
2. Group all mass features within a time group that have correlated responses across all chromatograms of an experimental data set, i.e., a cluster of mass features.
3. Store all mass features with group assignments within tabular data set for nontargeted data mining (*see* **Note 22**).

3.14 Response Normalization of Aligned Mass Features

Detected responses of mass features from mass spectrometric analyses need to be normalized prior to relative quantification to account for variations of sample amount and for variations due to differential losses in the course of the analytical process. Exact quantification requires in addition a metabolite-specific recovery factor and co-processed external calibration series to calculate metabolite concentrations.

1. Normalize responses of all remaining mass features to sample fresh weight and to the response of the internal standard, $^{13}C_6$-sorbitol.
2. Use normalized responses directly for statistical analyses or transform normalized responses according to the experimental design. For example, calculate ratios of each normalized mass

feature to the median of the non-treated control group or to the median normalized response of all chromatograms of an experimental set, subsequently calculate logarithms of the ratios to equally represent fold increases and fold decreases (Figs. 2 and 3).

3. Apply tools for data visualization, e.g., principal component analysis (PCA), hierarchical cluster analysis, or clustered heat map display of the tabular normalized data to analyze the variation between replicate samples representing identical conditions compared to the variation between conditions. Visualize data to analyze the groups of response patterns of mass features across all studied conditions.
4. Apply statistical procedures, such as ANOVA and post hoc tests or their nonparametric counterparts, with correction for multiple testing or use more elaborate approaches, e.g., ref. 11, to discover relevant temperature controlled mass features or clusters of mass features (*see* **Note 23**). During the discovery phase focus on clusters that comprise at least three mass features (Fig. 2b). Also consider mass features that are absent in control conditions and present in temperature stressed conditions or vice versa (*see* **Note 24**).
5. For the targeted exact quantification of specific metabolites, select at least one selective mass feature and use external calibration curves to determine the apparent concentration. Perform quantification within the range of the upper and lower limits of quantification of the calibration curve. Avoid peak overloading by reanalysis using split injection (*see* Subheading 3.5.2). Never extrapolate beyond the limits of quantification. Correct for the recovery of an isotopically labelled internal standard within each chromatogram or for the recovery factor determined in preceding recovery experiments to obtain the concentration within the sample.

3.15 Reconstitution and Matching of Mass Spectra for the Identification of Relevant Metabolites

The preceding discovery process and statistical evaluation provides information on relevant mass features and clusters of relevant mass features. These mass features need to be linked to the metabolite(s) that are represented by those features (Fig. 2b). In GC-MS profiling studies this identification process is most efficiently started with a mass spectrum.

1. Reconstitute mass spectrum of all mass features with overlapping retention index intervals, i.e., a reconstituted mass spectrum of a time group. Match reconstituted time group spectra to mass spectral libraries using reverse matching (*see* **Note 25**).
2. Reconstitute mass spectrum of all correlated mass features within a time group, i.e., a reconstituted mass spectrum of a cluster of mass features. Match reconstituted cluster spectra to mass spectral libraries using forward matching (*see* **Note 26**).

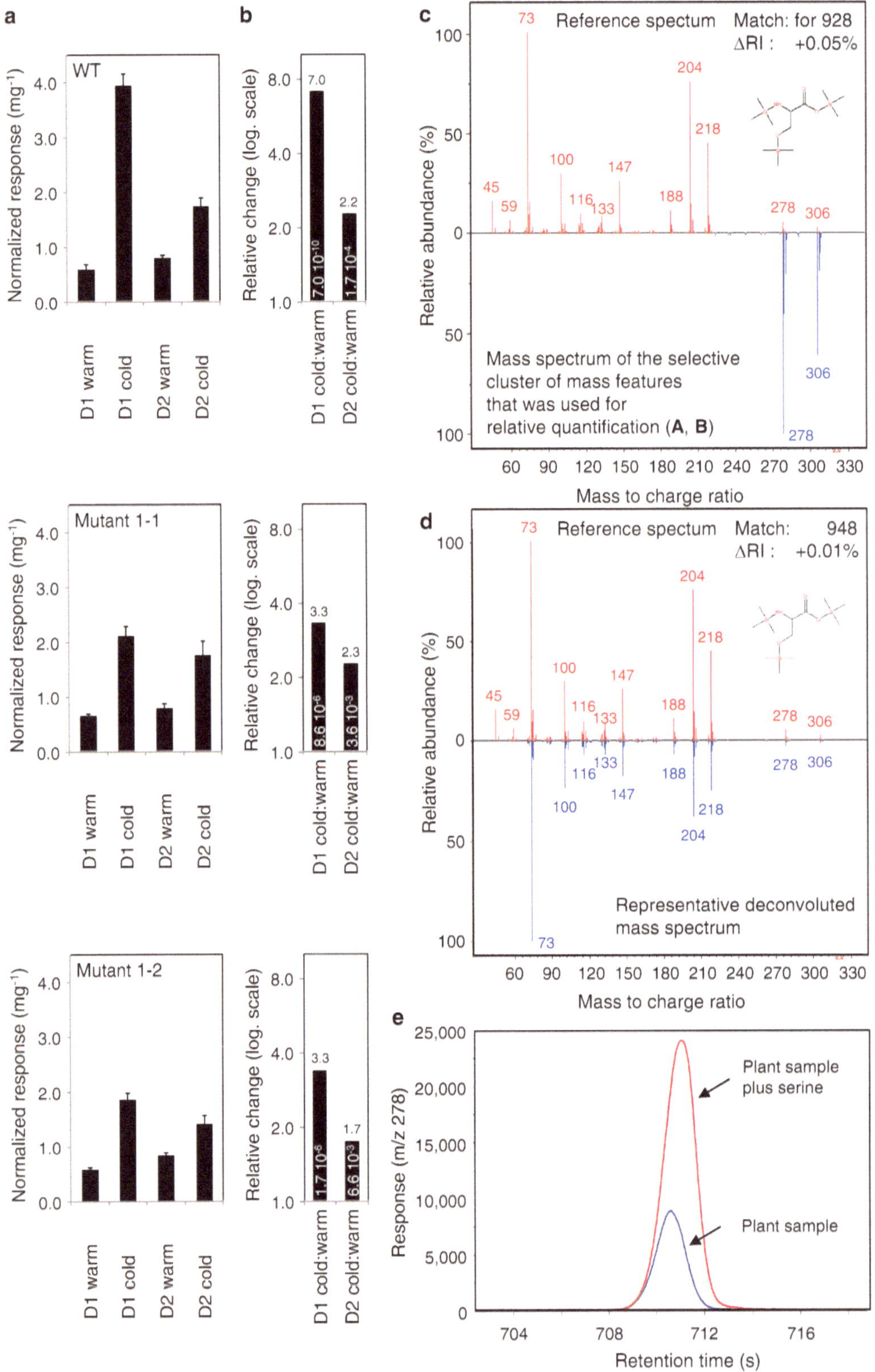

Fig. 3 Wild type (Col-0) and two allelic mutants of *Arabidopsis thaliana* plants were grown under optimal growth temperatures, 20 °C in the day 18 °C at night (warm), and under cold temperatures, 10 °C during the day 8 °C at night. Complete rosettes were harvested into liquid nitrogen at two developmental stages D1before appearance of the inflorescence and D2 after appearance of the florescence. The rosettes were

3. Retrieve representative full mass spectra. Apply automated mass spectral deconvolution algorithms to extract full mass spectra from the GC-MS chromatograms of the experimental data set that contain the relevant mass features at highest available responses and the least amounts of co-eluting compounds, both matching directions apply (*see* **Note 27**).
4. Export and archive all extracted and reconstituted mass spectra in a mass spectral interchange format, such as the MSP-format (Fig. 3).
5. Perform provisional identification manually (*see* **Note 28**), consider the best mass spectral match within an RI deviation window ≤1 %, when using identical capillary column types [17].
6. If no hit is obtained, the respective compound is an unknown metabolite that can be described by mass spectrum and retention index. To further characterize this unknown compound repeat mass spectral matching without RI constraint and interpret partially matching mass spectra, the fragmentation pattern of the available mass spectra of this unknown compound and attempt to deduce a possible structure or presence of likely substructures, e.g., ref. 7.
7. Archive mass spectrum and retention index of the unknown compound, for examples by submission to a public data base, such as the Golm Metabolome Database.

3.16 Verification of Provisional Metabolite Identifications

Mass spectral and retention index based identifications using respective library compendia need to be verified by authenticated reference substances, especially for the differentiation of co-eluting isomers.

1. Obtain commercially or synthesize candidate metabolites and alternative isomers.
2. Perform a standard addition experiment. For this purpose, process in parallel, a backup of the biological sample that contains the provisionally identified metabolite, a sample that contains

Fig. 3 (continued) homogenized and a representative aliquot of 60 ± 3 mg analyzed. The normalized responses (**a**) and the calculated relative pool size changes (**b**) indicate a robust and significant, *t*-test $p = 6.6 \times 10^{-3}$–7.0×10^{-10}, 1.7- to 7.0-fold increase of cluster 5082 (*see* Fig. 2) of mass features at two developmental stages and in wild type as well as in two allelic mutants (**c**). Cluster 5082 is deemed relevant for the cold response of wild type and of the mutants and the underlying metabolite should, therefore, be identified. (**d**) The mass spectrum and retention index of cluster 5082 and a representative deconvoluted mass spectrum were retrieved and matched to the Golm Metabolome Database, http://gmd.mpimp-golm.mpg.de/. The spectra matched to the reference entry of L-serine (3TMS) which is a chemical derivative of the metabolite, L-serine http://gmd.mpimp-golm.mpg.de/Metabolites/3ab40d3a-043a-488f-8361-d1bf309f842c.aspx. Note that the chemically correct identification is DL-serine, because a non-chiral capillary column was used. Subfigure (**e**) shows the typical result of a standard addition experiment shown at the expected retention time by a selective mass fragment, *m*/*z* 278. Note the slight shift of retention time that is caused by the increased amount of serine

the obtained and authenticated reference compound and an equal mixture. Adjust the amount of the reference compound to be approximately threefold higher than in the biological samples (Fig. 3).

3. Check metabolite identity by exact match of mass spectra and chromatographic retention.
4. If multiple isomers of the metabolite exist repeat the standard addition experiment using an alternative type of capillary column with changed polarity of the stationary phase.
5. Check metabolite identity again by exact match of mass spectra and chromatographic retention (*see* **Note 29**).
6. Archive the verification status of metabolite identifications or classifications, e.g., identification by standard addition of authenticated reference substance using one chromatographic system or two systems, preliminary identification by mass spectral and retention index match, or preliminary classification by mass spectral match.
7. Use the standard addition experiment to obtain a recovery value of the metabolite for the analyzed biological sample type.
8. If the identification fails, archive inferred classification, a structure hypothesis, if available, representative mass spectra and retention indices, as well as back up samples, if available, for later identification.

4 Notes

1. Metabolism has a rapid turnover which is stopped by immediate freezing. Samples need to stay frozen until extraction. Gravimetric determination of sample fresh weight by differential weighing while samples and containers are kept frozen is influenced by hoarfrost that forms from humid air during the weighing process and by residual droplets of liquid nitrogen. These influences are minimized if the container is controlled for residual liquid nitrogen before weighing and if the air humidity is kept low during differential weighing. Because humidity may change in the course of a day, determine the empty weight of the precooled micro vials directly before determining the differential weight with frozen sample. The tolerance of the scales must be 0.1 mg or better. The accuracy of the scales must be tested frequently by gauge weights (Fig. 1).
2. In the case that the single leaf is smaller than the 2.5 mg limit, pooling is required and single leaf analysis in a strict sense is impossible. Note that absolute quantification is delimited by the error of fresh weight determination. In the case that the leaf is larger than 125 mg or in the case that a complete rosette

is analyzed prepare a frozen homogenate and take a representative aliquot from the frozen homogenate. Surplus frozen homogenate of the sample can be used for analyses of other system components.

3. The duration and number of bursts required to obtain a fine powder may require adaptation to the physical properties of the plant material. Removal of the steel balls may lead to an uncontrolled loss of sample. Keep the time of storage to a minimum. Do not store at −80 °C for longer than 2 months without testing for storage effects.
4. Surface contaminations of all employed vials, flasks, bottles, pipetting devices, and steel balls can be avoided by careful washing with extraction solvents. Please consider that autoclaved material, while sterile, may nevertheless be chemically contaminated.
5. A recovery mixture of stable isotope labelled standards at concentrations that approximate the expected endogenous metabolite concentrations can be readily obtained by in vivo stable isotope labelling for example of *Arabidopsis thaliana*, *Saccharomyces cerevisiae*, or *Synechocystis* [22–24]. When performing standard addition experiments with non-labelled endogenous metabolites to determine recovery factors aim for an approximately threefold increased standard addition compared to the endogenous metabolite concentration.
6. Organic solvents can be pre-cooled to −20 °C. Water can be kept at 4 °C prior to extraction. Precooling is possible when using few internal standards at low concentrations. When using multiple internal standards or high concentrations of single standards, do not precool to avoid precipitation of added internal standards.
7. Leaves of some cold stressed and cold acclimated plants, such as *Arabidopsis thaliana*, accumulate soluble polar metabolites compared to non-stressed and non-acclimated leaves, e.g., refs. 8, 9. The sample fresh weight of cold treated *Arabidopsis thaliana* leaf samples can be reduced but the ratio of extract volume to sample fresh weight must be kept constant for all samples of an experimental set (*see* **Note 9**).
8. To increase metabolite recovery extract twice with 400 μL bi-distilled water. Combine the upper phases and dry a proportionally up-scaled aliquot in a vacuum concentrator. Should the bi-distilled water for the first extraction contain internal standards, use 400 μL bi-distilled water without internal standards for the second extraction step.
9. The fresh weight of single leaves changes during development. To compare small samples, e.g., 2.5–10.0 mg fresh weight, with larger samples the ratio of extract volume and fresh weight

should be kept approximately constant. When using 2.0 mL micro vials extract ≤10 mg with 400 μL methanol solvent, 10–20 mg with 800 μL, 20–30 mg with 1,200 μL, 30–40 mg with 1,600 μL, and 40–50 mg with 2,000 μL. Take a frozen aliquot from a homogenate to handle larger samples in 2.0 mL micro vials.

10. The protocol describes a two-step reaction. A methoxyamine reagent transfers carbonyl moieties into two products, an E- and a Z-methoxyamine. The two products are in most cases separable by gas chromatography. Methoxyamination transforms reducing sugars into open chain products and eliminates acetal and ketal formation prior to trimethylsilylation. Trimethylsilylation substitutes hydrogen atoms that are bound to heteroatoms, e.g., hydroxyl-, sulfhydryl-, or amino moieties, and thereby transforms most polar primary metabolites into volatile analytes. Note that amino moieties can be non-trimethylsilylated or trimethylsilylated once or twice, depending on the steric hindrance of the structure. The ratio between two trimethylsilylation products of an amine, e.g., an amino acid, can be used to assess the performance of the trimethylsilylation reaction. Non-methoxyaminated but trimethylsilylated hexoses can be used to monitor the efficiency of the methoxyamination reaction.

11. 4-(Dimethylamino)pyridine is added as catalyst in the presence of high amounts of reducing sugars. Do not change the sequence of dissolving steps. Dissolving methoxyamine hydrochloride first may cause precipitations when adding 4-(dimethylamino)pyridine.

12. Silylation reagents react with water, loose reactivity, and form polysiloxanes in the process which contribute to the chemical background of GC-MS profiling.

13. Check that solid residue is completely dissolved.

14. Erban and coauthors [16] describe an automated and miniaturized protocol for in-line chemical derivatization. This protocol allows continuous processing of large sample numbers and is exactly timed between chemical derivatization and injection into the GC-MS system.

15. Perform injections of pure *N*,*O*-bis(trimethylsilyl)trifluoroacetamide reagent between samples in regular intervals to control for carry over effects and to counter act build-up of semi-volatile chemical deposits within the injector system. Note that complex lipophilic and other non-volatile compounds accumulate in the GC injection port. An increased frequency of GC maintenance may be required when omitting phase separation using methanol extraction without liquid partitioning into chloroform. Use deactivated glass insert liners,

e.g., ref. 16. Change liner before each new experimental set. During analysis of each experimental set, change liner depending on the build-up of non-volatile chemical deposits within the injector system. Depending on the type of sample and applied extraction procedure liner may be changed every 12–60 injections. The concentration of chemically derivatized samples is adjusted to monitor a maximal number of peaks. The major metabolites are typically close to or at the upper detection limit when applying splitless injection. To also quantify these major metabolites perform a second analysis of the same samples using split injection. Use the split ratio to adjust the abundance of major metabolites to the linear range of quantification.

16. The ionization method determines the complexity of the recorded mass spectrum. Electron impact ionization generates highly repeatable mass spectral fingerprints that can be easily compared between different types of mass spectrometers and between systems of different manufacturers. The sensitivity of profiling analyses declines with aging mass detectors. Adjustments of detector voltage can be used within manufacturer defined limits. Note that increasing detector voltage may affect signal to noise and thereby the capability to monitor minor components in complex mixtures. Alternatives to electron impact ionization are (atmospheric pressure) chemical ionization approaches which generate a smaller number of fragments. Such systems are used to analyze the molecular ions of analytes and can have a higher sensitivity due to a lower degree of fragmentation [25].

17. The exact type and version of mass spectrometer must be documented. Similar to the influence of the type of capillary column on gas chromatographic separation, the type and build of a mass spectrometer may influence the characteristics and details of recorded mass spectra. Mass spectrometers with higher mass precision can be used to deduce the molecular formula of analytes or of mass fragments from accurate mass recordings. Mass spectrometers set to higher scan rates can be used for fast GC-MS but at a loss of sensitivity and/or separation. Mass spectrometers with lower scan rates should not be employed for fast GC-MS analysis, ≥10 scans per peak should be recorded.

18. The mentioned artifacts may frequently occur due to system aging. Most artifacts are caused by build-up of non-volatile deposits. We do not mention rare system's deviations that may also compromise analyses. All quantitative analyses and profiling approaches that are based on chromatography systems hyphenated to mass spectrometric detection are subject to slow drifts in chromatography, mass calibration, and quantitative response.

These aspects must be controlled using the quality control samples mentioned above and by system maintenance.

19. The TagFinder software offers all required options for the processing of GC-MS chromatogram files [20]. The handling of the TagFinder software and recommended parameter settings have been described elsewhere [3, 21]. Here we describe the essential steps of GC-MS data processing for metabolite profiling and quantification experiments. Similar analyses can also be performed by combination of other specialized software. Note that the initial discovery of metabolic markers requires nontargeted peak lists that contain all mass features (RI, m/z, response) observed in the chromatogram files comprising a profiling experiment. The verification of candidate marker metabolites can be performed by a predefined targeted list of selective and metabolite-specific mass features that can be either directly extracted from chromatogram files using vendor software or from the comprehensive peak lists.
20. Mass alignment according to equal nominal mass is a trivial process that is typically applied when using low mass resolution mass spectrometers. Nevertheless, this process needs to be described. With the availability of high-resolution mass spectrometers for GC-MS systems, alternative merging methods are required, for example, the creation of mass bins at decimal intervals that are in agreement with the mass resolution provided by the respective instrument. Minimal, average, maximal mass and mass range should be documented when using such instruments (*see* Subheading 3.12).
21. Chromatographic alignment can be achieved with and without use of internal standards. The alignment result and applied procedures must be reported and archived. Most chromatographic alignment procedures are confounded by random low-response peaks that are caused by electronic and chemical noise. These noise effects can be contained by applying a low-response threshold for analyzed mass features and by co-processing of a limited number of chromatograms, e.g., 50–200 chromatogram files. However, more confounding are peak broadening effects and retention index shifts which are caused by changes of metabolite concentrations, especially by increases approaching chromatographic overloading. The effects can be contained by accepting wider retention index intervals for mass feature alignment at the cost of a possible loss of isomer resolution. Especially closely eluting minor isomers which provide no differentiating mass fragments may get lost in the presence of large amounts of the abundant isomer. Note that alternative means of chromatographic alignment, such as chromatographic alignment algorithms and software, may be applied but need to be tested thoroughly for the mentioned challenges that may confound alignment.

22. Grouping of mass features and "de-isotoping" with the aim to remove redundancy can be achieved using diverse approaches. The TagFinder software uses the fact that co-eluting redundant mass features are highly correlated in GC-MS analyses, because natural mass isotopomers of elements occur in fixed ratios and because electron impact ionization generates mass fragments in stable ratios. Luedemann and coauthors [20, 21] describe the generation of a correlation network using Pearson's or Kendall's correlation coefficients and a core finding algorithm to assign clusters.
23. Mass spectrometry-based metabolite profiling experiments may contain large numbers of low-response mass features that are equally likely to be caused either by analytes or by electronic and chemical noise. The majority of irrelevant mass features can be removed before relative or absolute quantification by setting a minimal response threshold. If available through processing software use signal to noise values of each recorded mass feature to threshold.
24. Note that multiple statistical approaches may lead to the discovery of relevant regulated mass features. Try to combine the results of several tests and focus on those mass features that have high relative changes of normalized responses and that are part of clusters with a high number of correlated mass features. When checking mass features by analysis of absent/present calls avoid mass features with responses that are close to the lower limit of detection.
25. Note that reconstituted mass spectra of time groups are inherently composite mass spectra of co-eluting analytes. Therefore reverse matching of library spectra of pure authenticated reference compounds is the most appropriate matching approach.
26. Note that reconstituted mass spectra of clusters of mass features are inherently partial mass spectra. Mass fragments that are common to co-eluting analytes are eliminated because such mass fragments are not highly correlated. Also mass fragments with low-responses and mass fragments with an overloaded response are eliminated. Therefore forward matching to library spectra of pure authenticated reference compounds is the most appropriate matching approach.
27. Note that automatically deconvoluted mass spectra can still contain systematic errors, such as additional mass fragments of exactly co-eluting analytes, missing mass fragments due to false subtraction of fragments from neighboring analytes. If necessary, attempt manual extraction and correction of mass spectra.
28. Note that mass spectral and retention index matches alone represent a preliminary identification. Isomers are frequently present in biological samples, e.g., diverse saccharides. Such isomers can have almost identical mass spectra and highly similar chromatographic retention.

29. An elegant solution for this two column approach to compound identification is provided by two-dimensional GCxGC-MS systems. Note that some isomers cannot even be separated by changed column polarities. A non-separable set of metabolites should be reported in the sense of summarized information, e.g., the sum of two saccharides or, if non-chiral columns are used, the sum of D- and L-stereoisomers.

References

1. Fiehn O, Kopka J, Doermann P, Altmann T, Trethewey RN, Willmitzer L (2000) Metabolite profiling for plant functional genomics. Nat Biotechnol 18:1157–1161
2. Roessner U, Wagner C, Kopka J, Trethewey RN, Willmitzer L (2000) Simultaneous analysis of metabolites in potato tuber by gas chromatography-mass spectrometry. Plant J 23:131–142
3. Allwood JW, Erban A, de Koning S, Dunn WB, Luedemann A, Lommen A, Kay L, Löscher R, Kopka J, Goodacre R (2009) Inter-laboratory reproducibility of fast gas chromatography-electron impact-time of flight mass spectrometry (GC-EI-TOF/MS) based plant metabolomics. Metabolomics 5:479–496
4. Wagner C, Sefkow M, Kopka J (2003) Construction and application of a mass spectral and retention time index database generated from plant GC/EI-TOF-MS metabolite profiles. Phytochemistry 62:887–900
5. Schauer N, Steinhauser D, Strelkov S, Schomburg D, Allison G, Moritz T, Lundgren K, Roessner-Tunali U, Forbes MG, Willmitzer L, Fernie AR, Kopka J (2005) GC-MS libraries for the rapid identification of metabolites in complex biological samples. FEBS Lett 579:1332–1337
6. Kopka J, Schauer N, Krueger S, Birkemeyer C, Usadel B, Bergmueller E, Doermann P, Weckwerth W, Gibon Y, Stitt M, Willmitzer L, Fernie AR, Steinhauser D (2005) GMD@CSB.DB: the Golm Metabolome Database. Bioinformatics 21:1635–1638
7. Hummel J, Strehmel N, Selbig J, Walther D, Kopka J (2010) Decision tree supported substructure prediction of metabolites from GC-MS profiles. Metabolomics 6:322–333
8. Kaplan F, Kopka J, Haskell DW, Zhao W, Schiller KC, Gatzke N, Sung DY, Guy CL (2004) Exploring the temperature-stress metabolome of Arabidopsis. Plant Physiol 136:4159–4168
9. Kaplan F, Kopka J, Sung DY, Zhao W, Popp M, Porat R, Guy CL (2007) Transcript and metabolite profiling during cold acclimation of Arabidopsis reveals an intricate relationship of cold-regulated gene expression with modifications in metabolite content. Plant J 50: 967–981
10. Guy CL, Kaplan F, Kopka J, Selbig J, Hincha DK (2008) Metabolomics of temperature stress. Physiol Plant 132:220–235
11. Korn M, Gärtner T, Erban A, Kopka J, Selbig J, Hincha DK (2010) Predicting Arabidopsis freezing tolerance and heterosis in freezing tolerance from metabolite composition. Mol Plant 3:224–235
12. Dunn WB, Erban A, Weber RJM, Creek DJ, Brown M, Breitling R, Hankemeier T, Goodacre R, Neumann S, Kopka J, Viant MR (2013) Mass appeal: metabolite identification in mass spectrometry-focused untargeted metabolomics. Metabolomics 9:44–66
13. Sumner LW, Amberg A, Barrett D, Beale MH, Beger R, Daykin CA, Fan TWM, Fiehn O, Goodacre R, Griffin JL, Hankemeier T, Hardy N, Harnly J, Higashi R, Kopka J, Lane AN, Lindon JC, Marriott P, Nicholls AW, Reily MD, Thaden JJ, Viant MR (2007) Proposed minimum reporting standards for chemical analysis. Metabolomics 3:211–221
14. Fernie AR, Aharoni A, Willmitzer L, Stitt M, Tohge T, Kopka J, Carroll AJ, Saito K, Fraser PD, DeLuca V (2011) Recommendations for reporting metabolite data. Plant Cell 23:2477–2482
15. Lisec J, Schauer N, Kopka J, Willmitzer L, Fernie AR (2006) Gas chromatography mass spectrometry-based metabolite profiling in plants. Nat Protoc 1:387–396
16. Erban A, Schauer N, Fernie AR, Kopka J (2007) Non-supervised construction and application of mass spectral and retention time index libraries from time-of-flight GC-MS metabolite profiles. Methods Mol Biol 358:19–38
17. Strehmel N, Hummel J, Erban A, Strassburg K, Kopka J (2008) Retention index thresholds for compound matching in GC-MS metabolite profiling. J Chromatogr B 871:182–190
18. Van den Dool H, Kratz PD (1963) A generalization of the retention index system including linear

temperature programmed gas–liquid partition chromatography. J Chromatogr 11:463–471

19. Birkemeyer C, Kolasa A, Kopka J (2003) Comprehensive chemical derivatization for gas chromatography-mass spectrometry-based multi-targeted profiling of the major phytohormones. J Chromatogr A 993:89–102
20. Luedemann A, Strassburg K, Erban A, Kopka J (2008) TagFinder for the quantitative analysis of gas chromatography–mass spectrometry (GC-MS) based metabolite profiling experiments. Bioinformatics 24:732–737
21. Luedemann A, von Malotky L, Erban A, Kopka J (2012) TagFinder: preprocessing software for the fingerprinting and the profiling of gas chromatography–mass spectrometry based metabolome analyses. Methods Mol Biol 860:255–286
22. Huege J, Sulpice R, Gibon Y, Lisec J, Koehl K, Kopka J (2007) GC-EI-TOF-MS analysis of in vivo-carbon-partitioning into soluble metabolite pools of higher plants by monitoring isotope dilution after ($^{13}CO_2$)-labelling. Phytochemistry 68:2258–2272
23. Strassburg K, Walther D, Takahashi H, Kanaya S, Kopka J (2010) Dynamic transcriptional and metabolic responses in yeast adapting to temperature stress. OMICS J Integr Biol 14: 249–259
24. Huege J, Goetze J, Schwarz D, Bauwe H, Hagemann M, Kopka J (2011) Modulation of the major paths of carbon in photorespiratory mutants of Synechocystis. PLoS ONE 6:e16278
25. Strehmel N, Kopka J, Scheel D, Böttcher C (2013) Annotating unknown components from GC/EI-MS-based metabolite profiling experiments using GC/APCI(+)-QTOFMS. Metabolomics. doi:10.1007/s11306-013-0569-y

Chapter 15

A Lipidomic Approach to Identify Cold-Induced Changes in Arabidopsis Membrane Lipid Composition

Hieu Sy Vu, Sunitha Shiva, Aaron Smalter Hall, and Ruth Welti

Abstract

Lipidomic analysis using electrospray ionization triple quadrupole mass spectrometry can be employed to monitor lipid changes that occur during cold and freezing stress of plants. Here we describe the analysis of *Arabidopsis thaliana* polar glycerolipids with normal and oxidized acyl chains, sampled during cold and freezing treatments. Mass spectral data are processed using the online capabilities of LipidomeDB Data Calculation Environment.

Key words Cold acclimation, Freezing, Post-freezing recovery, Lipidomics, Galactolipid, Phospholipid, Oxidized lipid, Triple quadrupole, Mass spectrometry

1 Introduction

Cold and freezing stress can cause major crop losses. *Arabidopsis thaliana* can be employed as an experimental model for analysis of the biochemical changes that accompany the development of plant freezing tolerance by cold acclimation and the changes that occur in cold and freezing stress and during plant recovery from freezing stress. Cold acclimation, or the development of tolerance to freezing by exposure to cold, but non-freezing, temperature, is a capability of many plants originating in temperate climates. Cold acclimation of Arabidopsis (Columbia accession) for 1 day or more decreases the lethal temperature from approximately −2 to −8 °C [1, 2].

Lipid changes occur during cold acclimation, the freezing process, and recovery from freezing stress (e.g., refs. 2–5). One effect of lipid alterations in cold acclimation is to curb lipid phase changes that lead to membrane leakage [2]. Indeed, alterations in plant lipid metabolism before or during cold and freezing stress can modulate plant damage [2–5]. For example, fatty acid desaturases, such as *Ads-2*, act to increase fatty acid unsaturation during cold acclimation and have a positive effect on freezing tolerance [6]. Desaturases forming trienoic fatty acids also are required for effective

Dirk K. Hincha and Ellen Zuther (eds.), *Plant Cold Acclimation: Methods and Protocols*, Methods in Molecular Biology, vol. 1166, DOI 10.1007/978-1-4939-0844-8_15, © Springer Science+Business Media New York 2014

photosynthesis at low temperature [7]. Generally, acyl lipid levels increase during cold acclimation; indeed, fatty acid synthesis is critical for avoiding plant damage in low temperature stress [2, 3, 8]. The freezing-activated galactolipid:galactosyltransferase encoded by *SENSITIVE TO FREEZING 2* (*SFR2*) forms oligo-galatosyldiacylglycerols and also has a positive effect on survival [9]. Two phospholipase Ds, which hydrolyze phospholipids, act during freezing and post-freezing recovery, but one increases and the other decreases plant damage during the freezing and post-freezing recovery processes [3, 4, 10]. Cold- or freezing-induced lipid changes may alter binding of lipids to proteins, which may affect protein function and plant stress damage [11].

There is still much to be discovered about the role of lipids, lipid-metabolizing enzymes, and lipid-binding proteins in cold and freezing stress. Mass spectrometry-based lipidomics offers many advantages in monitoring cold- and freezing-induced lipid changes. A lipidomic method can be used to examine a large number of lipid molecular species and to perform the analysis in a relatively short time. Lipid extracts from cold-treated plants may be introduced to a mass spectrometer by direct infusion or by liquid chromatography, and both methods have been utilized in Arabidopsis cold stress studies [3, 12, 13]. Here we describe a direct infusion approach similar to that used to analyze lipids during cold acclimation and freezing stress in recent work by Vu et al. [13] and extend our previously described analytical approach [3, 14]. The procedure takes advantage of the automated capability for processing of triple quadrupole mass spectral data at LipidomeDB Data Calculation Environment [15].

The analytical procedure described here measures membrane phospholipids and galactolipids, with identification of mass spectral data in terms of lipid class, total acyl carbons, and total double bonds (i.e., total acyl carbons: total acyl double bonds of double bond equivalents beyond the carbonyl). For membrane lipids with normal acyl chains, compound identifications are based on the mass/charge ratio (*m/z*) of the intact ion and the mass or *m/z* of one fragment formed in the mass spectrometer. Typically, for polar lipids, this is a head group fragment. In addition, the analysis of phospholipids and galactolipids containing oxidized acyl chains is described; these lipids are specified by head group and acyl species.

Lipid amounts are determined as normalized mass spectral signal/plant dry mass. The intensities of peaks in each sample are compared to those of added internal standards. A value of 1 represents the same intensity as 1 nmol of a relevant internal standard (or an *m/z*-corrected intensity of two internal standards detected in the same mass spectral scan). For diacyl or monoacyl phospholipids, the response for each compound is very close (within 5 or 10 %) to the response of an internal standard of the same class. Thus, the normalized signal/dry mass for diacyl or monoacyl phospholipids can be considered to be equal to nmol/dry mass. On the other hand, the molar responses of galactolipids and the oxidized membrane lipids have not been carefully characterized. In these cases, comparison among samples of the normalized signal/dry mass for the same lipid is valid, but normalized signal/dry mass may not be an accurate indicator of the relative amount of each lipid compared to other lipids in the sample.

The protocol for cold and freezing described herein is for cold and freezing treatment of *Arabidopsis thaliana* (Columbia accession and accessions with similar freezing tolerance). More details on cold acclimation, freezing, and post-freezing treatments can be found elsewhere in this volume.

2 Materials (*See Note 1*)

2.1 For Cold Acclimation and Freezing Treatment

1. *Arabidopsis thaliana* plants, grown in a growth chamber and in soil, such as Pro-Mix "PGX" (Hummert International), in 3½″ Kord square pots (Hummert International), or 72-cell plug trays (International Greenhouse Company) (*see* **Note 2**).
2. Light meter.
3. Waxed paper, scissors.
4. Ice chips.
5. Growth chamber, such as a Conviron ATC26.
6. Walk-in cold room.
7. Light cart (Hummert International).
8. Freezing chamber.

2.2 For Sampling, Lipid Extraction, and Dry Mass Measurement

1. Scissors.
2. Isopropanol with 0.01 % BHT (w/v).
3. HPLC-grade chloroform.
4. HPLC-grade water.
5. Chloroform/methanol (2:1, v/v) with 0.01 % BHT (w/v).
6. 1 M KCl in water.
7. Glass tubes, 50 mL (25 × 150 mm) with Teflon-lined screw caps (Fisher).
8. Pasteur pipettes, 9-inch.
9. Dry block heater that accepts 50 ml tubes.
10. Vortex mixer.
11. Orbital shaker.
12. Vacuum concentrator (such as CentriVap), vented to hood, or nitrogen gas stream evaporator, in hood.

13. Low-speed (clinical-type) centrifuge.
14. Oven, vented to hood.
15. Balance that determines mass, preferably to micrograms.
16. Ionizer antistatic system (VWR) (*see* **Note 3**).

2.3 For Mass Spectrometry

1. Methanol/300 mM ammonium acetate in water (95:5, v/v).
2. Chloroform.
3. Internal standard mix, containing LPC(13:0) (*see* **Note 4**), LPE(14:0), LPE(18:0), PA(28:0) [di14:0], PA(40:0) [diphytanoyl], PA(40:0) [diphytanoyl], PC(24:0) [di12:0], PC(48:2) [di 24:1], PE(24:0) [di12:0], PE(46:0) [di23:0], PG(28:0) [di14:0], PG(40:0) [diphytanoyl], hydrogenated PI, PS(28:0) [di14:0], PS(40:0) [diphytanoyl], hydrogenated DGDG, and hydrogenated MGDG (14) (*see* **Note 5**).
4. Pre-slit, target Snap-it 11 mm Snap Caps (MicroLiter).
5. Amber vials, 12 × 32 mm (MicroLiter).
6. Autosampler, such as CTC Mini-PAL (LEAP), with 1 mL sample loop.
7. Sample trays to hold vials, such as VT54 (LEAP).
8. Large reservoir (e.g., 500 ml) syringe pump with pump controller to provide continuous infusion. Reservoir is filled with methanol or methanol:isopropanol (1:1, v/v).
9. Methanol/acetic acid (9:1, v/v) for washing between samples.
10. Methanol/chloroform/water (66.5:30:3.5, v/v/v) to fill the wash reservoirs on the autosampler for washing the syringe and injection port.
11. Triple quadrupole mass spectrometer, such as API 4000 (Applied Biosystems, Foster City, CA), with electrospray ionization source.

3 Methods

3.1 Cold Acclimation and Freezing Treatment

1. Transfer soil-grown Arabidopsis plants to the portable light cart. Put the light cart into cold room with desired temperature (1–4 °C) for cold acclimation. Light intensity and day/night cycle should be measured with a light meter and adjusted to match those of the growing condition.
2. Acclimate plants by placing in the cold room for the desired period (0–7 days).
3. For plants that will undergo freezing, cut pieces of waxed paper to cover half of soil around each plant. Gently place waxed paper under Arabidopsis rosettes and on top of soil as shown in Fig. 1.

Fig. 1 Arabidopsis plants prepared to undergo freezing. Two half circles of waxed paper have been placed under each rosette. The purpose of the waxed paper is to eliminate freezing of leaves to the soil, which makes it difficult to obtain clean leaf or rosette samples when the plants are frozen

4. Transfer plants to be frozen to the programmable freezing chamber. Program the freezing chamber so that the temperature drops from the cold acclimation point to −2 °C at 2 °C/h (*see* **Note 6**). Plants may be held at −2 °C for 1 h for ice crystal formation before the temperature is dropped at 2 °C/h to the final temperature. Ice chips may be added on soil (under or around waxed paper) at this step to prevent supercooling (*see* **Note 7**).
5. After the freezing treatment (typically 2 h), plants may be thawed at 4 °C or other desired temperature (*see* **Note 8**).

3.2 Sampling, Lipid Extraction, and Dry Mass Measurement

The method is modified from ref. 16. Additional recommendations and options for extraction have been provided previously in another article in this series [14].

1. Plants may be sampled at desired time points before cold acclimation, during acclimation, after freezing, or during a post-freezing recovery period (*see* **Note 9**). Cut leaves, rosettes, or other desired tissue and quickly submerge plant tissues in isopropanol (containing 0.01 % BHT) at 75 °C for 15 min (*see* **Note 10**).
2. For plant tissues with dry mass less than 30 mg, submerge in a 50-ml glass tube containing 3 ml of isopropanol (0.01 % BHT), add 1.5 ml chloroform and 0.6 ml water (*see* **Note 11**).
3. Shake the tube at 100 rpm for 1 h at room temperature. Transfer solvent to a different tube.

4. Add 4 ml of chloroform:methanol (2:1, v/v) to the tissues in the first tube. Shake for 30 min and transfer solvent to the same tube used in **step 2**.
5. Repeat **step 3** three additional times. Combine all extracts (*see* **Note 12**).
6. Evaporate the combined extract with a nitrogen stream or vacuum concentrator (i.e., centrifuging evaporator; CentriVap).
7. Dissolve lipids in 1 ml of chloroform and store at −20 °C or colder.
8. Extracted tissues should be dried overnight at 105 °C and cooled. The dry mass should be determined using the microgram balance.

3.3 Mass Spectrometry and Data Processing

1. Add 10 μl of internal standard mix to each 2-ml amber glass vial. From 1 ml of sample in chloroform, add a volume originating from 0.2 mg dry tissue mass. Bring the volume to 360 μl with chloroform. Add 840 μl of the mixture methanol/300 mM ammonium acetate in water (95:5, v/v). The final solvent composition should be chloroform/methanol/300 mM ammonium acetate in water (30:66.5:3.5, v/v/v) (*see* **Note 13**).
2. Make "standards-only" ("i.s.," internal standard) samples with 10 μl internal standard mix, 350 μl chloroform, and 840 μl methanol:300 mM ammonium acetate in water (95:5, v/v) (*see* **Note 14**).
3. Make a set of washing blank ("wb") vials (equal to number of sample vials) containing methanol:acetic acid (9:1, v/v) to wash the tubing and ion source system between samples (*see* **Note 15**).
4. On a VT-54 sample tray, arrange vials in order: wb, i.s.1, wb, sample1, wb, sample 2, wb, … sample 9, wb, sample10, wb, i.s.2, wb, … That is, a washing blank should be every other vial. "Standards-only" samples should be in the spot of a sample and should run after every ten samples.
5. Program pump, auto-sampler, and mass spectrometer to infuse each sample at 30 μl/min and acquire a combination of spectra using the parameters shown in Table 1. Use the multiple channel analyzer (MCA) feature and a scan speed between 50 and 100 mass unit/s (*see* **Note 16**).
6. After data acquisition, perform baseline subtraction (window width: 20 Da), smoothing (point weighting: 0.4, 1.0, 0.4), peak integration (centroiding) of the resultant spectra in mass spectrometer software.
7. Export these processed spectral data into Excel files in the format specified at LipidomeDB Data Calculation Environment (DCE) at http://lipidome.bcf.ku.edu:9000/Lipidomics/.

Table 1
Mass spectral acquisition parameters for analysis of polar plant membrane lipids

Class (target list in DCE)	Adduct	Scan mode (+/–), scan type, scan mass	*m/z* Range scanned	MCA cycles	Collision energy (V)	Depolarization potential (V)	Exit potential (V)	Collision exit potential (V)
DGDG (plant DGDG)	$[M+NH_4]^+$	+NL, 341.13	890–1,050	50	24	90	10	23
MGDG (plant MGDG)	$[M+NH_4]^+$	+NL, 179.08	700–900	50	21	90	10	23
PA (plant PA)	$[M+NH_4]^+$	+NL, 115.00	500–800	60	25	100	14	14
PC (plant PC), LPC (plant LPC)	$[M+H]^+$	+Pre, 184.07	450–690	20	40	100	14	14
PE (plant PE), LPE (plant LPE)	$[M+H]^+$	+NL, 141.02	420–920	50	28	100	15	11
PG (plant PG)	$[M+NH_4]^+$	+NL, 189.04	650–1,000	60	20	100	14	14
PI (plant PI)	$[M+NH_4]^+$	+NL, 277.06	790–950	150	25	100	14	14
PS (plant PS)	$[M+H]^+$	+NL, 185.01	600–920	80	26	100	14	14
ox-PC, ox-PE, ox-PG, ox-DGDG, ox-MGDG, acMGDG, and ox-acMGDG (*see* Table 2 for target list)	(*see* Table 2)	–Pre, 277.2 –Pre, 283.2 –Pre, 291.2 –Pre, 293.2 –Pre, 295.2	1,040–1,100 815–820 740–1,150 720–1,030 730–860	65	–45	–100	–10	–20

The first column, "Class (Target list in DCE)" indicates the lipid class and the name of the target lipid list at the online lipidomics data processing site, LipidomeDB Data Calculation Environment. Parameter settings refer to an API 4000 (Applied Biosystems, Foster City, CA) triple quadrupole mass spectrometer with electrospray ionization source

Abbreviations: *DGDG* digalactosyldiacylglycerol, *MCA* multiple channel analyzer, *MGDG* monogalactosyldiacylglycerol, *PA* phosphatidic acid, *PC* phosphatidylcholine, *PE* phosphatidylethanolamine, *PG* phosphatidylglycerol, *PI* phosphatidylinositol, *PS* phosphatidylserine, *acMGDG* acylated monogalactosyldiacylglycerol, *ox-* oxidized, i.e., containing an oxidized fatty acyl chain

This table is adapted and expanded from a table in ref. 14 (*see* **Note 1**)

8. Use LipidomeDB Data Calculation Environment (DCE) at http://lipidome.bcf.ku.edu:9000/Lipidomics/ for identification and quantification of lipids (detailed instruction is available at the website). *See* Tables 1 and 2 for *Target Compound Lists*, which are provided as preformulated lists in the DCE. Internal standards for normal-chain diacyl lipids are the two internal standards of the same class. The mass spectral signals for the compounds listed in Table 2 can be quantified in relation to MGDG(34:0), which is measured in the –Pre 283.2 scan (*see* **Note 17**). For normal-chain phospholipids, results produced by DCE can be interpreted as nmol of target compounds in the analyzed vial. For normal-chain galactolipids, oxidized lipids, and acylated lipids, results produced by DCE are intensities normalized to the internal standard(s). A normalized intensity of 1 is the same mass spectral signal as 1 nmol of internal standard (*see* **Note 18**).
9. Normalize amounts of lipid analyzed (nmol, for normal-chain phospholipids; normalized signal, for normal-chain galactolipids, oxidized and acylated lipids) to mg of dry tissue mass, using this formula: dry-mass-normalized amount (nmol/mg or normalized signal/mg) $= \frac{\text{Vt (ml)} \times \text{amount (nmol or normalized signal)}}{\text{Va (ml)} \times \text{dry mass (mg)}}$, where Vt is total original sample volume (1 ml); Va is analyzed volume (volume equivalent to the 0.2 mg dry mass that was used in **step 1**).

4 Notes

1. Many of the materials indicated here are the same as listed in ref. 14. The methods extend those described there and apply them to the analysis of lipids derived from cold and freezing experiments. Portions of Subheadings 2 and 3 and Table 1 are republished by permission (Springer license number 3207410145743).
2. We typically use 27-day-old Arabidopsis plants, from which we sample rosettes. However, plants at other developmental stages may be used. The current protocol is appropriate for any above-ground vegetative tissue, flowers, or siliques.
3. Using an antistatic system with a microgram-accurate balance (Mettler Toledo) will increase the stability of mass measurements.
4. Abbreviations are DGDG, digalactosyldiacylglycerol; LPC, lysophosphatidylcholine; LPE, lysophosphatidylethanolamine; LPG, lysophosphatidylglycerol; MGDG, monogalactosyldiacylglycerol; PA, phosphatidic acid; PC, phosphatidylcholine;

Table 2

Oxidized membrane lipids and acylated MGDG analyzed by the negative precursor scans 277.2, 291.2, 293.2, and 295.2 and target lists for their analysis

M mass	M formula	Compound designations	Scan mode and adduct	Target list in DCE
984.6	$C_{59}H_{100}O_{11}$	acMGDG(50:6)	–Pre 277.2, $[M+C_2H_3O_2]^-$	18:3-unox-acMGDG-1
990.6	$C_{59}H_{106}O_{11}$	acMGDG(50:3)	–Pre 277.2, $[M+C_2H_3O_2]^-$	18:3-unox-acMGDG-1
1,006.6	$C_{61}H_{98}O_{11}$	acMGDG(52:9)	–Pre 277.2, $[M+C_2H_3O_2]^-$	18:3-unox-acMGDG-1
1,012.6	$C_{61}H_{104}O_{11}$	acMGDG(52:6)	–Pre 277.2, $[M+C_2H_3O_2]^-$	18:3-unox-acMGDG-1
1,014.6	$C_{61}H_{106}O_{11}$	acMGDG(52:5)	–Pre 277.2, $[M+C_2H_3O_2]^-$	18:3-unox-acMGDG-1
1,034.6	$C_{63}H_{102}O_{11}$	acMGDG(54:9)	–Pre 277.2, $[M+C_2H_3O_2]^-$	18:3-unox-acMGDG-1
1,036.6	$C_{63}H_{104}O_{11}$	acMGDG(54:8)	–Pre 277.2, $[M+C_2H_3O_2]^-$	18:3-unox-acMGDG-1
1,038.6	$C_{63}H_{106}O_{11}$	acMGDG(54:7)	–Pre 277.2, $[M+C_2H_3O_2]^-$	18:3-unox-acMGDG-1
745.5	$C_{39}H_{72}O_{10}PN$	PE(34:3-2O)	–Pre 291.2, $[M-H]^-$	18:4-O-ox-lipid-1
756.5	$C_{40}H_{69}O_{11}P$	PG(34:5-O)	–Pre 291.2, $[M-H]^-$	18:4-O-ox-lipid-1
758.5	$C_{40}H_{71}O_{11}P$	PG(34:4-O)	–Pre 291.2, $[M-H]^-$	18:4-O-ox-lipid-1
760.5	$C_{43}H_{68}O_{11}$	MGDG(34:7-O)	–Pre 291.2, $[M-H]^-$	18:4-O-ox-lipid-1
767.5	$C_{41}H_{70}O_{10}PN$	PE(36:6-2O)	–Pre 291.2, $[M-H]^-$	18:4-O-ox-lipid-1
769.5	$C_{41}H_{72}O_{10}PN$	PE(36:5-2O)	–Pre 291.2, $[M-H]^-$	18:4-O-ox-lipid-1
774.5	$C_{43}H_{66}O_{12}$	MGDG(34:8-2O)	–Pre 291.2, $[M+C_2H_3O_2]^-$	18:4-O-ox-lipid-1
778.5	$C_{43}H_{70}O_{12}$	MGDG(34:6-2O)	–Pre 291.2, $[M+C_2H_3O_2]^-$	18:4-O-ox-lipid-1
787.5	$C_{42}H_{78}O_{10}PN$	PC(34:3-2O)	–Pre 291.2, $[M+C_2H_3O_2]^-$	18:4-O-ox-lipid-1
788.5	$C_{45}H_{72}O_{11}$	MGDG(36:7-O)	–Pre 291.2, $[M-H]^-$	18:4-O-ox-lipid-1

(continued)

Table 2 (continued)

M mass	M formula	Compound designations	Scan mode and adduct	Target list in DCE
792.5	$C_{43}H_{68}O_{13}$	MGDG(34:7-3O)	–Pre 291.2, $[M+C_2H_3O_2]^-$	18:4-O-ox-lipid-1
802.5	$C_{45}H_{70}O_{12}$	MGDG(36:8-2O)	–Pre 291.2, $[M+C_2H_3O_2]^-$	18:4-O-ox-lipid-1
806.5	$C_{45}H_{74}O_{12}$	MGDG(36:6-2O)	–Pre 291.2, $[M+C_2H_3O_2]^-$	18:4-O-ox-lipid-1
809.5	$C_{44}H_{76}O_{10}PN$	PC(36:6-2O)	–Pre 291.2, $[M+C_2H_3O_2]^-$	18:4-O-ox-lipid-1
811.5	$C4_4H_{78}O_{10}PN$	PC(36:5-2O)	–Pre 291.2, $[M+C_2H_3O_2]^-$	18:4-O-ox-lipid-1
820.5	$C_{45}H_{72}O_{13}$	MGDG(36:7-3O)	–Pre 291.2, $[M+C_2H_3O_2]^-$	18:4-O-ox-lipid-1
922.6	$C_{49}H_{78}O_{16}$	DGDG(34:7-O)	–Pre 291.2, $[M-H]^-$	18:4-O-ox-lipid-1
928.6	$C_{49}H_{84}O_{16}$	DGDG(34:4-O)	–Pre 291.2, $[M+C_2H_3O_2]^-$	18:4-O-ox-lipid-1
936.6	$C_{49}H_{76}O_{17}$	DGDG(34:8-2O)	–Pre 291.2, $[M+C_2H_3O_2]^-$	18:4-O-ox-lipid-1
940.6	$C_{49}H_{80}O_{17}$	DGDG(34:6-2O)	–Pre 291.2, $[M+C_2H_3O_2]^-$	18:4-O-ox-lipid-1
946.6	$C_{49}H_{86}O_{17}$	DGDG(34:3-2O)	–Pre 291.2, $[M-H]^-$	18:4-O-ox-lipid-1
950.6	$C_{51}H_{82}O_{16}$	DGDG(36:7-O)	–Pre 291.2, $[M+C_2H_3O_2]^-$	18:4-O-ox-lipid-1
954.6	$C_{49}H_{78}O_{18}$	DGDG(34:7-3O)	–Pre 291.2, $[M+C_2H_3O_2]^-$	18:4-O-ox-lipid-1
964.6	$C_{51}H_{80}O_{17}$	DGDG(36:8-2O)	–Pre 291.2, $[M+C_2H_3O_2]^-$	18:4-O-ox-lipid-1
968.6	$C_{51}H_{84}O_{17}$	DGDG(36:6-2O)	–Pre 291.2, $[M+C_2H_3O_2]^-$	18:4-O-ox-lipid-1
982.6	$C_{51}H_{82}O_{18}$	DGDG(36:7-2O)	–Pre 291.2, $[M+C_2H_3O_2]^-$	18:4-O-ox-lipid-1
992.6	$C_{59}H_{92}O_{12}$	acMGDG(50:10-O)	–Pre 291.2, $[M+C_2H_3O_2]^-$	18:4-O-acMGDG-3
998.6	$C_{59}H_{98}O_{12}$	acMGDG(50:7-O)	–Pre 291.2, $[M+C_2H_3O_2]^-$	18:4-O-acMGDG-3
1,006.6	$C_{59}H_{90}O_{13}$	acMGDG(50:11-2O)	–Pre 291.2, $[M+C_2H_3O_2]^-$	18:4-O-acMGDG-3

1,010.6	$C_{59}H_{94}O_{13}$	acMGDG(50:7-2O)	–Pre 291.2, $[M+C_2H_3O_2]^-$	18:4-O-acMGDG-3
1,012.6	$C_{59}H_{96}O_{13}$	acMGDG(50:8-2O)	–Pre 291.2, $[M+C_2H_3O_2]^-$	18:4-O-acMGDG-3
1,016.6	$C_{59}H_{100}O_{13}$	acMGDG(50:6-2O)	–Pre 291.2, $[M+C_2H_3O_2]^-$	18:4-O-acMGDG-3
1,020.6	$C_{59}H_{88}O_{14}$, $C_{61}H_{96}O_{12}$	acMGDG(50:12-3O), acMGDG(52:10-O)	–Pre 291.2, $[M+C_2H_3O_2]^-$	18:4-O-acMGDG-3
1,024.6	$C_{59}H_{92}O_{14}$	acMGDG(50:10-3O)	–Pre 291.2, $[M+C_2H_3O_2]^-$	18:4-O-acMGDG-3
1,026.6	$C_{61}H_{102}O_{12}$	acMGDG(52:7-O)	–Pre 291.2, $[M+C_2H_3O_2]^-$	18:4-O-acMGDG-3
1,030.6	$C_{59}H_{98}O_{14}$	acMGDG(50:7-3O)	–Pre 291.2, $[M+C_2H_3O_2]^-$	18:4-O-acMGDG-3
1,034.6	$C_{61}H_{94}O_{13}$	acMGDG(52:11-2O)	–Pre 291.2, $[M+C_2H_3O_2]^-$	18:4-O-acMGDG-3
1,038.6	$C_{61}H_{98}O_{13}$	acMGDG(52:9-2O)	–Pre 291.2, $[M+C_2H_3O_2]^-$	18:4-O-acMGDG-3
1,040.6	$C_{61}H_{100}O_{13}$	acMGDG(52:8-2O)	–Pre 291.2, $[M+C_2H_3O_2]^-$	18:4-O-acMGDG-3
1,044.6	$C_{61}H_{104}O_{13}$	acMGDG(52:6-2O)	–Pre 291.2, $[M+C_2H_3O_2]^-$	18:4-O-acMGDG-3
1,048.6	$C_{61}H_{92}O_{14}$, $C_{63}H_{100}O_{12}$	acMGDG(52:12-3O), acMGDG(54:10-O)	–Pre 291.2, $[M+C_2H_3O_2]^-$	18:4-O-acMGDG-3
1,052.7	$C_{61}H_{96}O_{14}$	acMGDG(52:10-3O)	–Pre 291.2, $[M+C_2H_3O_2]^-$	18:4-O-acMGDG-3
1,054.7	$C_{63}H_{106}O_{12}$	acMGDG(54:7-O)	–Pre 291.2, $[M+C_2H_3O_2]^-$	18:4-O-acMGDG-3
1,058.7	$C_{61}H_{102}O_{14}$	acMGDG(52:7-3O)	–Pre 291.2, $[M+C_2H_3O_2]^-$	18:4-O-acMGDG-3
1,062.7	$C_{63}H_{98}O_{13}$, $C_{61}H_{90}O_{15}$	acMGDG(54:11-2O), acMGDG(52:13-4O)	–Pre 291.2, $[M+C_2H_3O_2]^-$	18:4-O-acMGDG-3
1,066.7	$C_{61}H_{94}O_{15}$, $C_{63}H_{102}O_{13}$	acMGDG(52:11-4O), acMGDG(54:9-2O)	–Pre 291.2, $[M+C_2H_3O_2]^-$	18:4-O-acMGDG-3
1,068.7	$C_{63}H_{104}O_{13}$	acMGDG(54:8-2O)	–Pre 291.2, $[M+C_2H_3O_2]^-$	18:4-O-acMGDG-3
1,070.7	$C_{61}H_{98}O_{15}$, $C_{63}H_{106}O_{13}$	acMGDG(52:9-4O), acMGDG(54:7-2O)	–Pre 291.2, $[M+C_2H_3O_2]^-$	18:4-O-acMGDG-3

(continued)

Table 2 (continued)

M mass	M formula	Compound designations	Scan mode and adduct	Target list in DCE
1,072.7	$C_{63}H_{108}O_{13}$, $C_{61}H_{100}O_{15}$	acMGDG(54:6-2O), acMGDG(52:8-4O)	−Pre 291.2, $[M+C_2H_3O_2]^-$	18:4-O-acMGDG-3
1,076.7	$C_{63}H_{96}O_{14}$	acMGDG(54:12-3O)	−Pre 291.2, $[M+C_2H_3O_2]^-$	18:4-O-acMGDG-3
1,080.7	$C_{63}H_{100}O_{14}$	acMGDG(54:10-3O)	−Pre 291.2, $[M+C_2H_3O_2]^-$	18:4-O-acMGDG-3
1,084.7	$C_{61}H_{96}O_{16}$	acMGDG(52:10-5O)	−Pre 291.2, $[M+C_2H_3O_2]^-$	18:4-O-acMGDG-3
1,086.7	$C_{63}H_{106}O_{14}$	acMGDG(54:7-3O)	−Pre 291.2, $[M+C_2H_3O_2]^-$	18:4-O-acMGDG-3
729.5	$C_{39}H_{72}O_9PN$	PE(34:3-O)	−Pre 293.2, $[M-H]^-$	18:3-O-ox-lipid-1
747.5	$C_{39}H_{74}O_{10}PN$	PE(34:2-2O)	−Pre 293.2, $[M-H]^-$	18:3-O-ox-lipid-1
751.5	$C_{41}H_{70}O_9PN$	PE(36:6-O)	−Pre 293.2, $[M-H]^-$	18:3-O-ox-lipid-1
753.5	$C_{41}H_{72}O_9PN$	PE(36:5-O)	−Pre 293.2, $[M-H]^-$	18:3-O-ox-lipid-1
758.5	$C_{40}H_{71}O_{11}P$	PG(34:4-O)	−Pre 293.2, $[M-H]^-$	18:3-O-ox-lipid-1
760.5	$C_{40}H_{73}O_{11}P$	PG(34:3-O)	−Pre 293.2, $[M-H]^-$	18:3-O-ox-lipid-1
762.5	$C_{43}H_{70}O_{11}$	MGDG(34:6-O)	−Pre 293.2, $[M+C_2H_3O_2]^-$	18:3-O-ox-lipid-1
769.5	$C_{41}H_{72}O_{10}PN$	PE(36:5-2O)	−Pre 293.2, $[M-H]^-$	18:3-O-ox-lipid-1
771.5	$C_{41}H_{74}O_{10}PN$	PE(36:4-2O)	−Pre 293.2, $[M-H]^-$	18:3-O-ox-lipid-1
771.5	$C_{42}H_{78}O_9PN$	PC(34:3-O)	−Pre 293.2, $[M+C_2H_3O_2]^-$	18:3-O-ox-lipid-1
776.5	$C_{40}H_{73}O_{12}P$	PG(34:3-2O)	−Pre 293.2, $[M-H]^-$	18:3-O-ox-lipid-1
776.5	$C_{43}H_{68}O_{12}$	MGDG(34:7-2O)	−Pre 293.2, $[M+C_2H_3O_2]^-$	18:3-O-ox-lipid-1
778.5	$C_{40}H_{75}O_{12}P$	PG(34:2-2O)	−Pre 293.2, $[M-H]^-$	18:3-O-ox-lipid-1
789.5	$C_{42}H_{80}O_{10}PN$	PC(34:2-2O)	−Pre 293.2, $[M+C_2H_3O_2]^-$	18:3-O-ox-lipid-1

790.5	$C_{45}H_{74}O_{11}$	MGDG(36:6-O)	–Pre 293.2, $[M-H]^-$	18:3-O-ox-lipid-1
793.5	$C_{44}H_{76}O_{9}PN$	PC(36:6-O)	–Pre 293.2, $[M+C_2H_3O_2]^-$	18:3-O-ox-lipid-1
795.5	$C_{44}H_{78}O_{9}PN$	PC(36:5-O)	–Pre 293.2, $[M+C_2H_3O_2]^-$	18:3-O-ox-lipid-1
804.5	$C_{45}H_{72}O_{12}$	MGDG(36:7-2O)	–Pre 293.2, $[M+C_2H_3O_2]^-$	18:3-O-ox-lipid-1
811.5	$C_{44}H_{78}O_{10}PN$	PC(36:5-2O)	–Pre 293.2, $[M+C_2H_3O_2]^-$	18:3-O-ox-lipid-1
813.5	$C_{44}H_{80}O_{10}PN$	PC(36:4-2O)	–Pre 293.2, $[M+C_2H_3O_2]^-$	18:3-O-ox-lipid-1
924.6	$C_{49}H_{80}O_{16}$	DGDG(34:6-O)	–Pre 293.2, $[M+C_2H_3O_2]^-$	18:3-O-ox-lipid-1
930.6	$C_{49}H_{86}O_{16}$	DGDG(34:3-O)	–Pre 293.2, $[M+C_2H_3O_2]^-$	18:3-O-ox-lipid-1
938.6	$C_{49}H_{78}O_{17}$	DGDG(34:7-2O)	–Pre 293.2, $[M+C_2H_3O_2]^-$	18:3-O-ox-lipid-1
952.6	$C_{51}H8_{4}O_{16}$	DGDG(36:6-O)	–Pre 293.2, $[M+C_2H_3O_2]^-$	18:3-O-ox-lipid-1
966.6	$C_{51}H_{82}O_{17}$	DGDG(36:7-2O)	–Pre 293.2, $[M+C_2H_3O_2]^-$	18:3-O-ox-lipid-1
731.5	$C_{39}H_{74}O_{9}PN$	PE(36:2-O)	–Pre 295.2, $[M-H]^-$	18:2-O-ox-lipid-1
753.5	$C_{41}H7_{2}O_{9}PN$	PE(36:5-O)	–Pre 295.2, $[M-H]^-$	18:2-O-ox-lipid-1
755.5	$C_{41}H_{74}O_{9}PN$	PE(36:4-O)	–Pre 295.2, $[M-H]^-$	18:2-O-ox-lipid-1
760.5	$C_{40}H_{73}O_{11}P$	PG(34:3-O)	–Pre 295.2, $[M-H]^-$	18:2-O-ox-lipid-1
762.5	$C_{40}H_{75}O_{11}P$	PG(34:2-O)	–Pre 295.2, $[M-H]^-$	18:2-O-ox-lipid-1
773.5	$C_{42}H_{80}O_{9}PN$	PC(34:2-O)	–Pre 295.2, $[M+C_2H_3O_2]^-$	18:2-O-ox-lipid-1
795.5	$C_{44}H_{78}O_{9}PN$	PC(36:5-O)	–Pre 295.2, $[M+C_2H_3O_2]^-$	18:2-O-ox-lipid-1
797.5	$C_{44}H_{80}O_{9}PN$	PC(36:4-O)	–Pre 295.2, $[M+C_2H_3O_2]^-$	18:2-O-ox-lipid-1

"M mass" and "M formula" indicate the mass and formula of the uncharged lipid; ion *m/z* can be approximated by subtracting 1 from (for $[M-H]^-$ adduct) or adding 59 to (for $[M+C_2H_3O_2]^-$ adduct) molecular mass (M mass). "Compound designations" include class abbreviation (*acMGDG* acylated monogalactosyldiacylglycerol, *DGDG* digalactosyldiacylglycerol, *MGDG* monogalactosyldiacylglycerol, *PC* phosphatidylcholine, *PE* phosphatidylethanolamine, *PG* phosphatidylglycerol) and, in parentheses, total acyl carbons: total acyl double bonds or equivalents. "Scan mode and adduct" indicates the *m/z* of the negative precursor scan and the detected adduct. Entries with the same "Scan mode and adduct" (i.e., from the same mass spectrum) are listed together and the data are processed together. "Target list in DCE" indicates the title of the target lipid list in the online lipidomics data processing site, LipidomeDB Data Calculation Environment, for data from a particular scan mode. More details on chemical structures of the listed lipids are available in the Supplemental Data of ref. 13. The internal standard for these targets is MGDG(34:0), which is analyzed in the –Pre 283.2 scan (Table 1 and text)

PE, phosphatidylethanolamine; PG, phosphatidylglycerol; PI; phosphatidylinositol; PS, phosphatidylserine.

5. From 5 mM lipid stock solutions in chloroform or appropriate mixtures of chloroform, methanol, and water [14], mix 120 μl (600 nmol) of each LPC and PC; 60 μl (300 nmol) of each LPE, PA, PE, and PG; 80 μl (400 nmol) of PI; 40 μl (200 nmol) of each PS; 240 μl (1,200 nmol) of DGDG; and 480 μl (2,400 nmol) of MGDG. Bring this mixture to 10 ml by adding 8.16 ml chloroform. This produces 10 μl of a stock solution with 0.6 nmol of each LPC and PC, 0.3 nmol of each LPE, PA, PE, and PG, 0.4 nmol of total PI, 0.2 nmol of each PS; 1.2 nmol of total DGDG, and 2.4 nmol of total MGDG. It is best to determine the concentration of phospholipids for the stock solution by phosphate assay [17]. Concentrations of the total MGDG and individual MGDGs (i.e., MGDG(34:0) and MGDG(36:0)), total DGDG and individual DGDGs (i.e., DGDG(34:0) and DGDG(36:0)), and individual PIs (i.e., PI(34:0) and PI(36:0)) are best determined by gas chromatography of fatty acid methyl esters derived from these lipids.
6. Although going directly to the low freezing temperature may not perfectly mimic natural freezing, a freezing regimen without gradual temperature change can be employed. If plants are to be placed directly at the low freezing temperature, ice chips can be added right before placing the plants in the freezing chamber.
7. For any freezing regimen, soil should be saturated with water prior to adding of ice chips. An alternative approach to placing ice chips on the soil is to partly submerge the 3½″ square pots or the 72-well plug tray in an ice slurry (1.5 kg of ice plus tap water to make 4 L).
8. Plants may be thawed at 4 °C or at the growing temperature. Although plants may sustain more damage with recovery at the growing temperature, recovery characteristics of acclimated plants are clearly distinguishable from those of non-acclimated plants (Vu, Hieusy and Welti, Ruth, Kansas State University, 2014).
9. Depending on the particular experimental goal, plants can be sampled early or late in cold acclimation (to measure early or late cold-induced molecular changes), right after freezing treatment (to measure freezing-induced changes), and/or during the recovery phase (to measure thawing-related changes). During the cold acclimation period, it is best to sample inside the cold room, and the temperature of the heating block may need to be closely monitored to maintain 75 °C. To sample right after freezing, it is critical to collect the plant tissues quickly without allowing them to thaw. Especially

when handling a large number of plants with a reach-in freezing chamber, avoid letting plants wait outside of the chamber; instead, pull out only the number of plants that can be sampled in less than 30 s. Within 30 s, two workers typically can sample four Arabidopsis rosettes. If using a 72-well plug tray, the tray can be cut, before planting, into sections of four plants for sampling by two workers.

10. It is critical to drop harvested leaves into isopropanol at 75 °C immediately to prevent activation of phospholipase D, a wound-induced enzyme, which will degrade membrane lipids and produce phosphatidic acids.
11. Should a different volume of isopropanol be required (to fully submerge plant tissues when harvesting), the volumes of chloroform and water can be varied accordingly.
12. For Arabidopsis leaves, five rounds of chloroform:methanol extraction are usually sufficient. Leaves should be completely white. For the last extraction, samples may be shaken overnight.
13. 1.2 ml total volume is required when using the indicated amber vial together with the indicated autosampler to ensure complete filling of the 1-ml sample loop without introducing an air bubble. When using a different type of sample vial or a different autosampler, a test filling should be performed to determine optimal total volume.
14. "Standards-only" spectra are used to correct instrument background signal and assess sample carryover. Internal standard peaks in "standards-only" spectra will likely have higher intensities than those in other spectra, because of low ion suppression. Intensities of plant analyte peaks in "standards-only" spectra should be very low, and analyte peak intensities from "standards-only" spectra may be subtracted from the intensities of the same peak from plant lipid spectra to remove background signal in the plant spectra.
15. The "washing blank" has a high concentration of acetic acid to wash the sample loop, the tubing between autosampler and the ion source, and the ion source needle to prevent carryover of acidic lipids such as PA and PS. Right after the sample loop is filled, the sample syringe and the injection port are washed with methanol/chloroform/water (66.5:30:3.5, v/v/v) contained in the two wash reservoirs of the CTC Mini-PAL autosampler.
16. Parameters, including collision energy, source temperature, source voltages, collision gas pressure, and scanning time should be optimized for each system. It is recommended NOT to use the first and the last 1.5 min of the total run time allowed by the injected volume (i.e., an injected volume of 1,000 μl allows ~33 min of run time at flow rate of 30 μl/min) because of the instability of the ion flow during these periods and because it takes some time (depending on tubing length and

diameter) for sample to reach the electrospray source. Instead, for the first 1.5 min, an MS scan for a wide mass range (*m/z* 200–800) can be acquired and monitored to ensure a continuous ion stream is reaching the detector.

17. On the target list page in LipidomeDB Data Calculation Environment, there is a check box at the top of the page, with the phrase, "Check here if the standards are in a separate spectrum from target compounds." This feature is described in the "advanced users" part of the tutorial. You should not check this box when processing data for normal-chain lipids. You should check this box when processing data for the compounds in Table 2. You will need to load the spectra for the internal standard (the –Pre 283.2 scan) separately from the spectra of the target compounds.
18. Normal-chain phospholipids and their internal standards of the same class have very similar response factors (the amount of mass spectral intensity per mol), and this allows accurate quantification of these lipids. On the other hand, the mass spectral response factors of normal-chain, oxidized, and acylated galactolipids may differ somewhat from those of their internal standard(s). In particular, for analysis of the compounds listed in Table 2, many compounds differ in structure from the internal standard (MGDG(34:0)).

Acknowledgements

The authors would like to thank lab members Mary Roth, Pamela Tamura, Thilani Samarakoon, Sam Honey, Drew Roach, and Kaleb Lowe for their contributions to plant stress experiments in our laboratory. This work was funded by National Science Foundation MCB 0920663. Contribution no. 14-036-J from the Kansas Agricultural Experiment Station.

References

1. Gilmour SJ, Hajela RK, Thomashow MF (1988) Cold acclimation in *Arabidopsis thaliana*. Plant Physiol 87:745–750
2. Uemura M, Joseph RA, Steponkus PL (1995) Cold acclimation of *Arabidopsis thaliana* (effect on plasma membrane lipid composition and freeze-induced lesions). Plant Physiol 109:15–30
3. Welti R, Li W, Li M, Sang Y, Biesiada H, Zhou HE, Rajashekar CB, Williams TD, Wang X (2002) Profiling membrane lipids in plant stress responses. Role of phospholipase Dα in freezing-induced lipid changes in Arabidopsis. J Biol Chem 277:31994–32002
4. Li W, Wang R, Li M, Li L, Wang C, Welti R, Wang X (2008) Differential degradation of extraplastidic and plastidic lipids during freezing and post-freezing recovery in *Arabidopsis thaliana*. J Biol Chem 283:461–468
5. Degenkolbe T, Giavalisco P, Zuther E, Seiwert B, Hincha DK, Willmitzer L (2012) Differential remodeling of the lipidome during cold acclimation in natural accessions of *Arabidopsis thaliana*. Plant J 72:972–982
6. Chen M, Thelen JJ (2013) *ACYL-LIPID DESATURASE2* is required for chilling and freezing tolerance in Arabidopsis. Plant Cell 25:1430–1444

7. Routaboul JM, Fischer SF, Browse J (2000) Trienoic fatty acids are required to maintain chloroplast function at low temperatures. Plant Physiol 124:1697–1705
8. Moellering ER, Muthan B, Benning C (2010) Freezing tolerance in plants requires lipid remodeling at the outer chloroplast membrane. Science 330:226–228
9. Takami T, Shibata M, Kobayashi Y, Shikanai T (2010) De novo biosynthesis of fatty acids plays critical roles in the response of the photosynthetic machinery to low temperature in Arabidopsis. Plant Cell Physiol 51:1265–1275
10. Li W, Li M, Zhang W, Welti R, Wang X (2004) The plasma membrane-bound *PHOSPHOLIPASE* Dδ enhances freezing tolerance in *Arabidopsis thaliana*. Nat Biotechnol 22:427–433
11. Chen QF, Xiao S, Chye ML (2008) Overexpression of the Arabidopsis 10-kilodalton acyl-coenzyme A-binding protein ACBP6 enhances freezing tolerance. Plant Physiol 148: 304–315
12. Burgos A, Szymanski J, Seiwert B, Degenkolbe T, Hannah MA, Giavalisco P, Willmitzer L (2011) Analysis of short-term changes in the *Arabidopsis thaliana* glycerolipidome in response to temperature and light. Plant J 66:656–668
13. Vu HS, Tamura P, Galeva NA, Chaturvedi R, Roth MR, Williams TD, Wang X, Shah J, Welti R (2012) Direct infusion mass spectrometry of oxylipin-containing Arabidopsis membrane lipids reveals varied patterns in different stress responses. Plant Physiol 158:324–339
14. Shiva S, Vu HS, Roth MR, Zhou Z, Marepally SR, Nune DS, Lushington GH, Visvanathan M, Welti R (2013) Lipidomic analysis of plant membrane lipids by direct infusion tandem mass spectrometry. Meth Mol Biol 1009:79–91
15. Zhou Z, Marepally SR, Nune DS, Pallakollu P, Ragan G, Roth MR, Wang L, Lushington GH, Visvanathan M, Welti R (2011) LipidomeDB data calculation environment: online processing of direct-infusion mass spectral data for lipid profiles. Lipids 46:879–884
16. Bligh EG, Dyer WJ (1959) A rapid method of total lipid extraction and purification. Can J Biochem Physiol 37:911–917
17. Ames BN (1966) Assay of inorganic phosphate, total phosphate and phosphatases. In: Neufeld E, Ginsburg V (eds) Methods in enzymology, vol. VIII: complex carbohydrates. Academic Press, New York, NY, pp 115–118

Chapter 16

Quantification of Superoxide and Hydrogen Peroxide in Leaves

Ilona Juszczak and Margarete Baier

Abstract

Reactive oxygen species (ROS) are produced in plants under both non-stressful and stressful conditions. Various histochemical staining methods have been developed and are widely used to visualize ROS accumulation sites. In contrast to qualitative analysis, quantification of ROS has been time- and labor consuming. As a consequence, the number of samples, which could be analyzed in parallel, has been limited. To overcome this problem, we introduce an improved semiquantitative method, in which ROS levels are quantified after histochemical staining in plant organs with the digital image analysis package ImageJ.

Key words ROS, Chloroplast, H_2O_2 staining, Superoxide staining, Digital imaging, Quantification

1 Introduction

Exposure to unfavorable conditions, such as low temperatures and excess light, can cause severe impairment of photosynthetic electron transport and cellular metabolism and often leads to enhanced production of reactive oxygen species [1–3]. ROS, such as superoxide radical anions ($O_2^{\bullet-}$), hydrogen peroxide (H_2O_2), and hydroxyl radicals ($OH^{\bullet}$) are highly reactive molecules. They are formed enzymatically, e.g., in oxidative bursts in response to pathogen infection or by peroxidases upon cell wall maturation [4–6] and non-enzymatically by reduction of oxygen in the Mehler reaction [7]. Due to their high reactivity, they can cause damage to a wide range of biomolecules, such as lipids, proteins, and DNA, and eventually they lead to cell death [8, 9]. Accumulation of ROS is antagonized by a highly efficient antioxidative defense system consisting of antioxidant enzymes and low molecular weight antioxidants [10–12]. The capacity of this system responds to ROS accumulation: The levels of low-molecular weight antioxidants increase and the genes for antioxidant enzymes are induced in

Dirk K. Hincha and Ellen Zuther (eds.), *Plant Cold Acclimation: Methods and Protocols*, Methods in Molecular Biology, vol. 1166, DOI 10.1007/978-1-4939-0844-8_16, © Springer Science+Business Media New York 2014

response to oxidative stress [13]. The steady state ROS level and the rates by which ROS can be trapped by dyes in tissues reflect the balance of ROS production and ROS detoxification.

To detect ROS in plant cells and organs various in situ and in vitro methods have been developed. For in situ pattern analysis preference has been given to histochemical staining [14–16] and, for quantification, to ROS-trapping by fluorophors [17, 18].

Here, we present a method for quantitative analysis of ROS accumulation in leaves based on histochemical ROS staining. ROS levels are quantified with the help of digital image analysis. Our method is cheap, fast, and easy. Special preparation and illumination of stained plant material enable visualization of ROS production sites not only in the external cell layer but also in the inner parts of these organs. Therefore, the method permits determination of ROS-accumulating leaf areas as well as quantification of ROS accumulation within particular leaf areas. The method has been optimized with respect to quantification. It can be used for both, detection of $O_2^{\bullet-}$ by NBT staining and the analysis of H_2O_2 accumulation by DAB staining, with minimal variation.

2 Materials

All solutions are prepared using distilled water and analytical grade reagents (*see* **Note 1**). All reagents and solutions are stored at 4 °C (unless indicated otherwise). For disposal of waste material carefully follow your waste disposal regulations (remember that sodium azide is acutely toxic).

1. 10× Phosphate-buffered saline (PBS) stock solution: Dissolve 80 g of NaCl, 2 g of KCl, 14.4 g of Na_2HPO_4, and 2.4 g of KH_2PO_4 in 800 mL of distilled water. Mix and adjust pH to 7.4. Adjust volume with distilled water to 1 L. Sterilize by autoclaving and store at 4 °C.
2. 1× PBS: To obtain 1 L of 1× PBS dilute 100 mL of 10× PBS with 900 mL of sterile water. Store at 4 °C.
3. 10 mM sodium azide: Dissolve 0.65 g of sodium azide (NaN_3) in 950 mL of 1× PBS. Adjust volume to 1 L. Caution! Sodium azide is acutely toxic. Do not prepare more solution than you really need.
4. 1 mg/mL NBT staining solution: Dissolve 1 g of nitro blue tetrazolium chloride (NBT) in 1 L of 1× PBS (*see* Subheading 2, **item 1** for recipe). Store in the darkness at 4 °C.
5. 1 mg/mL DAB staining solution: Dissolve 1 g of 3,3′-diaminobenzidine (DAB) in 1 L of 1× PBS (*see* Subheading 2, **item 1** for recipe). Store in the darkness at 4 °C.
6. Sample containers (bottles/beakers): Prepare an appropriate number of sample containers (each sample should be placed in

a separate sample container; the size of sample container depends on the size of stained plant tissue). Fill them with NBT or DAB staining solution for NBT and DAB staining, respectively.

7. Desiccators and vacuum pump: The number of desiccators needed depends on the number of samples. The same desiccators can be used for both NBT and DAB stainings.
8. De-staining solution: Mix 100 mL of 100 % acetic acid, 100 mL of glycerol, and 300 mL of 96 % ethanol. Right before starting the de-staining procedure, heat the mixture in a water bath up to 60–80 °C.
9. Photos: You can take photos using either a camera with a high resolution or a light microscope with an integrated camera. If not using a light microscope, an additional light source might be helpful, which lights up your samples from the bottom (e.g., White-light plate, INTAS, Germany). For quantification it is absolutely essential that the light is evenly applied and the object is not shaded or shading.
10. Quantification: Download and install the freeware package ImageJ on your computer (http://rsbweb.nih.gov/ij/download.html) [19].

3 Methods

All staining reactions are light sensitive. Therefore, the whole staining procedure should be carried out in darkness or at very low light intensities. Samples subjected to staining (e.g., leaves, seedlings) should be treated with care. Injuries promote ROS formation and lead to false-positive results.

3.1 NBT Staining for $O_2^{\cdot-}$ Detection

1. Place samples (e.g., detached leaves, whole seedlings, or rosettes) immediately after the harvest in 10 mM sodium azide. The azide inhibits superoxide dismutases and heme-type peroxidases at the active sites. The azide solution should fully cover the plant material. Put the sample containers to the desiccator and apply a vacuum (*see* **Note 2**) to remove air from the intercellular spaces and to flood the tissue with the azide solution. Incubate the plant material for 10 min under vacuum. Afterwards, slowly release the vacuum. Take the sample containers out of the desiccator. Discard the sodium azide solution.
2. Fill the sample containers (containing your samples) with NBT staining solution, place them back into the desiccator and apply a vacuum (*see* **Note 3**) to remove the oxygen in the desiccator and to protect the NBT from auto-oxidation.
3. Keep the samples under vacuum for 6–24 h (*see* **Note 4**). Release the vacuum and subject the samples to the de-staining procedure (Subheading 3.3).

3.2 DAB Staining for H_2O_2 Detection

1. Place the samples in containers filled with DAB staining solution immediately after the harvest. Put them into a desiccator and apply a vacuum (*see* **Note 3**).
2. Incubate the samples in the DAB solution for 10–24 h in darkness until they are optimally stained.
3. For clearing of the background from chlorophylls and other pigments, continue with (Subheading 3.3).

3.3 De-staining Procedure

1. Discard the staining solutions.
2. Fill the sample containers (containing samples) with hot de-staining solution, which will remove chlorophylls, carotenoids, and other plant pigments, but stabilize the dye. Incubate the samples in the de-staining solution until the chlorophyll is completely removed (*see* **Note 5**). Non-stained tissue (parts) should be as white as possible. Due to the high glycerol content the samples get soft and luminescent. Caution! Acetic acid has a pungent smell (especially when hot). Perform de-staining under fume hood!

3.4 Photographing of Stained Samples

1. Take the samples out of the de-staining solution (be careful not to destroy the sample with forceps while taking out).
2. Remove the traces of de-staining solution by placing the samples on a paper towel.
3. Place the samples under the microscope or on the external light device.
4. Illuminate the samples with white light from the bottom and take photos (*see* **Note 6**).
5. Using your favorite software (e.g., Adobe Photoshop or the freeware GIMP) subject the photos to digital processing (e.g., background subtraction, contrast improvement, color correction). For taking photos from NBT-stained plants, decrease the yellow background to its minimum. For DAB-stained samples remove the yellow and the blue colors to minimize background effects (*see* **Note 7**). The photos should be of similar quality to those in Fig. 1 after such processing.

3.5 Quantification of Staining Results

1. Open the photo, which you want to analyze, with ImageJ.
2. Change the color scale of the photo to a grey-scale. To do that, choose from ImageJ main menu "Image" (Fig. 2a) and from the sub-menus "Type" and "32-bit." Store the grey picture in a new file.
3. To segment the grey picture into features of interest (area stained stronger than a defined threshold) and background, adjust the threshold for the intensity of grey color. From ImageJ main menu use the following options: "Image"→

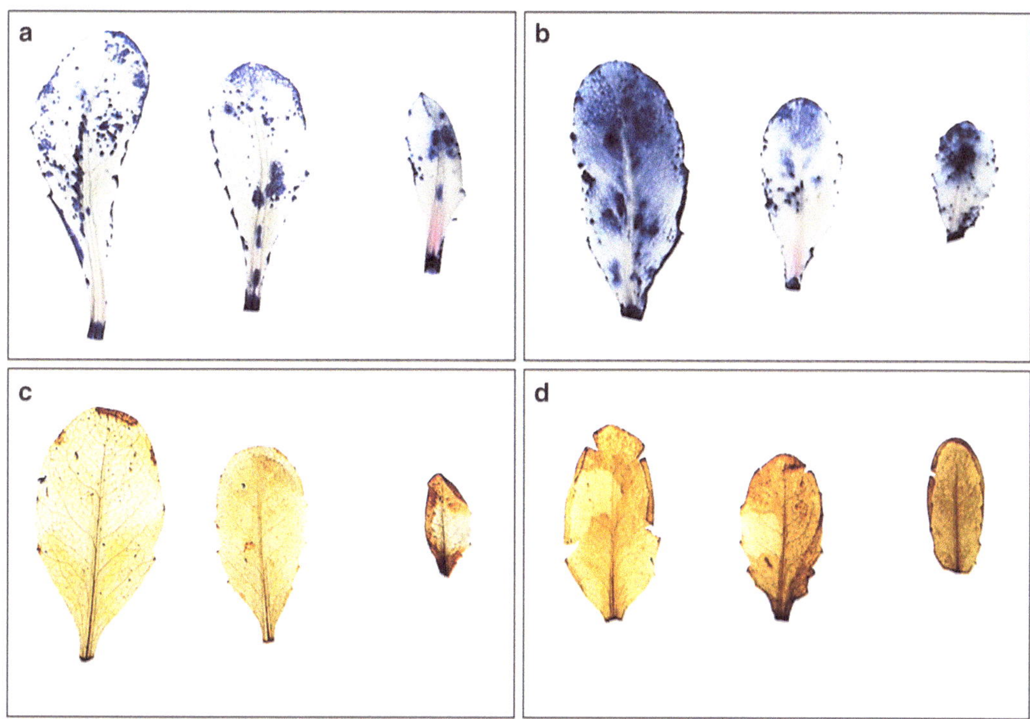

Fig. 1 *Arabidopsis thaliana* leaves (middle-age and young) stained with NBT (**a**, **b**) and DAB (**c**, **d**). The samples were harvested before (**a**, **c**) and after (**b**, **d**) 14 days of cold treatment [8]

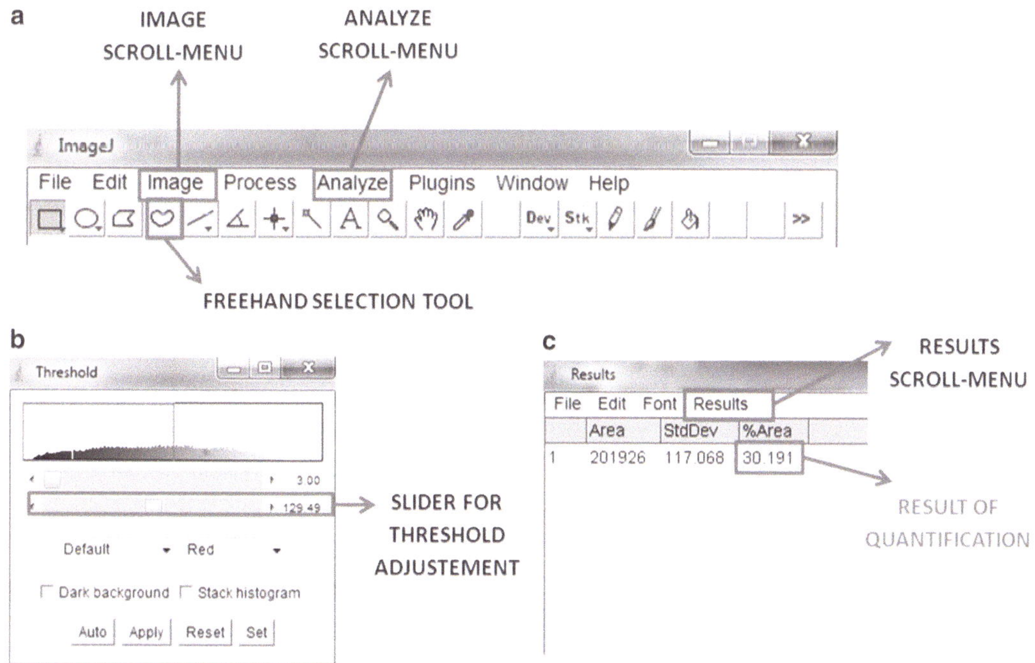

Fig. 2 ImageJ main menu (**a**), Threshold window (**b**), and Results window (**c**) with the most important options marked with *red squares* (Color figure online)

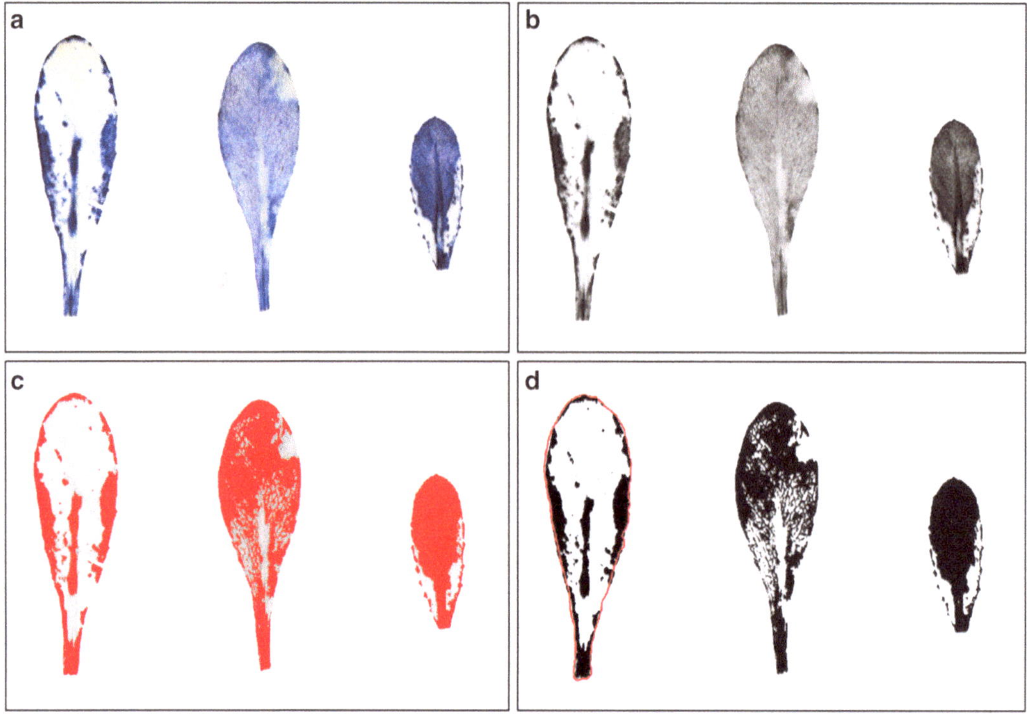

Fig. 3 Quantification workflow. (**a**) Original picture; (**b**) Picture in *grey* (32-bit) scale; (**c**) Picture before threshold setting, in which the stained area is marked in *red*; (**d**) Picture after threshold setting with the area of the first leaf surrounded by the *red line* (using Freehand selection tool)

"Adjust" → "Threshold" (Fig. 2a). In the "Threshold window" (Fig. 2b) move the lower slider to the set threshold level. Make sure that only the stained area is marked in red (Fig. 3c).

4. Surround the area of the first leaf using the "Freehand" selection tool (Fig. 2a).
5. Choose "Measure" from the ImageJ main menu "Analyze."
6. To choose the type of measurement, here in [%] of stained leaf area, use the following options in the "Results" window: Results → Set measurements → Area fraction (Fig. 2c). The results are shown in [%] as %Area (Fig. 2c).
7. To determine the staining intensity within defined leaf areas take the grey picture. Select the area of interest using the rectangular or oval selection tool. Measure the staining intensity by choosing "Measure" from the ImageJ main menu "Analyze." The type of the measurement should be set as Mean grey value (from the Results window choose: Results → Set measurements → Mean grey value). The result is shown as mean of the intensities within the selected area.
8. Repeat **steps 4–7** to quantify the staining in further leaves.

4 Notes

1. Use of freshly prepared staining solutions improves the quality of staining. PBS can be prepared in advance and stored at 4 °C.
2. To enable the comparison between samples (e.g., control and stress treatment), put all of them to the same desiccator and incubate them for the same time. It ensures that the same strength of vacuum is applied to all of them.
3. To improve penetration of the staining solution into your sample, apply and release the vacuum at least three times (each for 5 min) prior to the incubation.
4. Longer staining periods usually give better results, but can also result in over-staining. Therefore, carefully optimize the duration of staining.
5. To speed up de-staining, the incubation can be performed in a thermoblock (approximately 15–20 min at 60 °C). It is recommended to change the de-staining solution at least twice for optimal de-staining of young tissues and three times for older leaves, which are more difficult to de-stain.
6. Avoid any light from the side. If possible use a soft ring light for taking the photos.
7. Take care that all digital modifications are identical for all samples treated in parallel. If samples should be compared which are harvested on different days, prepare extra samples for comparison of staining intensities and de-staining efficiency and maintain half of them in the dark in the staining solution and half in the de-staining solution until the next set of plants is prepared.

References

1. Thomashow MF (1999) Plant cold acclimation: freezing tolerance genes and regulatory mechanisms. Annu Rev Plant Physiol Plant Mol Biol 50:571–599
2. Heidervand L, Amiri LM (2010) What happens in plant molecular responses to cold stress? Acta Physiol Plant 32:419–431
3. Juszczak I, Rudnik R, Pietzenuk B, Baier M (2012) Natural genetic variation in the expression regulation of the chloroplast antioxidant system among *Arabidopsis thaliana* accessions. Physiol Plant 146:53–70
4. Torres MA, Dangl JL, Jones JD (2002) Arabidopsis gp91phox homologues AtrbohD and AtrbohF are required for accumulation of reactive oxygen intermediates in the plant defense response. Proc Natl Acad Sci U S A 99:517–522
5. Bolwell GP, Bindschedler LV, Blee KA, Butt VS, Davies DR, Gardner SL, Gerrish C, Minibayeva F (2002) The apoplastic oxidative burst in response to biotic stress in plants: a three-component system. J Exp Bot 53:1367–1376
6. Bindschedler LV, Dewdney J, Blee KA, Stone JM, Asai T, Plotnikov J, Denoux C, Hayes T, Gerrish C, Davies DR, Ausubel FM, Bolwell GP (2006) Peroxidase-dependent apoplastic oxidative burst in Arabidopsis required for pathogen resistance. Plant J 47:851–863

7. Mehler AH (1951) Studies on reactivities of illuminated chloroplasts. I. Mechanism of the reduction of oxygen and other Hill reagents. Acta Biochem Biophys 33:65–77
8. Mittler R (2002) Oxidative stress, antioxidants and stress tolerance. Trends Plant Sci 7: 405–410
9. Apel K, Hirt H (2004) Reactive oxygen species: metabolism, oxidative stress, and signal transduction. Annu Rev Plant Biol 55:373–399
10. Mittler R, Vanderauwera S, Gollery M, Van Breusegem F (2004) Reactive oxygen gene network of plants. Trends Plant Sci 9:490–498
11. Baier M, Dietz KJ (1997) The plant 2-Cys peroxiredoxin BAS1 is a nuclear-encoded chloroplast protein: its expressional regulation, phylogenetic origin, and implications for its specific physiological function in plants. Plant J 12:179–190
12. Baier M, Pitsch NT, Mellenthin M, Guo W (2010) Regulation of genes encoding antioxidant enzymes in comparison to regulation of the extra-plastidic antioxidant defense system. In: Anjum NA, Chan MT, Umar S (eds) Ascorbate-glutathione pathway and stress tolerance in plants. Springer, Dordrecht, Heidelberg, London, New York
13. Foyer CH, Shigeoka S (2011) Understanding oxidative stress and antioxidant functions to enhance photosynthesis. Plant Physiol 155: 93–100
14. Vanacker H, Carver TL, Foyer CH (2000) Early H_2O_2 accumulation in mesophyll cells leads to induction of glutathione during the hyper-sensitive response in the barley-powdery mildew interaction. Plant Physiol 123:1289–1300
15. Kawai-Yamada M, Ohori Y, Uchimiya H (2004) Dissection of Arabidopsis Bax inhibitor-1 suppressing Bax-, hydrogen peroxide-, and salicylic acid-induced cell death. Plant Cell 16:21–32
16. Bournonville CF, Diaz-Ricci JC (2010) Quantitative determination of superoxide in plant leaves using a modified NBT staining method. Phytochem Anal 22:268–271
17. Hideg E, Barta C, Kalai T, Vass I, Hideg K, Asada K (2002) Detection of singlet oxygen and superoxide with fluorescent sensors in leaves under stress by photoinhibition or UV radiation. Plant Cell Physiol 43: 1154–1164
18. Fryer MJ, Oxborough K, Mullineaux PM, Baker NR (2002) Imaging of photo-oxidative stress responses in leaves. J Exp Bot 53:1249–1254
19. Schneider CA, Rasband WS, Eliceiri KW (2012) NIH Image to ImageJ: 25 years of image analysis. Nat Methods 9:671–675

Chapter 17

Estimating Ice Encasement Tolerance of Herbage Plants

Bjarni E. Gudleifsson and Brynhildur Bjarnadottir

Abstract

One of the key stresses acting on herbage plants during winter is ice encasement, when plants are enclosed in compact ice and turn from aerobic to anaerobic respiration. The cause of cell death is related to the accumulation of metabolites to toxic levels during winter and perhaps also to production of reactive oxygen species (ROS) when plants escape from long-lasting ice cover. The process of ice encasement damage has been studied by sampling studies, indirect measurements of ice tolerance, field tests and provocation methods by increasing stress in the field artificially, thus increasing the ice stress. Here we describe a laboratory method to measure ice encasement tolerance. This is the most common and effective way to measure ice encasement tolerance of large plant material. Plants are raised from seeds (or taken from the field), cold acclimated, usually at +2 °C under short day conditions, in a greenhouse or growth chamber (or in the field during fall). Plants are submerged in cold water in beakers and frozen encased in ice, usually at −2 °C. Plants are kept enclosed in ice at this temperature. Samples are taken at intervals, depending on species and tolerance of plant material, and put smoothly to regrowth. Damage is then evaluated after a suitable time of regeneration.

Key words Herbage plants, Ice cover, Ice damage, Ice encasement, Ice tolerance, Testing methods

1 Introduction

Plants are subject to many kinds of stresses during winter, usually related to the direct or indirect impact of low winter temperatures. Plants can be stressed by stressors of physical, chemical, and biotic origin, with one or more stressors acting simultaneously. Usually one type of stress dominates and causes damage. The winter stresses of herbage plants can be freezing, frost heave, drought, energy starvation, flooding, ice encasement, photoinhibition, attacks of snow molds, wind stress, and other stresses [1]. Some of the stresses mainly apply to certain plant groups, e.g., snow molds to grasses or winter cereals and wind stress to trees. Plants have different tolerance levels to the different stresses and it is of interest to measure the tolerance of species and cultivars in order to select or breed plants with increased tolerance. If the stress intensity exceeds the tolerance of the plant it can be damaged or even killed.

Dirk K. Hincha and Ellen Zuther (eds.), *Plant Cold Acclimation: Methods and Protocols*, Methods in Molecular Biology, vol. 1166, DOI 10.1007/978-1-4939-0844-8_17, © Springer Science+Business Media New York 2014

Fig. 1 Ice cover on hayfields in northern Iceland in February 1994 (B.E. Gudleifsson)

Tolerance to ice encasement is named ice encasement tolerance or simply ice tolerance. The combined ability of plants to tolerate all kinds of winter stresses is expressed as winter tolerance or winter hardiness. Freezing tolerance is frequently expressed as LT_{50}, the temperature killing 50 % of the plant population. Similarly, ice tolerance is expressed as LD_{50}, i.e., the number of days under anoxia that kills 50 % of the plant population. This paper focuses on ice damage on herbage plants, mainly perennial grasses and winter cereals.

Ice damage is most frequently found on herbage plants in certain maritime areas in the Northern hemisphere. It has repeatedly been reported in the coastal areas of the United States, Canada, Iceland, Scandinavia, Russia, and Japan [2]. During winter, plants use carbohydrate reserves to respire, but the respiration rate is low because of the low temperatures [3, 4]. The ice damage is caused by a compact ice layer on the field, in most cases formed in shorter or longer thaw periods during winter [2]. The melted snow and rain water freeze into a more or less continuous ice cover on the ground and the plants may become partly or completely encased in ice (Fig. 1) and then turn from aerobic to anaerobic respiration, thus producing a particular mixture of metabolites. The ice is almost completely impermeable to gases [5] and the plants have little or no access to O_2 and no possibility of releasing metabolites into the surroundings. In some cases the anoxia damage has been related directly to lack of O_2, but later to the combined action of plant metabolites [6–8]. Under ice cover the plants cannot release the metabolites [9] and the metabolites therefore accumulate in and around the plant cells; if the ice encasement lasts for a sufficiently long time, the concentration of these metabolites reaches toxic levels, resulting in ice damage. The metabolites accumulated are mainly CO_2, ethanol, lactate, and malate [3, 7], of which CO_2 is considered the most toxic [7, 9]. Some other metabolites have been detected from meltwater after long-lasting ice cover, such as citrate, propionate, pyruvate, fumarate, and shikimate [10–12] and also increased concentrations of alanine and tyrosine have been detected [8] as well as acetate, butyrate, ethylbutyrate, and butanol [11–13]. In spring, a strong odor has been reported from the meltwater after long-lasting ice cover, most likely related to the release of encased metabolites, mainly volatile fatty acids [11]. Some of the metabolites are supposed to damage cell membranes, resulting in increased microviscosity, increased levels of fatty acids, decreased ion uptake, and electrolytic leakage [14–16]. It has been proposed that higher ice tolerance is related to slower glycolytic metabolism [17, 18].

When plants escape long-lasting ice cover they often look green and healthy at first but then they may wilt and die within a few days [2, 19]. It has therefore been proposed that plants are not only killed by metabolic toxicity under the ice, but rather by the shock exerted by the transition from long-lasting anaerobic conditions under the ice to the effect of open air after the ice melts. The transition from anoxia to air is a great shock to the plant and reactive oxygen species (ROS) might develop and subsequently kill the plant cells [20]. This theory is supported by the fact that antioxidants added to the water before icing in a laboratory test have increased plant survival [11]. The ROS are supposed to damage cell membranes, as do the accumulated metabolites. Thus, toxicity of metabolites under ice and formation of ROS during melting may both participate in cell membrane damage during ice encasement. This results in damaged or dead plants appearing after ice melt in spring (Fig. 2).

2 Materials

Details in the laboratory methods used to measure ice tolerance vary from author to author. The tests are based on keeping the plants in an anoxic environment, usually compact ice, for different lengths of time. Here we describe a laboratory method in detail that originated from studies on winter cereals in Canada [9, 21, 22] and has later been modified for use in ice tolerance testing of grasses [23] and legumes [24], Brassica species (winter rape and stubble turnip), alpine plants and arctic shrubs [3, 25]. The method has been described in detail before [3]. This is the most reliable and common way to measure ice tolerance and the method fits better for timothy than perennial ryegrass and red clover [26]. In addition to cold acclimated plant material, growth chambers or a greenhouse and controlled freezers are needed.

Fig. 2 Dead grasses in depressions in a permanent hayfield in Iceland (B.E. Gudleifsson)

2.1 Plant Material

The plant material should be as homogeneous as possible and plants should be cold acclimated before treatment. Plants may be artificially raised and acclimated in a growth chamber or acclimated naturally. The acclimation and management practices, such as different cutting regimes or fertilizing, are usually performed under field conditions.

2.2 Anoxic Simulation

Plants can be encased in ice in beakers or boxes, with or without soil, or they can be enclosed in sealed plastic bags in an anoxic atmosphere with or without soil [13, 18]. The oxygen is believed to be quickly used up in respiration (*see* **Note 1**). The replicates, beakers, boxes, or bags can contain single plants or a bundle of plants.

2.3 Freezing Equipment

During treatment plants should be kept at natural winter temperatures. In nature the temperatures during the winter in the plant layer under ice or snow usually fluctuates around 0 °C [27]. Therefore plants should be tested at subzero temperatures and programmed freezers are needed to secure freezing of external water to ice. Usually a temperature of −2 °C has been used. The freezers should be equipped with a hot gas bypass to minimize temperature fluctuations [28] (*see* **Note 2**).

2.4 Recovery

When plants have been treated by ice encasement, they are removed from the freezer and put in a growth chamber or refrigerator at 4 °C to ensure a smooth transition from frost to growth conditions. Then a regeneration area is needed, a greenhouse or growth chamber, with suitable temperature, light intensity, soil, and humidity. Plants are transplanted into a suitable growth medium and kept under conditions that permit growth.

3 Methods

As pointed out by Baadshaug [29] studies of ice damage may have the aim to:

(a) Study the impact of climate and plant environment on the intensity of damage [3, 29].

(b) Sort out the impact of management on plant tolerance and damage [30, 31].

(c) Explore the genetic background of plant tolerance to ice damage [32].

(d) Study physiological and biochemical processes taking place in the plant during cold acclimation and ice damage [3, 9].

Different types of studies have been used to approach these goals [33–35]. These include sampling studies, indirect measurement of tolerance, field tests, provocation methods, and laboratory experiments.

3.1 Sampling Studies

Information that is collected on farmland is useful to clarify the impact of the environment on ice damage. In spring plant survival is evaluated and subsequently related to climatic data during winter, soil characteristics, and topographical differences. By comparing field survival records with climatic and soil data, the relationship between climate, soil, and the type of winter damage in Norway and Iceland have been illustrated [30, 36–38]. Sampling of information on topography, soil conditions, and management can provide a basis for investigating the main effects of such factors on survival, but uncontrolled conditions and the confounding effects of many factors may lead to difficulties in interpretation.

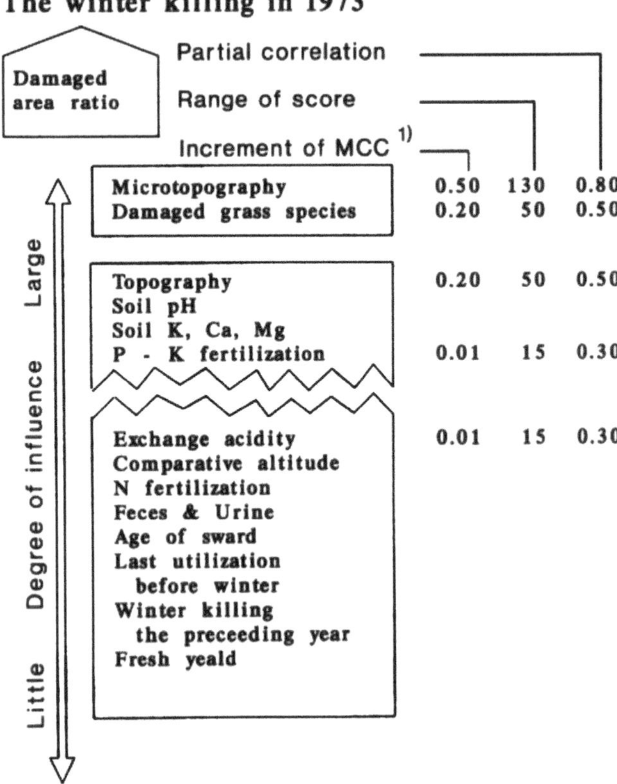

Fig. 3 Rank of factors affecting survival of grasses in Japanese pastures after ice encasement (from [2] originating from [39] with permission)

Hakamata et al. [39] used a sampling method to demonstrate the impact of soil conditions and management on ice damage in Japan and registered the greatest impact of microtopography on the ice encasement injury. These results are illustrated in Fig. 3. Field sampling can help in clarifying the main factors in the impact of environment and management on winter damage, but more precise experiments are needed to explain the details.

3.2 Indirect Measurements of Tolerance

Such measurements may be used to evaluate ice tolerance. The measurements of chemical, physiological, or morphological characters that are related to ice tolerance may indicate the level of ice tolerance. Indirect measurements are rapid, sensitive, repeatable, and usually nondestructive of the plant material. Many plant characteristics have been related to winter survival or freezing tolerance [34, 40, 41], but no such tests have yet been related directly to ice encasement tolerance. Although freezing and ice encasement are two different stresses, there is a general link between tolerance to these two different stress types and LD_{50} and

LT_{50} are highly correlated [2, 4, 26]. Therefore, estimates on freezing tolerance partly indicate the ice tolerance. Fowler et al. [42] and Gusta et al. [41] tested the relationship between a number of chemical, biochemical, physiological, and morphological characteristics and field survival in winter wheat under prairie conditions. Most of the characteristics were associated with field survival but were of little practical use. The crown LT_{50} was best correlated and the best predictor of field survival. Because ice tolerance is related to anoxic conditions, the measurement of respiration rate or production of specific metabolites and free radicals may indicate the ice tolerance of plants and provide valuable information on ice tolerance.

3.3 Field Tests

Field tests of ice encasement often indicate differences between species and cultivars of plants or the effect of management on survival [37, 43]. The intensity of winter stresses is often too low or too high to separate the treatments and the experiments may therefore be of limited value. Small differences in topography will affect the snow and ice cover, and may result in variable intensity of ice stress in the experiment and increase experimental error. The stress levels experienced in field trials have been shown to change dramatically over distances as short as a few meters and ice encasement damage often exhibits a patchy character (*see* Fig. 2) and is unevenly distributed over the experimental area [44–46]. To minimize these errors, comparisons should be restricted to plots that are in close proximity to one another and topographically equal. The results in field experiments are often not presented as percent of damage or survival but as yield at first cut or plant cover. It is quite understandable that the yield of permanent grassland is not a good criterion for estimating damage because weeds, taking over the space of killed plants, can partly compensate for the yield lost by killed grasses. Therefore direct evaluation of the damage is preferred. Ordinary field experiments can be of use in ice encasement studies, but these experiments should be backed up by climatic data and field observations during winter to establish the type of winter damage. Experiments should be placed on well-leveled fields to ensure as equal a stress intensity as possible.

3.4 The Provocation Method

This includes monitoring the winter climate and artificially increasing the ice stress to test plant tolerance in the field. The stress is increased by producing compact ice to kill plants and measure the ice tolerance [27, 29]. This might be performed on a spot with several plant species or cultivars to compare their tolerance, where ice cover is established artificially by flooding the plots during frost. To do this, water is poured over framed experimental areas. Such experiments have the advantage, compared to laboratory experiments, that the plant material is in natural conditions. Such experiments are of course highly

Fig. 4 Ice cover protected with styrofoam in a provocation test in Iceland in early spring 1979 (B.E. Gudleifsson)

dependent on the local winter climate conditions and can in certain cases be endangered or completely destroyed by mild winters and long-lasting thaws. The ice cover can be protected from melting (or from extremely low temperatures) by snow cover or insulating materials such as styrofoam (Fig. 4). The provocation method has been successfully used in many locations and in many years to test ice encasement tolerance, but because of difficulties in monitoring the artificial ice cover in the field, this method is used only sporadically [7, 27, 47].

3.5 Laboratory Experiments

Laboratory experiments are the most common and most effective way to study the impact of ice encasement. This method is therefore discussed here in detail. Conditions are much more exactly controlled in the laboratory than in the field. Testing the ice tolerance of herbage plants involves five steps: plant growth, cold acclimation, ice encasement stress, recovery and assessment of damage. The plants are cultivated and exposed to the actual stress factor in controlled environments. In many cases good agreement between the results of ice encasement in the laboratory and field survival have been demonstrated [26, 48–50], but in some cases the relationship was poor [51], as could be expected where stress factors other than ice encasement are involved in the field. The laboratory tests have the advantage that extensive plant material can be tested for a specific stress factor in a relatively short time. However, for all laboratory tests there are many crucial questions. How old should the plants be when tested, at what stage of acclimation should the test be executed, at what temperature should plants be tested, and how should the damage be evaluated?

3.5.1 Plant Growth

It is important to use homogenous plant material. Plants can be raised from seeds in the greenhouse or growth chambers or alternatively they can be taken from the field. Plants are frequently raised in a growth chamber or greenhouse at 18–20/15–17 °C day/night with a 16 h photoperiod [9, 18, 23, 49]. For winter annual plants, like the winter cereals, it is natural to test young seedlings because they enter winter in the field as seedlings. For perennial plants such as grasses and legumes the age of plants might affect the results, but in most cases in these species also rather young seedlings are preferred in controlled environments in order to save time and space. In addition, seedlings are easier to handle than older plants with their extensive root systems. Grass plants of first year seedlings often survive ice encasement better in the field than plants from later production years [10, 30]. In a laboratory study there was no difference between seedlings and old plants of timothy [52], indicating that the difference in the field may be related to soil compaction that leads to poor plant growth, poor cold acclimation, and probably greater anaerobic stress.

3.5.2 Cold Acclimation

Plants are cold acclimated in the laboratory for several weeks at low temperatures and under short day conditions. Also, frequently plants are taken from the field after natural cold acclimation and then tested for ice encasement tolerance in controlled environments [24, 51–59]. They can only be taken from the field before freezing of the soil in fall. In other seasons field plants must be acclimated artificially or collected with a concrete chipper during the winter [51, 57, 60]. Sometimes field plants are transferred to flats in the fall and moved to the laboratory after acclimation [4]. This of course involves a slight modification of the environment. Usually plant material from the field is not as homogenous as plants from growth chambers and the transport from field to laboratory may increase that variability. Plants cold acclimated in controlled environments usually do not reach the same level of tolerance as field derived plants [48, 58, 61]. Plants seem to increase their ice encasement tolerance under the same acclimation conditions that induce freezing tolerance [49], but timothy and winter wheat seem to increase ice encasement tolerance faster than freezing tolerance [58]. In all experiments the environmental conditions during growth and cold acclimation should be standardized. Plants are frequently acclimated at 2/0 °C day/night at a 16 h photoperiod for 2–7 weeks [9, 51]. However, there is still the question at what level of acclimation plants should be tested for ice encasement tolerance. As pointed out by Larsen [33] plants in coastal regions with a variable climate do not always reach maximum acclimation before winter, and in addition ice encasement stress often occurs in late winter and early spring when plants have reduced tolerance. Therefore, it is not necessary to attain maximal cold acclimation before ice encasement. The level of acclimation is not as critical when LD_{50} is determined using variable duration of icing as when survival is measured at a single stress level [61, 62].

3.5.3 Ice Encasement Stress

Stressing includes keeping the plants under anoxic conditions for different lengths of time. Anoxic stress during ice encasement has been simulated by placing the plants in sealed bags in an N_2 environment at low temperatures [5, 63–65]. This simulation has the disadvantage that gaseous substances produced by the plants during anoxia will evaporate from the plants into the gaseous surroundings. Thus the toxic stress of anaerobic respiration products is not as intense in an N_2 atmosphere as in ice. Most laboratory experiments with ice encasement have therefore been performed in ice at subzero temperatures. The ice encasement tests have been executed in boxes or flats where the plants are covered with ice [48, 49, 54, 66] or the whole pot with soil and plants is submerged in bigger boxes [12, 67, 68] (*see* **Note 3**). In some cases field or greenhouse grown plants are encased with the root system in the soil medium in order to reduce labor and to simulate the natural conditions [19, 55, 56]. This method increases the requirement for freezing space and has not reduced the variability between parallels [68]. There is a possibility of variable air inclusions in the soil, thus reducing the anaerobic stress and increasing the variability. The removal of soil from plants before ice exposure increases the uniformity of stress but adds to the labor needed [9, 68]. The roots and tops of plants are trimmed to 2–3 cm and plants or crown segments are submerged in water in beakers that are frozen to solid ice [5, 9, 21, 23, 68–70] (*see* **Note 4**). Plants are put into cold tap water and ice cubes may be added to speed up freezing. The temperature during encasement has varied between different studies:

-1.0 °C [21, 70, 71]

-2.0 °C [4, 32, 59, 66–69]

-2.5 °C [12, 54, 72, 73]

-3.0 °C [52, 67]

-4.0 °C [47, 74, 75]

-5.0 °C [5, 24]

The temperature should secure complete freezing of the surrounding water without freezing the plant cells. Cold acclimated winter cereal plants start freezing at -6 °C [76]. It has been demonstrated that survival decreases as the ice encasement temperature is lowered [21, 77, 78]. Because the exact ice tolerance of the plant material is never known beforehand, samples are taken at intervals after different lengths of icing. This secures a suitable discrimination of treatments. In experiments with winter cereals, samples are taken at daily intervals for up to 2 weeks and for grasses at weekly intervals for up to 7 weeks. This method offers the possibility to express the results as LD_{50}, i.e., the number of days required to kill 50 % of the plant population [61].

3.5.4 Recovery from Stress

After stressing plants, one or more replicates are removed from the freezer at intervals. To reduce the shock and to secure slow melting of the ice, the boxes are first transferred to +4 °C and kept there for 2–4 days before plants are removed from the boxes and transplanted into soil for regeneration. The different sampling times increase the accuracy of the survival determination but also increase the need for plant material and ice encasement space. Even though these methods demonstrate differences between plant species in ice encasement tolerance, the variability within replicates and parallels have been considerable and have limited their use in a breeding program [69, 79]. Efforts have been made to reduce this variability [80] (*see* **Notes 1–5**).

3.5.5 Assessment of Damage

When the plants have recovered from the stress, the impact of stress is evaluated. The impact can be evaluated in several ways, including plant survival, cell damage, and metabolite production. This makes it possible to separate the tolerance of different species and cultivars and even the impact of different management practices. Several methods have been developed to determine the injury quickly after stressing. Smillie & Hetherington [63] used chlorophyll fluorescence of wheat leaves and demonstrated that FR, the maximal rate of the induced rise in chlorophyll fluorescence, decreased more in sensitive than in tolerant cultivars during anaerobic stress. Tanino & McKersie [19] determined cell viability after ice encasement of winter wheat by tetrazolium staining and observed that crown cells were viable immediately after ice encasement, which was lethal to the plants, and that cell viability was lost during the first days of regrowth. Many scientists have used plant regrowth or dry matter yield as a measure of ice encasement injury [24, 29, 81]. The dry matter yield is not a good criterion of damage when plants of different species or different ages are compared [24] but, as pointed out by Larsen [33], when comparing material with little genetic variation in tolerance, the scoring of vigor or amount of regrowth seems to express plant differences best. The percent surviving plants is probably the most reliable determination of injury and it is also most widely used (*see* **Note 5**). It should be noted that bacteria might be involved in the damage to plants during regeneration after freezing [82] or ice encasement [83, 84], but their role is still unknown. When plants have recovered, the damage is registered, and each plant is evaluated as dead or alive. LD_{50} is then calculated for each treatment based on the regression line between days in ice and survival.

Fig. 5 Scoring of damage on single plants after freezing [67]

As the plants may be partly harmed and may gradually loose vigor during the regeneration, scoring of the vigor of each plant on a scale of 0 (dead) to 9 (unharmed) has been used, as illustrated in Fig. 5. This has increased the accuracy of the test [33, 67].

4 Notes

1. Plants may be tested with the roots incorporated in soil or alternatively the roots can be washed clean of soil and the testing performed on more or less clean plants. As the soil will in most cases include some air bubbles, testing of clean plants is expected to involve a stronger stress and to be more reliable. Cold tap water must be used to wash the plants to prevent loss of tolerance.
2. Since many samples are tested at the same time, it is important that samples are equally spread in the freezer. Care should be taken that samples do not touch the sides of the freezer. To ensure equal temperature a net bottom should be inserted 10–20 cm above the bottom of the freezer with a fan blowing continuously underneath to equalize the temperature.
3. When plants are kept encased in boxes for a long time, as is needed for grasses, the ice can sublimate from the ice surface. Therefore the boxes should be closed with caps or alternatively water should be added occasionally in small doses to the ice surface.

4. When plants without soil are submerged in water they may float up and stay on the water surface until the water freezes to ice. To sink them a small spike may be attached to the plant bundle with a rubber band.
5. The timing of evaluation varies from one experiment to another, depending partly on growth conditions. If each plant is evaluated on a scale of 0–10, the evaluation is undertaken before complete death, but if the evaluation is based on dead/alive plants, it is performed later.

References

1. Griffith M, Gudleifsson BE, Fukuta N (2001) Abiotic stresses in overwintering crops. In: Iriki N, Gaudet DA, Tronsmo AM et al (eds) Low temperature plant microbe interactions under snow. Hokkaido National Agricultural Experiment Station, Hitsujigaoka, Toytrira, Sapporo, Japan, pp 101–114
2. Gudleifsson BE, Larsen A (1993) Ice encasement as a component of winter kill in herbage plants. In: Christersson L, Li PH (eds) Advances in plant cold hardiness. CRC Press, Boca Raton, pp 229–249
3. Gudleifsson BE (2010) Ice tolerance and metabolite accumulation in herbage crops in Iceland and impact of climate change. Icel Agr Sci 23:111–122
4. Höglind M, Bakken AK, Jörgensen M et al (2010) Tolerance to frost and ice encasement in cultivars of timothy and perennial ryegrass during winter. Grass Forage Sci 65:431–445
5. Rakitina ZG (1965) The permeability of ice for O_2 and CO_2 in connection with a study of the reason for winter cereal mortality under the ice crust. Sov Plant Physiol 12:795–803
6. Rakitina ZG (1970) Effect of an ice crust on gas composition of the internal atmosphere in winter wheat. Sov Plant Physiol 17:755–759
7. Andrews CJ, Pomeroy MK (1990) Low temperature anaerobiosis in ice encasement damage to winter cereals. In: Jackson MB, Davies DD, Lambers H (eds) Plant life under oxygen deprivation. SPS Academic Publishing, The Hague, The Netherlands, pp 85–99
8. Castonguay Y, Thibault C, Rochette P et al (2009) Physiological responses of annual bluegrass to contrasted levels of O_2 and CO_2 at low temperatures. Crop Sci 49:671–689
9. Andrews CJ (1977) Accumulation of ethanol in ice-encased winter cereals. Crop Sci 17: 157–161
10. Gudleifsson BE (1997) Estimating ice encasement tolerance in the laboratory. In: Molecular and physiological aspects of cold and chilling tolerance of northern plants. Proceedings of the Finish-Japanese Workshop, Agricultural Research Center of Finland, Jokioinen, pp 14–15
11. Gudleifsson BE (2009) Ice encasement damage on grass crops and alpine plants in Iceland – impact of climate change. In: Wisniewski M, Tanino K, Gusta L (eds) Plant cold hardiness. From the laboratory to the field. CAB International, Wallingford, UK, pp 163–172
12. Brandsæter LO, Haugland E, Helgheim E et al (2004) Identification of phytotoxic substances in soil following winter injury of grasses as estimated by a bioassay. Can J Plant Sci 85: 115–123
13. Aamlid TS, Landschoot PJ, Huff DR (2009) Tolerance to simulated ice encasement and *Microdochium nivale* in USA selections of greens-type *Poa annua*. Acta Agric Scand Sect B Soil Plant Sci 59:170–178
14. Hetherington PR, McKersie BD, Borochov A (1987) Ice encasement injury to microsomal membranes from winter wheat crowns. I. Comparison of membrane properties after lethal ice encasement and during a post-thaw period. Plant Physiol 85:1068–1072
15. Hetherington PR, Broughton HL, McKersie BD (1988) Ice encasement injury to microsomal membranes from winter wheat crowns. II. Changes in membrane lipids during ice encasement. Plant Physiol 86:740–743
16. Andrews CJ, Pomeroy MK (1989) Physiological properties of plants affecting ice encasement tolerance. Icel Agr Sci 2:41–51
17. Andrews CJ (1996) How do plants survive ice? Ann Bot 78:529–536
18. Bertrand A, Gastonguay Y, Nadeau P et al (2001) Molecular and biochemical responses of perennial forage crops to oxygen deprivation at low temperatures. Plant Cell Environ 24: 1085–1093

19. Tanino KK, McKersie BD (1985) Injury within the crown of winter wheat seedlings after freezing and icing stress. Can J Bot 63:432–436
20. McKersie BD, Leshem YY (1994) Stress and stress coping in cultivated plants. Kluwer Academic, Dordrecht
21. Andrews CJ, Pomeroy MK (1975) Survival and cold hardiness of winter wheats during partial and total ice immersion. Crop Sci 15:561–566
22. Andrews CJ, Pomeroy MK (1981) The effect of flooding pretreatment on cold hardiness and survival of winter cereals in ice encasement. Can J Plant Sci 61:507–513
23. Gudleifsson BE, Andrews CJ, Bjornsson H (1986) Cold hardiness and ice tolerance of pasture grasses grown and tested in controlled environments. Can J Plant Sci 66:601–608
24. Bowley SR, McKersie BD (1990) Relationships among freezing, low temperature flooding, and ice encasement tolerance in alfalfa. Can J Plant Sci 70:227–235
25. Preece C, Callaghan TV, Phoenix GK (2012) Impacts of winter icing events on the growth, phenology and physiology of sub-arctic dwarf shrubs. Physiol Plant 146:460–473
26. Pulli S, Hjortsholm K, Larsen A et al (1996) Development and evaluation of laboratory testing methods for winter hardiness breeding, vol 32. Nordic Gene Bank, Alnarp, Sweden, p 98
27. Baadshaug OH (1973) En vurdering av forskjellige metoder for overvintringsundersökelser i eng og beitevekster. [An evaluation of various methods for investigating wintering of meadow and pasture plants]. Forskn fors Landbr 24:221–234, In Norwegian with English Summary
28. Voisey PW, Moulton F (1968) Precise temperature control for a domestic freezer. Can J Plant Sci 49:107–110
29. Baadshaug OH (1973) Effects of soil type and soil compaction on the wintering of three grass species under different wintering conditions. Acta Agric Scand 23:77–86
30. Gudleifsson BE (1971) Overvintringsskadar i grasmark på Island, omfang og årsaker. [Extent and causes of winter-damages in Icelandic grasslands]. Lisensiatoppgave, Norges landbrukshögskole, p 130, [In Norwegian with English Summary]
31. Andrews CJ, Seaman WL, Pomeroy MK (1984) Changes in cold hardiness, ice tolerance and total carbohydrates of winter wheat under various cutting regimes. Can J Plant Sci 64:547–558
32. McKersie BD, Hunt LA (1987) Genotypic differences in tolerance of ice encasement, low temperature flooding, and freezing in winter wheat. Crop Sci 27:860–863
33. Larsen A (1986) Test methods for wintering characters. NJF seminar Nr.84 Lantbruksväxternas övervintring, pp 149–166
34. Sakai A, Larcher W (1987) Frost survival of plants. Responses and adaptation to freezing stress. Springer, Berlin, Heidelberg
35. Larcher W (1985) Schädigung der Pflanzen durch Bodenfrost und Schneebedeckung. In: Larcher W, Häckel H, Sakai A (eds) Handbuch der Pflanzenkrankheiten. I, 5. Die Nichtparasitären Krankheiten. Verlag Paul Parey, Berlin, pp 274–287, (In German)
36. Fridriksson S (1954) Rannsóknir á kali túna árin 1951 og 1952. [Winter injury of plants in Icelandic hayfields]. Rit Landb.Deild. B-7, 72s, In Icelandic with English Summary
37. Andersen IL (1963) Overvintringsundersøkelser i eng i Nord-Norge. II. Noen undersøkelser over is og vannskader i eng. [Investigations on the wintering of meadow plants in northern Norway. II. Some investigations on damages caused by ice and water choking on meadows]. Forsk Fors Landbr 14:639–669, In Norwegian with English Summary
38. Årsvoll K (1973) Winter damage in Norwegian grasslands, 1968–1971. Meld Norges Landbrukshögskole 52(3):21
39. Hakamata T, Noshiro M, Hirashima T, Nose I (1978) Investigation of actual condition on the winter-killing of pasture species in the Nemuro-Kushiro district. Exploration of factors by the quantification No 1. J Jpn Grassl Sci 23: 280–288
40. Levitt J (1980) Responses of plants to environmental stresses. I. Chilling, freezing, and high temperature stresses. Academic Press, New York
41. Gusta LV, Fowler DB, Tyler MJ (1983) An evaluation of several chemical tests as possible selection measures for winter hardiness in wheat. Can J Plant Sci 63:115–119
42. Fowler DB, Gusta LV, Tyler NJ (1981) Selection for winter hardiness in wheat. III. Screening methods. Crop Sci 21:896–901
43. Andersen IL (1960) Overvintringsundersökelser i eng i Nord Norge. I. [Investigation on the wintering of meadow plants in Northern-Norway]. Forskn fors Landbr 11:635–660, In Norwegian with English Summary
44. Brink RA, Keller W, Eisenhart C (1939) Differential survival of alfalfa under an ice sheet. J Agr Res 59:59–71

45. Gudleifsson BE, Sigvaldason J (1972) Um kal og kalskemmdir. II. Kalskemmdir í tilraunareitum Tilraunastöðvarinnar á Akureyri. [On winter damage. II. Winter damage on experimental plots in Akureyri]. Ársrit Ræktunarf Norðurlands 69:84–101, In Icelandic with English Summary
46. Fowler DB (1979) Selection for winter hardiness in wheat. II. Variation in field trials. Crop Sci 19:773–775
47. Beard JB (1965) Effects of ice covers in the field on two perennial grasses. Crop Sci 5:139–140
48. Sjøseth H (1959) Studies on ice encasement in strains of red clover (*Trifolium pratense*) and timothy (*Phleum pratense*). Acta Agric Scand 9:292–298
49. Sjøseth H (1971) Vinterhardførhet hos ulike eng- og beitevekster. [Winter hardiness in different meadow and pasture plants]. Meld Norges Landbrukshögskole 50:39, In Norwegian with English Summary
50. Andrews CJ, Pomeroy MK (1977) Changes in survival of winter cereals due to ice cover and other simulated winter conditions. Can J Plant Sci 57:1141–1149
51. Andrews CJ, Pomeroy MK, Seaman WL (1986) The response of fall-sown cereals to winter stress in eastern Ontario. Can J Plant Sci 66:25–37
52. Gudleifsson BE (1997) Survival and metabolite accumulation by seedlings and mature plants of timothy grass during ice encasement. Ann Bot 79(Suppl A):93–96
53. Sprague MA, Graber LF (1943) Ice sheet injury to alfalfa. J Am Soc Agron 35:881–894
54. Smith D (1952) The survival of winter-hardened legumes encased in ice. Agron J 44:469–473
55. Beard JB (1964) Effects of ice, snow and water covers on Kentucky bluegrass, annual bluegrass and creeping bentgrass. Crop Sci 4:638–640
56. Freyman S, Brink VC (1967) Nature of ice-sheet injury to alfalfa. Agron J 59:557–560
57. Andrews CJ, Pomeroy MK, de la Roche IA (1974) Changes in cold hardiness of overwintering winter wheat. Can J Plant Sci 54:9–15
58. Andrews CJ, Gudleifsson BE (1983) A comparison of cold hardiness and ice encasement tolerance of timothy grass and winter wheat. Can J Plant Sci 63:429–435
59. Gudleifsson BE (1986) Måling av isdekketoleranse hos gras i laboratoriet. [Evaluating ice tolerance of plants in the laboratory]. NJF seminar Nr. 84 Lantbruksväxternas övervintring. pp 171–175, [In Norwegian]
60. Bolduc R (1976) Tecnique pour echantillonner les racines de plantes dans le sol gelé et enneige. [Method to collect samples of plants under snow and ice]. Can J Plant Sci 56:633–638, In French with English Summary
61. Pomeroy MK, Fowler DB (1973) Use of lethal dose temperature estimates as indices of frost tolerance for wheat cold acclimated under natural and controlled environments. Can J Plant Sci 53:489–494
62. Brule-Babel AL, Fowler DB (1989) Use of controlled environments for winter cereal cold hardiness evaluation: controlled freeze tests and tissue water content as prediction tests. Can J Plant Sci 69:355–366
63. Smillie RM, Hetherington SE (1983) Stress tolerance and stress-induced injury in crop plants measured by chlorophyll fluorescence in vivo. Chilling, freezing, ice cover, heat and high light. Plant Physiol 72:1043–1050
64. Samygin GA, Rakitina ZG, Livshin AZ (1971) Resistance of winter wheat tissues to freezing and desiccation under adverse gas regime. Dokl Bot Sci 196–198:77–80
65. Samygin GA, Rakitina ZG, Livshin AZ (1973) Influence of an adverse gaseous regime and ice-cold bark on the water-retaining capacity of the leaves and tillering nodes of hardened winter wheat. Dokl Bot Sci 209:60–62
66. Hope HJ, Comeau A, Hasty P (1984) Ice encasement tolerance of prairieland wild ryegrass, orbit tall wheatgrass and puma rye grown under controlled environments. Cereal Res Com 12:101–103
67. Larsen A (1978) Freezing tolerance in grasses. Methods for testing in controlled environments. Meldinger fra Norges Landbrukshögskole 57(23):56
68. Gudleifsson BE, Björnsson H (1989) Methods for estimating ice-encasement tolerance of grasses in the laboratory. Icel Agr Sci 2: 99–103
69. Tronsmo AM, Svendsen S (1989) Ice encasement in timothy and cocksfoot – a possible screening method for application in breeding programs. Icel Agr Sci 2:105–107
70. McKersie BD, McDermott BM, Hunt LA et al (1982) Changes in carbohydrate levels during ice encasement and flooding of winter cereals. Can J Bot 60:1822–1826
71. Poysa VW (1984) The genetic control of low temperature, ice-encasement, and flooding tolerances by chromosomes 5A, 5B and 5D in wheat. Cereal Res Comm 12:135–141
72. McKersie BD (1983) Types of winter stresses – do winter wheat cultivars respond differently?

In: Fowler DB, Gusta LV, Slinkard AE et al (eds) New frontiers in winter wheat production, Western Canada winter wheat conference., pp 26–38
73. Sjøseth H (1957) Undersøkelser over frostherdighet hos engvekster. [Studies on frost hardiness in meadow plants]. Forskn fors Landbr 8:77–98, In Norwegian with English Summary
74. Freyman S (1969) Role of stubble in the survival of certain ice-covered forages. Agron J 61:105–107
75. Tompkins DK, Ross JB, Moroz DL (2004) Effects of ice cover on annual bluegrass and creeping bentgrass putting greens. Crop Sci 44:2157–2179
76. Pukacki PM, McKersie BD (1990) Supercooling and ice nucleation events in the crown of winter wheat seedlings. Can J Plant Sci 70:1179–1182
77. Pomeroy MK, Andrews CJ (1985) Effect of low temperature and calcium on survival and membrane properties of isolated winter wheat cells. Plant Physiol 78:484–488
78. Gudleifsson BE (1989) Laboratoriemetoder til testing av isdekketoleranse hjå gras. [Laboratory methods to study ice tolerance of grasses]. Nord Jordbruksforskn 71:81, In Norwegian
79. Björnsson H (1986) Statistisk analys och planläggning av laboratorieførsøk med frost- och isresistens. Statistical analysis and planning of laboratory testing of freezing and ice tolerance]. NJF seminar Nr. 84 Lantbruksväxternas övervintring, pp 176–179, [In Swedish]
80. Gudleifsson BE (1993) Methods for testing ice encasement tolerance in herbage plants. [Methods for testing ice tolerance of herbage plants]. Röbäcksdalen Meddelar 1993:83–94, In Norwegian
81. Rakitina ZG (1977) Effect of a surrounding ice crust on winter wheat plant as a function of their flooding before freezing. Sov Plant Physiol 24:317–324
82. Olien CR, Smith MN (1981) Extension of localized freeze injury in barley by acute post-thaw bacterial disease. Cryobiology 18:404–409
83. Gudleifsson BE (1983) Isdekkeresistens og frostherdsle hjå enggras. [Ice tolerance and frost tolerance of meadow grasses]. Nord Jordbruksforskn 65, 3, [In Norwegian]
84. Gudleifsson BE (1997) Microbes active under snow and ice cover in hayfields in Iceland. In: International workshop on plant-microbe interactions at low temperatures under snow. Hokkaido National Agricultural Experiment Station, Hitujioaoka, pp 109–118

Chapter 18

Characterization of Ice Binding Proteins from Sea Ice Algae

Maddalena Bayer-Giraldi, EonSeon Jin, and Peter W. Wilson

Abstract

Several polar microalgae are able to live and thrive in the extreme environment found within sea ice, where growing ice crystals may cause mechanical damage to the cells and reduce the organisms' living space. Among the strategies adopted by these organisms to cope with the harsh conditions in their environment, ice binding proteins (IBPs) seem to play a key role and possibly contribute to their success in sea ice. IBPs have the ability to control ice crystal growth. In nature they are widespread among sea ice microalgae, and their mechanism of function is of interest for manifold potential applications. Here we describe methods for a classical determination of the IBP activity (thermal hysteresis, recrystallization inhibition) and further methods for protein characterization (ice pitting assay, determination of the nucleating temperature).

Key words Sea ice microalgae, Diatoms, Ice binding proteins, Antifreeze, Thermal hysteresis (TH), Nanoliter osmometer, Recrystallization inhibition (RI), Recrystallometer, Pitting assay, Nucleation, Supercooling, Lag time

1 Introduction

Sea ice is mainly a two-phase system, and its porous structure is largely determinant for biological activity within ice. During ice formation, solutes in the seawater are excluded from the ice matrix and segregate into brine droplets or brine channels, generally defined as brine inclusions inside sea ice [1]. Outflow of high salinity brine and inflow of seawater of lower salinity, as well as further cooling, cause brine inclusions to narrow and eventually separate into individual pockets divided by ice bridges.

Despite the harsh conditions that govern the conditions within sea ice, where temperatures range from about −1.8 °C on the bottom to −20 °C or less on the top [2, 3], and brine salinities can be as high as 200 on the Practical Salinity Scale, corresponding to roughly 200 g/kg [4], brine inclusions offer a habitat for a variety of microalgae. These algae play a crucial role for the ecology of the

Dirk K. Hincha and Ellen Zuther (eds.), *Plant Cold Acclimation: Methods and Protocols*, Methods in Molecular Biology, vol. 1166, DOI 10.1007/978-1-4939-0844-8_18, © Springer Science+Business Media New York 2014

Polar Oceans, since they represent a concentrated food source in the low-productivity ice-covered sea and in the months of melting they initiate blooms by seeding the water column [5]. Algae have been found distributed within brine inclusions throughout the entire thickness of the ice column.

The strategies adopted by ice microorganisms to cope with conditions in sea ice remain to be unraveled. Recent studies showed that several organisms that populate sea ice, spreading from bacteria [6] to diatoms [7–12] and a crustacean species [13], have ice binding proteins (IBPs). These proteins are common in polar species, but lack in temperate organisms [14], suggesting that IBPs play a key role in adaptation to subzero conditions. The nomenclature of these proteins varies, depending on authors, from ice binding to antifreeze or ice structuring (for reviews *see* refs. 15, 16). In the generally accepted adsorption–inhibition model describing the mechanism of action of IBPs, proteins bind to the ice lattice and locally inhibit ice growth by the Gibbs-Thomson effect [17, 18]. Recent publications showed that some IBPs organize water molecules into an ice-like structure that matches defined planes of the ice crystal, and is then gradually frozen into the ice lattice [19, 20]. One of the most prominent and best described effects of IBPs is thermal hysteresis, which describes the lowering of the freezing point of a solution below the melting point. Another effect which defines IBPs is inhibition of recrystallization, the grain boundary migration resulting in a growth of larger crystals at the expense of small grains.

The biological role of IBPs from sea ice microalgae remains an open question. The importance of some IBP families, as observed in fishes or insects, lies in lowering the freezing point below environmental temperature, in order to avoid ice formation in cells or organs. Other IBPs have the function to inhibit recrystallization, as it has been suggested for plant IBPs. In the context of sea ice, it seems unlikely that the biological role of IBPs may be thermal hysteresis (measured in the order of 1 °C) or recrystallization inhibition. Most of the IBPs from sea ice algae are active extracellularly. It has been suggested that they are trapped and accumulate within a layer of extracellular polysaccharide substances (EPS) secreted by several sea ice organisms [11]. Microalgal IBPs produced recombinantly or collected from spent growth medium affect the structure of the ice surface, causing pitting and characteristic microstructural features [7, 8, 11]. This suggests that the proteins shape their frozen environment in order to increase their habitable space within sea ice.

However, the characterization of IBPs is of relevance not only to understand their functional role in sea ice but also in the frame of possible applications of IBPs in the medical field, in the food industry, and in other fields related to a control of ice crystals.

In the following we present some standard techniques to determine the protein activity in terms of thermal hysteresis (TH) and recrystallization inhibition (RI), which define the proteins as ice binding. Also, we present further methods (ice pitting assay, determination of the nucleating temperature) to characterize the activity of IBPs.

2 Materials

2.1 Measurement of Thermal Hysteresis Using a Nanoliter Osmometer

1. Naoliter Osmometer.
2. Glass micropipettes (pre-pulled or pulled individually), rubber tubes fitting the micropipettes.
3. Nanoliter osmometer: cooling stage, sample holder plate with sample wells [6–8], controlling unit.
4. Tap water if colder than 18 °C or refrigerated circulating cooling fluid, dry air (nitrogen gas), stereo microscope, coverglass.
5. Immersion oil A (viscosity 150 cSt(lit)), B (viscosity 1,250 cSt(lit)), thermal heatsink paste, fine needles, chloroform.

2.2 Measurement of Ice Recrystallization Using an Optical Recrystallometer

1. Optical recrystallometer (Otago Osmometers Ltd, Dunedin, New Zealand).
2. Refrigerated circulating cooling fluid (ethylene glycol), dry air (nitrogen gas).
3. Multimeter.
4. Sample glass tubes (dimensions 8 mm outer diameter, 0.45 mm wall thickness, 8 cm high).
5. A beaker with 100 % ethanol cooled over night to −40 °C or −80 °C.

2.3 Ice Pitting Activity of IBPs

1. Ice-pitting instrument (Figs. 1 and 2) composed of a sample holder with the IBP solution set in a temperature-controlled bath with refrigerated circulating cooling fluid.
2. Optical microscope.

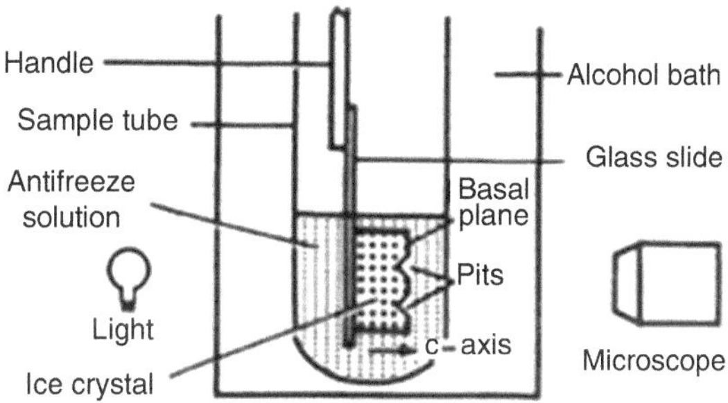

Fig. 1 Schematic of experimental instrument used to observe growth of ice crystals [27]

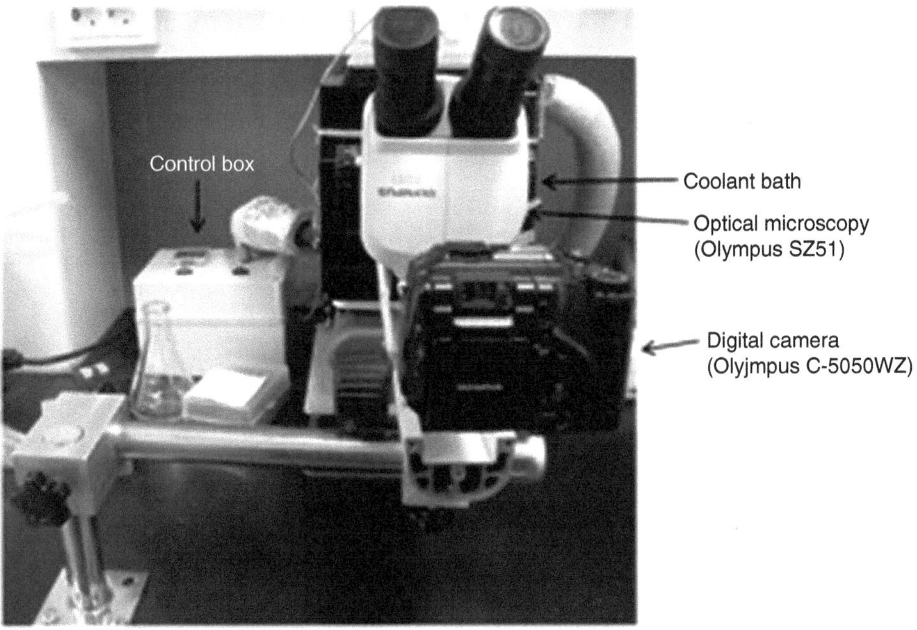

Fig. 2 Instrument for ice-pitting assay in the laboratory

3. Digital camera or movie-clip recording instruments.
4. Glass Petri dishes.
5. Glass slides (Dimension of slides: 0.5 cm (width) × 4 cm (height) × 0.1 cm (depth)).
6. Glass sample tubes (Dimension of rectangular part: 1 cm (width) × 0.5 cm (depth) × 2 cm (height).

2.4 Measuring the Ice Nucleation Properties of IBPs from Sea Ice Algae

1. Automatice lag time apparatus (purpose built by Otago Osmometers Ltd, Dunedin, New Zealand).
2. Dry air source.
3. Refrigerated circulating cooling fluid (isopropanol/water mix).
4. Data recorder.
5. Purpose built cold stage to sit DSC pan on or modified commercial cold stage capable of reaching –30 °C.
6. Aluminum DSC pans.

3 Methods

3.1 Measurement of Thermal Hysteresis Using a Nanoliter Osmometer

One characteristic of IBPs is their ability to lower the freezing point of a solution and separate it from the melting point [21–23]. This effect, known as thermal hysteresis (TH), is non-colligative and differs from equilibrium freezing point depression, where freezing and melting temperatures still coincide. Thermal hysteresis is easily determined with the Nanoliter Osmometer [24],

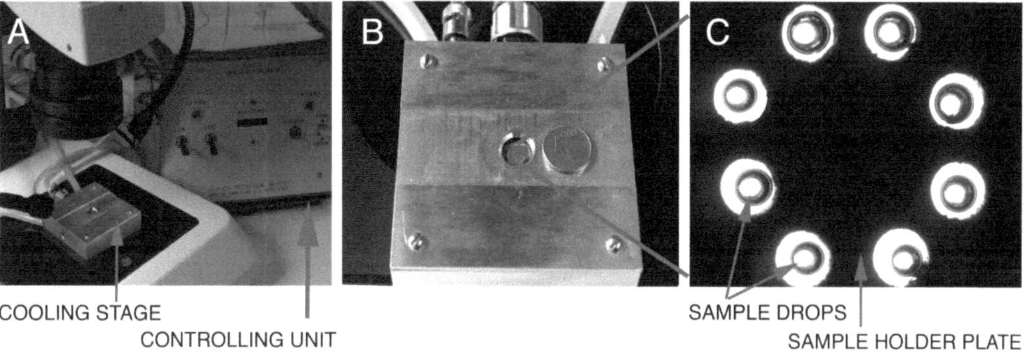

Fig. 3 Clifton Nanoliter Osmometer and loaded samples. (**a**) Controlling unit and cooling stage positioned on a stereo microscope; (**b**) Cooling stage with central sample holder plate, one cent coin for size comparison; (**c**) Sample holder plate viewed with stereo microscope, with samples within each of the 8 oil-loaded wells

which measures the melting and the freezing points of small volume samples (1–10 nl). We describe a measurement performed with the Clifton Nanoliter Osmometer (Clifton Technical Physics, Hartford, NY), but other osmometers follow the same principle (Otago Osmometer Ltd, Dunedin, New Zealand or LabVIEW operated devices, [25]). The Clifton Nanoliter Osmometer (Fig. 3) consists of a cooling stage, operated by a Peltier element, in contact with a sample holder plate. The cooling stage is regulated by a controller unit, which permits temperature adjustment in the range of approximately 1mOs = 0.00186 °C. Samples are viewed through a stereo microscope. Sample drops are loaded into wells in the sample holder plate, which are filled with oil to reduce sample evaporation, the melting and freezing points are determined visually.

1. Pull the glass micropipettes with a capillary puller in order to have a sharp end. If the fine end of the micropipette is closed, open it by gently scratching it (*see* **Note 1**).
2. Fill the glass micropipettes and the rubber tubes by setting them over night in immersion oil A.
3. Assemble the sample loading tube connecting each tube to two micropipettes, one on each end (*see* **Note 2**).
4. Connect the cooling fluid device to the osmometer.
5. Set the sample holder plate (cleaned in chloroform and dried) on the cooling stage using thermal paste (*see* **Note 3**). Fill the sample wells on the sample holder plate with immersion oil B using a fine needle.
6. Cover the sample with immersion oil A to avoid evaporation.
7. Load the sample into a sample loading tube by gently pressing the rubber part of the tube and then releasing it when submerged in the sample. Clean the loading tip with a tissue to remove remains of sample and obtain a clean tip.

8. Carefully insert the clean tip of the sample loading tube into a sample well. Release a drop of liquid by gently pressing the rubber part of the tube (*see* **Note 4**). Fill in all wells.
9. Turn on the dry air to avoid condensation during measurement and cover the sample plate with a coverglass.
10. Set the temperature of the osmometer slightly below melting temperature. Turn on the osmometer, shock-freeze samples at –40 °C. A sudden color change of the samples, from clear to dark, indicates their freezing. Release the Freeze switch and adjust temperature to the melting point.
11. Melt sample until only one crystal is left (*see* **Note 5**). Adjust the volume of the crystal to be as small as possible. Observe the crystal and note melting temperature (shrinking of the grain).
12. Slowly lower temperature and note freezing temperature (grain growth). The presence of IBPs becomes evident in a "hysteresis gap," a temperature range between freezing and melting point. Within this range the crystal will neither melt nor grow. The freezing of the sample can be observed as a "burst," if protein concentration is not too low (*see* **Note 6**).

3.2 Measurement of Ice Recrystallization Using an Optical Recrystallometer

Ice grains, especially small grains after shock-freezing, are subjected to instabilities causing migration of grain boundaries. This effect, called ice recrystallization, leads to the growth of the larger ice crystals, at the expenses of the smaller ones. Temperatures close to the melting point accelerate the recrystallization. A characteristic for IBPs is their ice recrystallization inhibition (RI), as they stabilize the original grain dimensions [26]. This effect can be measured with an Optical Recrystallometer. After shock-freezing, a sample appears optically thick, due to the several, small ice grains, whereas a recrystallized sample with few, large grains is clear. The recrystallometer detects the light intensity of a beam passing through the sample and gives an estimate of the recrystallization process (Fig. 4).

1. Connect the recrystallometer to the cooling fluid device and to the dry air supply. Connect the multimeter to the recrystallometer.
2. Turn on the recrystallometer and set the temperature to an annealing temperature close to the melting point, e.g., –4 °C (*see* **Note 7**).
3. Cool the sample tubes for at least 10 min in the cooled ethanol (*see* **Note 8**).
4. Load the sample (150 μl) into the sample tubes using a Pasteur pipette. The solution will freeze immediately (*see* **Note 9**). Put the tube back into the cool ethanol, incubate for 30 min.
5. Take the sample and quickly dry the tube with a tissue to remove the ethanol (*see* **Note 10**). Position the tube into the recrystallometer.

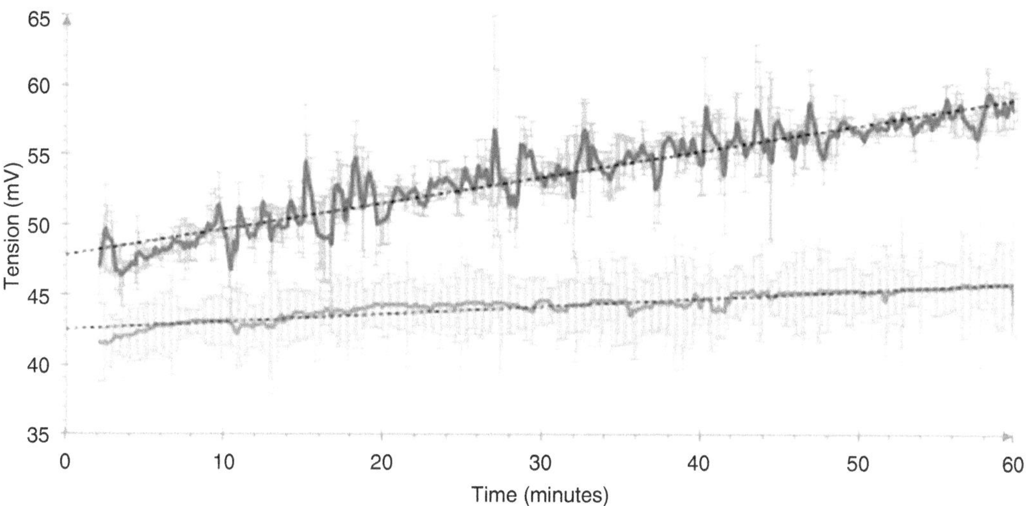

Fig. 4 Recrystallization measured using the Optical Recrystallometer. The negative control (*red*) with a buffer (phosphate-buffered saline) recrystallizes over time, which can be seen as an increase in voltage due to higher light intensities passing through the sample with larger ice grains. In the presence of IBPs the signal does not change over time, grains maintain their small dimensions (Color figure online)

6. Turn on the recrystallometer's light beam.
7. Start the multimeter. Measure for 1 h or the time more appropriate for your needs.
8. Download the multimeter results expressed as tension (mV). Recrystallization causes an increase in tension over time, which is less marked in samples with IBPs (Fig. 4).

3.3 Ice Pitting Activity of IBPs

IBPs have the ability to bind to ice surfaces and to prevent the growth of ice. The proteins usually bind specific planes of an ice crystal, typical for each protein family. Within the hysteresis gap between the melting and the freezing points (*see* Subheading 3.1) the crystal will only grow in the directions not affected by the proteins, developing characteristic pitting patterns (Fig. 5). For a simple determination of ice binding activity from samples, the depth and degree of the pitting deformations on small ice plates can be observed and the ice binding activity can be determined qualitatively [27, 28].

1. Prior to the experiment, check the osmolality of the samples. The temperature of the cooling liquid in the temperature-controlled bath should be slightly below the freezing point. In case of osmolality of approximately 1,000 mOsm/kg, set the temperature to −1.9 °C.
2. Fill half of the glass Petri dish with distilled water previously degassed by vacuum pump, to create the ice plates for the assay.

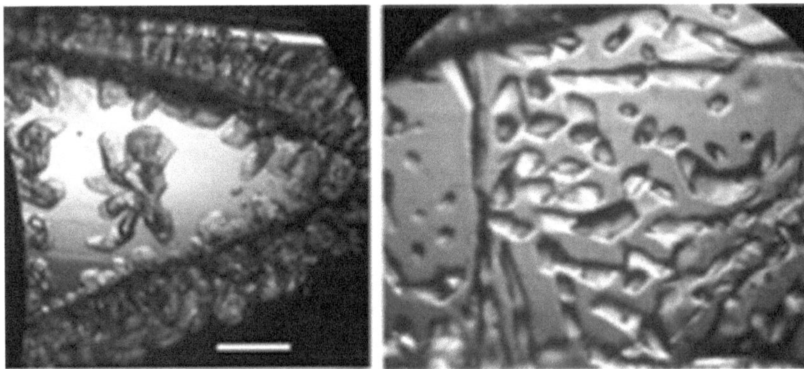

Fig. 5 Images of the deformation on the ice surface as seen with the ice-pitting assay. Scale bars indicate 1 mm [28]

3. Shock-freeze the water by incubating at −20 °C for approximately 40 min, in order to make a flat ice sheet of approximately 0.5 mm thickness (*see* **Note 11**).
4. Cool a glass slide at −20 °C.
5. Prepare a 0.5 cm × 0.5 cm piece of ice from the Petri dish and attach it to the chilled glass slide.
6. Put the sample tube filled with approximately 1.0 ml of the solution to be tested into the temperature-controlled bath. Adjust the position for optimal observation of the samples.
7. Put the glass slide with the ice piece into the cooled sample tube with great care (*see* **Note 12**).
8. Adjust the focus and magnification of the optical microscope and take pictures of the deformation on the surface of the ice sheet.

3.4 Measuring the Ice Nucleation Properties of IBPs from Sea Ice Algae

Determination of the ice nucleation properties of proteins from polar algae requires multiple measurements on the same batch, in the same container, since the temperature of nucleation always differs with successive runs, even when all other factors are kept constant. Polar diatoms synthesize IBPs, but it remains unclear what the actual purpose of these proteins is during the life cycle of the diatoms [11]. One possibility is that they use them to bind to the sea ice and remain in the photo-layer. In order to determine if diatoms enhance or inhibit ice nucleation we must first know the average nucleation temperature of the culture medium, in its container. It is often the container which causes nucleation of supercooled solutions—a scratch on the wall, a piece of dirt, etc. We describe here two methods for analyzing the nucleation characteristics of diatoms, the first requires a purpose-built device known as an automatic lag time apparatus (ALTA), while the second is simpler, requiring only a cooling stage and differential scanning calorimeter (DSC) pans.

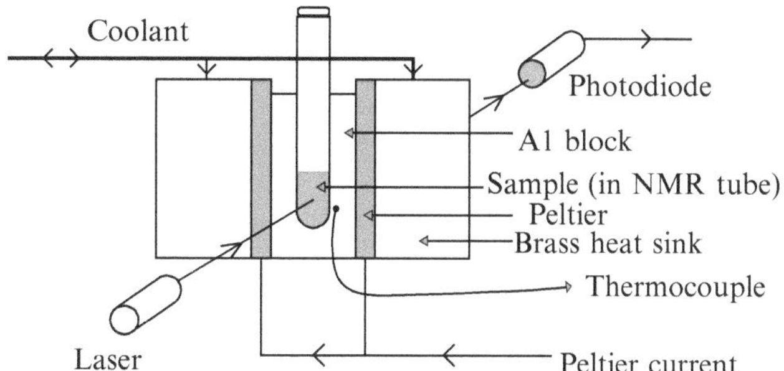

Fig. 6 Schematic of the automatic lag time apparatus (ALTA)

3.4.1 Measurements with an Automatic Lag Time Apparatus

ALTA repeatedly supercools a single liquid volume until it freezes [29]. The sample is cooled linearly, freezing is detected optically, the sample warmed and the process repeated around 200 times, as shown in Fig. 6 (*see* **Notes 13** and **14** for details).

1. A volume of typically 200 μl is placed in a glass tube which resides snugly in a hollowed-out 10 mm thick aluminum block. A thermocouple rests outside the tube to prevent unwanted nucleation sites within the liquid.
2. The metal sample holder is sandwiched between two Peltier modules which are used to heat and cool the sample by computer control. Excess heat is removed by heat sinks cooled by flowing isopropanol and water.
3. The freezing of the sample is monitored by the (interrupted) transmission of a low power diode laser, which causes the computer to switch direction of the current to the Peltiers. The sample is then heated to 283 K, or more, for some time to ensure melting of all residual ice crystals prior to commencing another run.

3.4.2 DSC Pan Type Measurement

A simpler method is to use a DSC pan sitting on a cold stage, with a typical sample volume of 10 μl of water, which will supercool and freeze at about −23 °C [30] (Fig. 7). What temperature will the water/solution freeze at (in that pan) if a sample of diatoms is added?

1. Sample is added to the DSC pan, a cover slip placed on top and cooling of the pan begun.
2. The freezing event is detected optically and temperature of freezing recorded.
3. The current to the cold stage peltiers is reversed and the DSC pan is heated to above 10 °C in order to melt the ice.
4. The process is then repeated, at least 50 times.

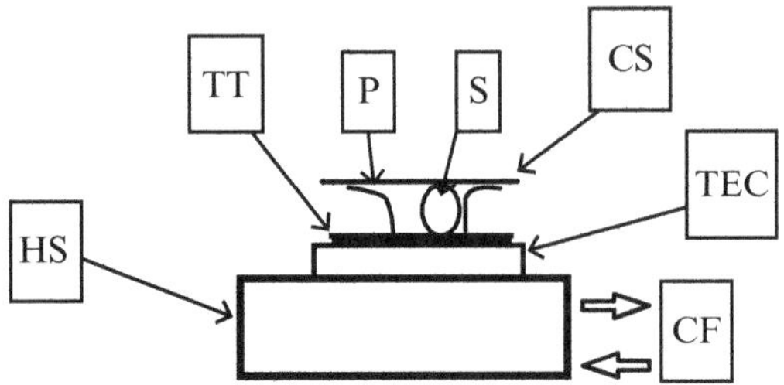

Fig. 7 Schematic of the second experimental arrangement: heat sink (HS), thermoelectric cooling (TEC), cooling fluid (CF), thermal transfer material (TT), DSC pan (P) with liquid sample (S) inside, and cover slip (CS). Freezing is detected optically from above

4 Notes

1. You can control the opening size of the micropipette tip by immersing it into water and passing air though the micropipette. The size of the air bubbles will be indicative for the tip opening.
2. Take care to remove air bubbles from the sample loading tubes by gently pressing the rubber part.
3. Use only a small amount of paste. Too much paste will mix with the immersion oil and disturb the measurement. Distribute the paste homogeneously on the sample holder and press it carefully on the cooling stage, in order to ensure maximal heat transfer between the cooling stage and the sample holder plate.
4. The diameter of the liquid sample drop should be around half the diameter of the well. You can control the size by pressing or releasing the rubber part of the sample loading tube. The sample should float in the oil and be positioned centrally within the sample well, without any direct contact with the plate. The sample can be carefully moved with a fine needle.
5. The bias on the freezing point due to the supercooling effect is avoided by observing one individual crystal.
6. Temperature changes should be performed carefully and not too fast, since the temperature response of the osmometer is slow. For time dependency of TH measurements see [25].
7. The system needs approximately 45 min to adjust to the set temperature.
8. The tubes should be inserted by approximately 2/3 into ethanol.
9. Set the tip of the pipette on the bottom of the sample tube, release the sample and quickly withdraw the pipette before the sample freezes.

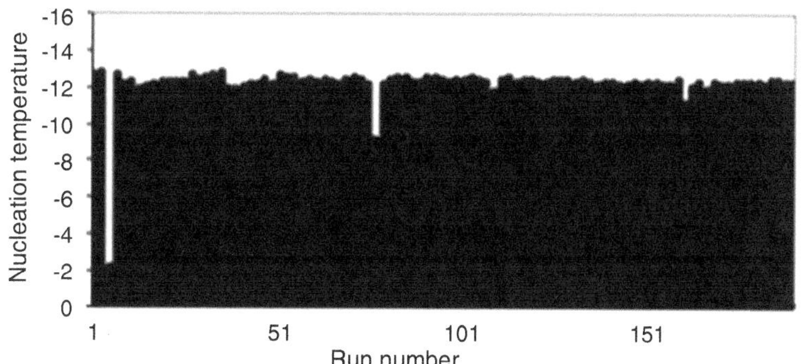

Fig. 8 Manhattan for a typical set of runs on ALTA, showing the stochastic nature of nucleation, where each run on the same sample freezes at a different temperature

10. Dry the glass carefully but do not hesitate to quickly set the tube into the recrystallometer in order to avoid condensation. Water on the glass surface may freeze the tube to the recrystallometer, causing errors in measurement and breaking of the tube.
11. In order to obtain ice with a flat surface, any disturbance or shock during the freezing process should be avoided.
12. When attaching the ice on the chilled glass slide, small drops of water can help to bind the ice to the cooled glass. After the attachment of the ice on the slide, residual water on the glass slide should be removed by precooled clean tissues.
13. Figure 8 shows these data collected from 200 consecutive heating/cooling cycles on a sample which was cooled at 0.018 K/s. We call this type of plot a "Manhattan" and it neatly demonstrates the stochastic nature of the nucleation process. The lack of slope of the Manhattan indicates that the sample has not changed during these many hours of recording freeze/thaw cycles [31].
14. The natural definition of the supercooling point (SCP) is the temperature at which the curve crosses the 50 % unfrozen mark, called here a T50. For the data shown in Fig. 9, the proposed SCP, or T50, is 8.17 K below the melting point. Adding substances to the sample tube can shift the S-curve to warmer T50s if they enhance nucleation, or to colder regimes if they inhibit nucleation, probably by masking potential nucleation sites inside the existing sample of liquid and/or container. To determine if sea ice microalgae or their IBPs have any effect on the nucleation temperature, the T50 must first be determined for a sample volume of the culture medium, in the container in which they will be measured [32]. The level of supercooling afforded by the osmolality of the solution

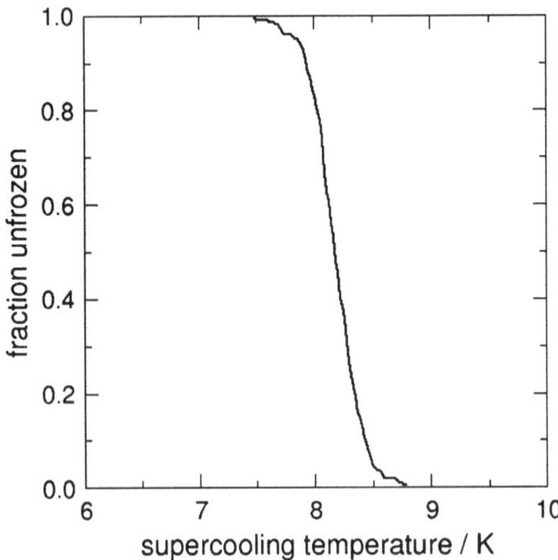

Fig. 9 Data for an ALTA sample set showing the spread of nucleation temperatures

is 2.0 times the melting point depression [33], i.e., if water in a given pan has a T50 of −20 °C then sea water in the same pan will have a T50 of (2.0×−1.9) = 3.8 °C colder, i.e., −23.8 °C. If the T50 of a given pan with culture medium is −24 °C and the new T50 is −19 °C the IBPs or algal cells have enhanced nucleation, if it is −28 °C the algae have produced some substance which is helping to inhibit nucleation. It is also critical that the addition of diatoms does not cause any slope to the Manhattan, which would mean that the sample is changing, almost certainly due to breaking up of the diatoms into smaller pieces—with rougher edges or in some way better ability to nucleate ice.

References

1. Weeks WF, Ackley SF (1982) The growth, structure, and properties of sea ice. In: CRREL monograph. U.S. Army ed., Hanover, p 82
2. Maykut GA (1986) The surface heat and mass balance. In: Untersteiner N (ed) NATO ASI series. Plenum Press, New York, NY, USA, pp 396–463
3. Eicken H (1992) The role of sea ice in structuring Antarctic ecosystems. Polar Biol 12:3–13
4. Cox GFN, Weeks WF (1983) Equation for determining the gas and brine volumes in sea-ice samples. J Glaciol 29:306–316
5. Lizotte MP (2001) The contributions of sea ice algae to Antarctic marine primary production. Am Zool 41:57–73
6. Raymond JA, Fritsen C, Shen K (2007) An ice-binding protein from an Antarctic sea ice bacterium. FEMS Microbiol Ecol 61:214–221
7. Raymond JA, Sullivan CW, DeVries AL (1994) Release of an ice-active substance by Antarctic sea ice diatoms. Polar Biol 14:71–75
8. Janech MG, Krell A, Mock T, Kang J-S, Raymond JA (2006) Ice-binding proteins from sea ice diatoms (*Bacillariophyceae*). J Phycol 42:410–416

9. Bayer-Giraldi M, Uhlig C, John U, Mock T, Valentin K (2010) Antifreeze proteins in polar sea ice diatoms: diversity and gene expression in the genus *Fragilariopsis*. Environ Microbiol 12:1041–1062
10. Gwak IG, Jung WS, Kim HJ, Kang S-H, Jin E (2010) Antifreeze protein in Antarctic marine diatom, *Chaetoceros neogracile*. Mar Biotechnol 12:630–639
11. Bayer-Giraldi M, Weikusat I, Besir H, Dieckmann G (2011) Characterization of an antifreeze protein from the polar diatom *Fragilariopsis cylindrus* and its relevance in sea ice. Cryobiology 63:210–219
12. Uhlig C, Kabisch J, Palm GJ, Valentin KU, Schweder T, Krell A (2011) Heterologous expression, refolding and functional characterization of two antifreeze proteins from *Fragilariopsis cylindrus* (Bacillariophyceae). Cryobiology 63:220–228
13. Kiko R (2010) Acquisition of freeze protection in a sea-ice crustacean through horizontal gene transfer? Polar Biol 33:543–556
14. Uhlig C (2011) Living inside sea ice – distribution and functional characterisation of antifreeze proteins in polar diatoms. Ph.D. thesis, University of Bremen
15. Barrett J (2001) Thermal hysteresis proteins. Int J Biochem Cell Biol 33:105–117
16. Venketesh S, Dayananda C (2008) Properties, potentials, and prospects of antifreeze proteins. Crit Rev Biotechnol 28:57–82
17. Raymond JA, DeVries AL (1977) Adsorption inhibition as a mechanism of freezing resistance in polar fishes. Proc Natl Acad Sci U S A 74:2589–2593
18. Kristiansen E, Zachariassen KE (2005) The mechanism by which fish antifreeze proteins cause thermal hysteresis. Cryobiology 51: 262–280
19. Nutt D, Smith JC (2008) Dual function of the hydration layer around an antifreeze protein revealed by atomistic molecular dynamics simulations. J Am Chem Soc 130: 13066–13073
20. Garnham CP, Campbell RL, Davies PL (2011) Anchored clathrate waters bind antifreeze proteins to ice. Proc Natl Acad Sci U S A 108:7363–7367
21. DeVries AL, Wohlschlag DE (1969) Freezing resistance in some Antarctic fishes. Science 163:1073–1075
22. Knight C (2000) Adding to the antifreeze agenda. Nature 406:249–250
23. Celik Y, Graham LA, Mok Y-F, Bar M, Davies PL, Braslavsky I (2010) Superheating of ice crystals in antifreeze protein solutions. Proc Natl Acad Sci U S A 107:5423–5428
24. Chakrabartti A, Hew CL (1991) The effect of enhanced α-helicity on the activity of a winter flounder antifreeze polypeptide. Eur J Biochem 202:1057–1063
25. Braslavsky I, Drori R (2013) LabVIEW-operated novel nanoliter osmometer for ice binding protein investigations. J Vis Exp 72:4189
26. Knight CA, Wierzbicki A (2001) Adsorption of biomolecules to ice and their effects upon ice growth. 2. A discussion of the basic mechanism of "antifreeze" phenomena. Crystal Growth Design 1:439–446
27. Raymond J, Wilson P, DeVries AL (1989) Inhibition of growth of nonbasal planes in ice by fish antifreezes. Proc Natl Acad Sci U S A 86:881–885
28. Raymond JA, Janech MG, Fritsen C (2009) Novel ice-binding proteins from a psychrophilic antarctic alga (Chlamydomonadaceae, Chlorophyceae). J Phycol 45:130–136
29. Barlow TW, Haymet ADJ (1995) ALTA: An automated lag-time apparatus for studying nucleation of supercooled liquids. Rev Sci Instrum 66:2996–3007
30. Wilson PW, Lu W, Xu H, Kim P et al (2012) Inhibition of ice nucleation by slippery liquid-infused porous surfaces (SLIPS). Phys Chem Chem Phys 15:581–585
31. Wilson PW, Heneghan AF, Haymet ADJ (2003) Ice nucleation in Nature: supercooling point measurement and the role of heterogeneous nucleation. Cryobiology 46:88–98
32. Wilson PW, Osterday KE, Heneghan AF et al (2010) Type I antifreeze proteins enhance ice nucleation above certain concentrations. J Biol Chem 285:34741–34745
33. Wilson PW, Haymet ADJ (2009) Effect of solutes on the heterogeneous nucleation temperature of supercooled water: an experimental determination. Phys Chem Chem Phys 11:2679–2682

Chapter 19

Isolation and Characterization of Ice-Binding Proteins from Higher Plants

Adam J. Middleton, Barbara Vanderbeld, Melissa Bredow, Heather Tomalty, Peter L. Davies, and Virginia K. Walker

Abstract

The characterization of ice-binding proteins from plants can involve many techniques, only a few of which are presented here. Chief among these methods are tests for ice recrystallization inhibition activity. Two distinct procedures are described; neither is normally used for precise quantitative assays. Thermal hysteresis assays are used for quantitative studies but are also useful for ice crystal morphologies, which are important for the understanding of ice-plane binding. Once the sequence of interest is cloned, recombinant expression, necessary to verify ice-binding protein identity can present challenges, and a strategy for recovery of soluble, active protein is described. Lastly, verification of function *in planta* borrows from standard protocols, but with an additional screen applicable to ice-binding proteins. Here we have attempted to assist researchers wishing to isolate and characterize ice-binding proteins from plants with a few methods critical to success.

Key words Ice-binding proteins, Antifreeze proteins, Ice-recrystallization inhibition, Thermal hysteresis, Ice crystals, Ice affinity purification, Disulphide bonds, Floral dip, Guttation fluid

1 Introduction

Ice-binding proteins (IBPs) have variously been named thermal hysteresis proteins, ice recrystallization inhibition (IRI) proteins, ice structuring proteins, and antifreeze proteins. Whatever the name, the study of IBPs from higher plants does not appear to be as popular as the isolation and characterization of those from other organisms. Many examples of IBPs are from freeze-intolerant fish and insects. These, in contrast to many plant IBPs described to date, show more thermal hysteresis (TH) activity, defined as the depression of the freezing point relative to the melting point. The strategy of plants we have studied is to survive low environmental temperatures by actually freezing with no substantial supercooling. Freezing initiates where the solute concentration is low, in the vascular tissues and the apoplast [1]. The resulting elevated

Dirk K. Hincha and Ellen Zuther (eds.), *Plant Cold Acclimation: Methods and Protocols*, Methods in Molecular Biology, vol. 1166, DOI 10.1007/978-1-4939-0844-8_19, © Springer Science+Business Media New York 2014

intracellular osmolarity, as well as the production of cold shock proteins, chaperones, and other molecules helps confer cellular low temperature protection. Presumably membranes are vulnerable to the advance of ice crystal spicules coming from the extracellular space and IBPs could control that growth. At temperatures close to the melting point, they inhibit the typical growth progression of large ice crystals at the expense of small ones. This phenomenon is referred to as Ostwald ripening or ice recrystallization [2]. It was the observation that grass had high IRI activity that helped spur more research into these proteins derived from higher plants [3]. Ice recrystallization is a major challenge in the transport and storage of frozen foods, and industry turned to the high IRI found in plants for a solution to this problem.

As a consequence of commercial interest, it is likely that much of the research on plant IBPs has not been released. Even a quick appreciation of the wide-spread distribution of IBPs in higher plants can best be obtained by perusing the patent literature. References can be accessed at: tinyurl.com/k5bmebd, tinyurl.com/m2olwo3, tinyurl.com/kkvyrk9, tinyurl.com/jw8omup, tinyurl.com/kza2pky, as well as the academic literature (e.g., [4–17], among others). As these citations indicate, IBPs have been reported in a variety of gymnosperms (e.g., spruce and Ginko), angiosperms (e.g., brassicas, carrot, onion, asters, plantain, potato, dandelion, night shade, maple, legumes, cotton wood, privet, sea buckthorn, and oak), and a variety of monocots, mostly grasses and cereals (e.g., onion, rye grass, fescue, bamboo, brome, bentgrass, daylily, triticale, spring oats, barley, and rye). In our own labs, undergraduate students have identified numerous additional vascular plants with IRI activity, but IRI or even TH activity does not ensure that there is an associated IBP, and thus these are not reported here. Our goal in this paper was to present the following methods designed to be useful for those moving from "prospecting" for new plant IBPs to expression of these *in planta*.

2 Materials

2.1 IRI (Capillary)

1. 10 μL capillaries (e.g., Drummond Microcaps).
2. Buffer (e.g., 50 mM Tris–HCl, pH 7.5 or phosphate-buffered saline; generally keep the salt concentration <100 mM).
3. Vacuum grease.
4. Packing tape (e.g., duct tape).
5. 95 % Ethanol.
6. Dry ice.
7. A double-walled glass chamber containing clear ethylene glycol and attached to a low temperature circulating bath filled

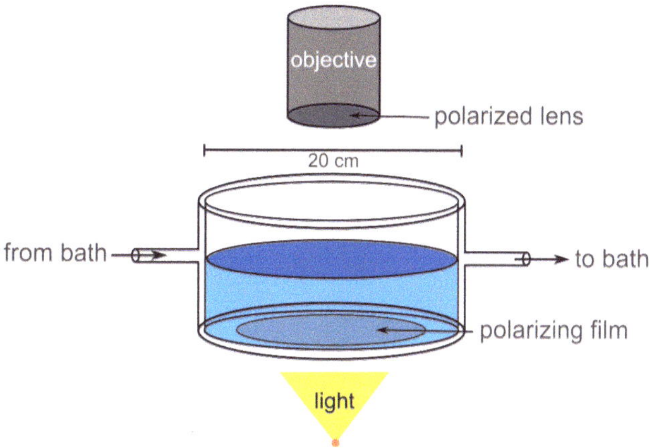

Fig. 1 Assembly of the double-walled glass chamber filled with clear antifreeze, hexane or isooctane (2,2,4-trimethyl pentane) and used for annealing at −4 °C and the subsequent examination of ice crystals for the ice recrystallization inhibition assays (both capillary and splat methods)

with clear ethylene glycol, surrounded by Styrofoam insulation (Fig. 1).

8. Dissecting microscope, capable of 40× magnification and fitted with a camera port.
9. Two polarizing filters or a section cut from a polarizing film; with one film placed below the objective and the other resting at the bottom of the insulated chamber (Fig. 1).

2.2 IRI (Splat)

1. Buffer: as suggested for the capillary assays.
2. Tube with a diameter of at least 5 cm and a minimum length of 1 m.
3. Retort stand and carpenter's level or plumb line.
4. Metal block (e.g., an inverted heating block).
5. Polystyrene container (e.g., a chemical shipping box).
6. Dry ice.
7. Glass microscope cover slides.
8. A double-walled glass chamber attached to a low temperature circulating bath filled with clear ethylene glycol, surrounded by Styrofoam insulation (Fig. 1; also used for the capillary method).
9. Hexane or isooctane (2,2,4-trimethyl pentane) for use inside chamber.
10. Desiccation beads (*see* **Note 1**).

2.3 TH and Ice Crystal Morphology

1. Nanolitre osmometer (see Bayer-Giraldi et al., this volume).
2. Sample of protein or a plant extract at a suitable concentration to allow measurement of TH (*see* **Note 2**).

2.4 Insoluble Recombinant Plant IBPs

1. A putative open reading frame from an IBP cloned into an expression vector (such as pET24a) and transformed into T7 RNA polymerase-compatible cells. Stop codons must be excluded from the 3′ end and the sequence will need to be cloned into the appropriate restriction sites to incorporate the poly-histidine tag (6× His; e.g., NdeI and XhoI for pET24a).
2. Lysogeny Broth (LB): 10 g/L tryptone, 5 g/L yeast extract, 10 g/L NaCl, pH adjusted to 7.5 with NaOH. Autoclave and allow to cool. Add appropriate amount of antibiotic just prior to inoculation (e.g., for pET24a use filter sterilized 50 μg/mL kanamycin).
3. Filter sterilized isopropyl β-D-1-thiogalactopyranoside (IPTG: dissolved in 10 mL of dH_2O to use at a final concentration of 0.5 mM) prepared fresh before use (*see* **Note 3**).
4. 8 M urea stock solution: deionize by incubating with mixed resin ion exchange beads (5 g/ 100 mL of sample) for 2 h at room temperature on a shaker (100 rpm). Beads are subsequently removed using a vacuum filter and the solution is degassed by placing the liquid in a sidearm flask while applying a vacuum. Adjust to pH 8.0 (*see* **Note 4**).
5. Buffers will vary, depending on the properties of the IBP. For purification under denaturing conditions buffers should be prepared fresh on the day of use (*see* **Note 5**).

   Buffer A (Lysis Buffer): 10 mM Tris–HCl, 100 mM NaCl, pH 8.0.

   Buffer B (Denaturing Buffer): 8 M urea, 100 mM NaH_2PO_4, 10 mM Tris–HCl, 10 mM β-mercaptoethanol, 2 % glycerol (v/v), pH 8.0.

   Buffer C (Wash Buffer): 8 M urea, 100 mM NaH_2PO_4, 10 mM Tris–HCl, 10 mM β-mercaptoethanol, pH 6.3.

   Buffers D (Elution Buffer): 8 M urea, 100 mM NaH_2PO_4, 10 mM Tris–HCl, 10 mM β-mercaptoethanol, pH 5.9.

   Buffer E (Elution Buffer): 8 M urea, 100 mM NaH_2PO_4, 10 mM Tris–HCl, 10 mM β-mercaptoethanol, pH 4.5.

   Buffers F-I (Dialysis Buffers): Four separate buffers containing either 6 M urea (F), 4 M urea (G), 2 M urea (H) or no urea (I) plus 100 mM NaCl, 15 mM Tris–HCl, 2 mM β-mercaptoethanol, 2 % glycerol (v/v), pH 8.0.
6. Mini EDTA-free protease inhibitor cocktail tablets (e.g., from Sigma-Aldrich Co.).
7. Centrifuge bottles.
8. Ni-NTA beads and gravity flow column.
9. Dialysis tubing.
10. Incubator shakers set at 37 °C and 24 °C.

2.5 Ice Affinity Purification

1. Programmable circulating water bath filled with ethylene glycol or antifreeze (plumbing or automotive). Attach the brass cooled probe to the inlet and outlet ports of the bath.
2. Stir plate and small stir bar (~15 mm × 7 mm).
3. Lab jack.
4. Insulated container to fit a 100 mL beaker.

2.6 Verification In Planta

1. A stationary phase liquid culture of a suitable *Agrobacterium* strain (e.g., GV3101, LBA4404, EHA105) confirmed to be carrying a binary vector (e.g., pCambia1305). This must include the desired promoter (e.g., the constitutive promoter CamV 35S or an inducible promoter such as the plant-specific cold-inducible RD29A promoter) followed by the IBP-encoding DNA sequence of interest.
2. Several 10 cm pots with approximately 30–50 healthy and well watered 3–4 week old *Arabidopsis* plants per pot in a tray that can be covered with a lid or with plastic wrap, plus additional pots and soil for subsequent steps.
3. A growth chamber or greenhouse set to appropriate conditions (e.g., 150 μE/m^2/s light intensity on a 16 h/8 h light/dark cycle at 22 °C and 70 % relative humidity).
4. LB: 10 g/L tryptone, 5 g/L yeast extract, 10 g/L NaCl. Autoclave 300 mL in a 1 L flask. Add appropriate antibiotic once solution has cooled (e.g., 50 μg/mL kanamycin for pCambia1305).
5. Sterile dH_2O (autoclave 150 mL aliquots).
6. Inoculation medium: 1.2 g Murashige and Skoog (MS) basal salts, 25 g sucrose in 500 mL dH_2O, pH 7.0. Prepare fresh or autoclave in advance. Just prior to inoculation, 0.5 μL of a 10 mg/mL stock of benzylamino purine in dimethyl sulfoxide (DMSO) and 90 μL Silwet L-77 (Lehle Seeds, Texas) are added to the inoculation medium plus *Agrobacterium* in a shallow container (at least 12.5 cm long and wide and ideally ~5–5.5 cm deep).
7. Incubator shaker set at 28 °C.
8. Centrifuge bottles and a centrifuge.
9. Metal sieve with very fine mesh (e.g., as would be used for sifting flour).
10. Small tubes (e.g., microcentrifuge tubes) for seed collection.
11. Sealable glass vessel (e.g., desiccator jar) located in a fume hood. This may require vacuum grease to make a complete seal.
12. Glass beaker containing 100 mL of household bleach.
13. Concentrated HCl.
14. Transfer pipet.

15. MS-agar plates: 0.5× MS salts (pH 5.8), 8 g/L agar. Autoclave and, in a laminar flow hood, add appropriate antibiotic (e.g., 40 μg/mL hygromycin for pCambia1305) once solution has cooled to ~55 °C, just prior to pouring.
16. Laminar flow hood.
17. Micropore surgical tape (3 M) and fine forceps.
18. Liquid nitrogen.
19. Mortar and pestle.
20. Extraction buffer: 10 mM Tris–HCl (pH 7.0), 25 mM NaCl.
21. Heating block.

3 Methods

3.1 IRI (Capillary)

Surveys of various plants and in a variety of tissues (leaves, stems, roots etc.) can be efficiently accomplished using the capillary method for IRI. Normally, plant material is collected after low temperature exposure and ground in liquid nitrogen using a mortar and pestle with suitable buffers (e.g., [3, 18]; see below). After brief centrifugation (12,000 × *g*) the extract supernatants can be assayed. The capillary method allows for efficient processing and comparisons and when used with twofold serial dilutions, permits an estimate of IRI activity [19].

1. Prepare the samples in buffer. If doing a semi-quantitative assay, prepare 1:2 serial dilutions (20 μL each) of the plant extract(s) in buffer. Each dilution must be mixed thoroughly and pipette tips must be changed after each dilution. We routinely use two capillary tubes at each dilution (*see* **Note 6**).
2. Adjust the temperature of the circulating bath so that the ethylene glycol in the double-walled glass chamber is at −4 °C (*see* **Note 7**).
3. Using forceps, dip one end of a 10 μL capillary into the diluted samples one at a time, allowing the sample to be drawn into the sample via capillary action (*see* **Note 8**). Seal the ends with vacuum grease and place each capillary into a custom-made holder or, alternatively, packing tape. Be sure to load buffer alone as a negative control and a sample of a known IBP (if available) as a positive control (Fig. 2). Seal the tape or assemble the holder (*see* **Note 9**).
4. Flash freeze the tubes in ethanol chilled with dry ice. The temperature of the ethanol should be <−55 °C before immersion of the capillaries. When they freeze, they will become translucent (*see* **Notes 10** and **11**).
5. Rapidly move the capillaries into the double-walled chamber. Observe them under a microscope at ~40× magnification

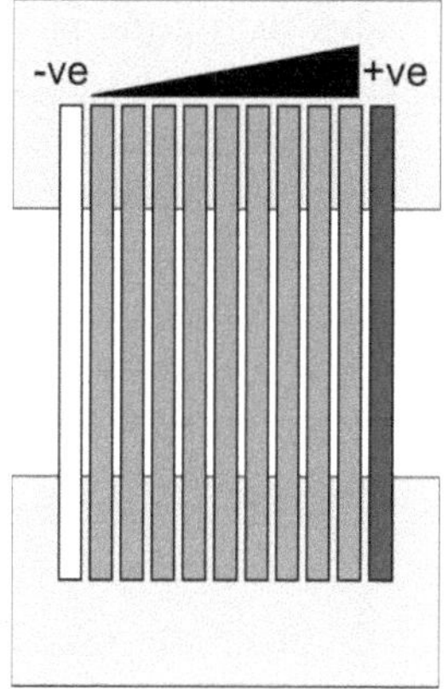

Fig. 2 Suggested method for the assembly of the samples in capillaries for ice recrystallization inhibition assays showing the placement of the dilution series (increasing concentrations from *left* to *right*). Also shown is the position of the positive and negative control samples. Shown in *gray* at the *top* and *bottom* of the capillaries is the packing tape, which is folded over to keep the tubes aligned

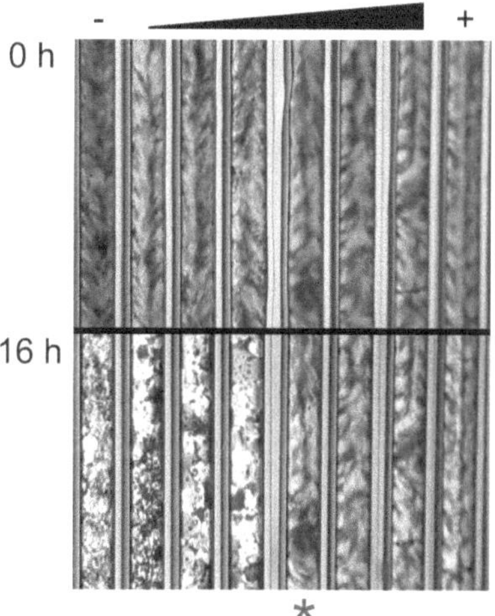

Fig. 3 Ice recrystallization inhibition (capillary assay) using recombinantly produced IBP originally isolated from *Lolium perenne* (perennial rye grass). Negative control (buffer alone: 50 mM Tris–HCl (pH 7.4), 100 mM NaCl, 1 mM EDTA) is indicated by "–"; positive control (purified type III AFP at 3 μM) by "+." 1:2 serial dilutions of purified *L. perenne* IBP were loaded into single capillary tubes. Concentrations increase from *left* to *right*. The endpoint is the last tube that shows no change in comparison with time 0 and is indicated with an *asterisk*

under crossed-polarized light. The ice crystals should be readily apparent. Take a photograph (time = 0) (*see* **Note 12**).

6. After annealing for 16–24 h, observe the capillaries again. Take a picture and compare the crystal shapes in the capillaries with those at time = 0 (Fig. 3).

3.2 IRI (Splat)

This technique is applicable to plant extracts as described for the capillary assays and is also suitable for monitoring the progress of IBP purification. The splat technique is more time-consuming than the capillary method when processing multiple samples, but photographic visualization of ice crystal size is more easily captured. Rough estimates of IRI activity can be made by determining mean ice crystal diameters or edge lengths, but this must be done within an appropriate dilution range.

1. This procedure uses the same chamber as the capillary method (*see* Subheading 3.1), but the method of freezing is different.
2. Place the metal block into a polystyrene container along with retort stand. Attach the tube to the retort stand so that it hangs slightly above the metal block (Fig. 4). Ensure that the tube is perpendicular to the ground using a carpenter's level or a plumb line. Fill container with dry ice and let the metal block cool for ~45 min.
3. Prepare samples for analysis. Make 1:10 dilutions of the plant extract (100 μL) and a positive and negative control as described for the IRI (capillary) method (*see* **Note 13**).
4. Label glass cover slides for each dilution and control. Place the slides on the cooled metal block and let their temperature equilibrate.
5. Starting with the negative control, place the cover slide directly underneath the tube. Using an automatic pipette, sample a drop (10 μL) of the diluted extract. Carefully dispense the drop of solution at the top center of the tube (*see* **Notes 14** and **15**).
6. Repeat procedure for all samples.

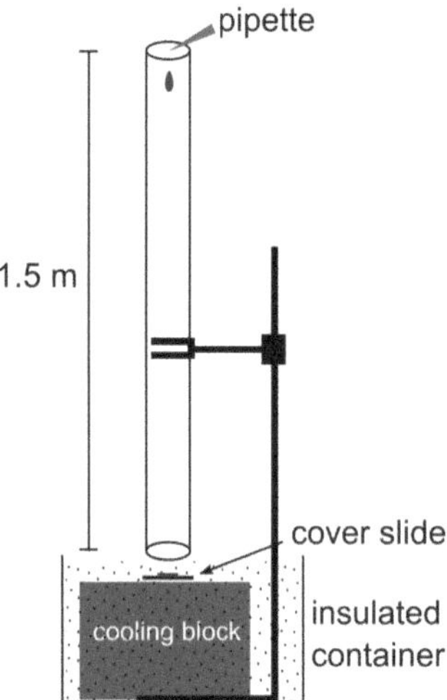

Fig. 4 Assembly showing the equipment for the generation of a thin layer of sample solution necessary for ice recrystallization inhibition assays (splat method)

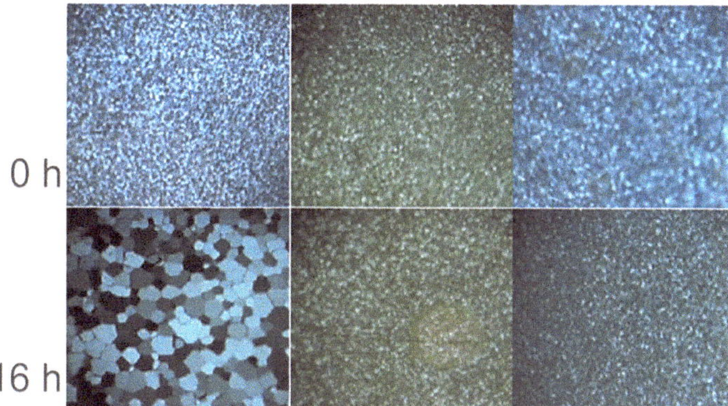

Fig. 5 Ice recrystallization inhibition (splat assay) in plant extracts and for purified plant IBPs. Samples include (from *left* to *right*): 50 mM Tris–HCl buffer, extract of *Solidago sp.* (goldenrod) leaves, and ice affinity purified recombinant IBP from *L. perenne* (perennial rye grass; 1 mg/ml). Images show the ice crystals at the beginning of the experiment (0 h) and after annealing for 16 h at −4 °C

7. Quickly move each slide into the incubation chamber and photograph under crossed-polarized light (as done with the capillary IRI procedure).
8. Repeat after ≥16 h and compare appearance of ice crystals with earlier time point (Fig. 5) (*see* **Note 16**).

3.3 TH and Ice Crystal Morphology

TH assay has been described elsewhere in this volume (Bayer-Giraldi et al.). The TH reading (°C) for a given IBP solution depends on several variables that need to be defined to facilitate comparison between laboratories and samples. This is an important consideration given the modest TH levels found in many plants. An obvious variable is IBP concentration, with TH typically showing a hyperbolic relationship to this factor [20]. However, by slowing sample cooling, higher TH values for an IBP can also be obtained [21]. A third variable is the size of the starting ice crystal. The larger the seed ice crystal, the smaller the TH value. Here we describe ways of changing the cooling rate and ice crystal size to show the dependence of the TH gap on these variables. In particular, a slower cooling rate may be necessary for dilute plant extracts in order to measure TH. These experiments will also provide an ice crystal growth habit that reflects IBP adsorption to particular ice planes [22].

1. Following the protocol outlined in Bayer-Giraldi et al. (this volume), load the sample into the grid of a nanolitre osmometer.
2. Flash cool the sample so that it freezes.
3. Slowly melt the drop until it contains only one ice crystal. Be careful that the sample does not warm too much since the ice crystal will rapidly melt as the melting point is approached. Stabilize the ice crystal by finely controlling the temperature of

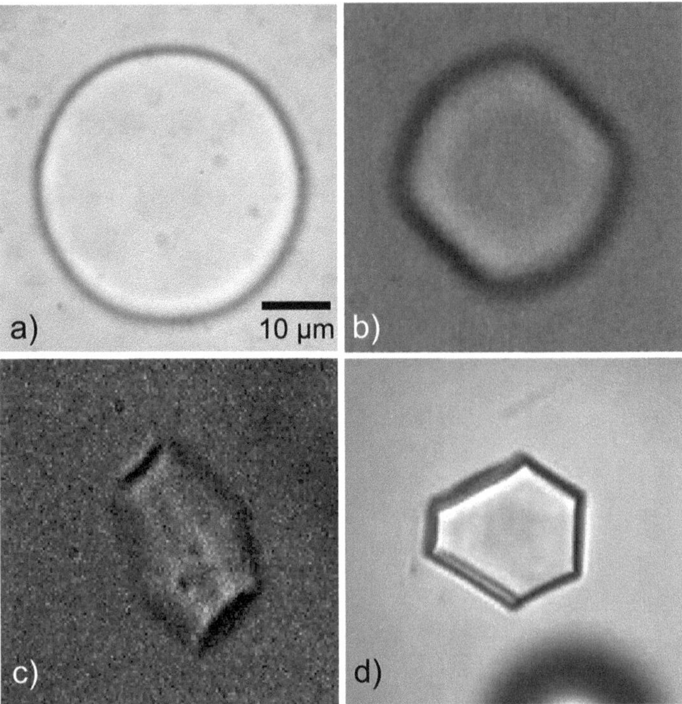

Fig. 6 Ice crystal morphologies formed in the presence of several plant IBPs. Samples include (**a**) buffer, (**b**) recombinant *Brachypodium distachyon* (purple false brome) IBP, (**c**) an extract of *Solidago sp.* (goldenrod) leaves, and (**d**) recombinant *L. perenne* (perennial rye grass) IBP

the sample (*see* **Note 17**). The temperature at which the ice crystal is stable is the melting point.

4. Lower the temperature of the sample by 0.01 °C every 10 s until the ice crystal begins to grow. Depending on the concentration and activity level of the IBP, this can range from 10–15 s to 5–10 min. As explained in Bayer-Giraldi et al. (this volume), the gap between the melting and freezing points is the TH reading.
5. Repeat **steps 2–4**, but this time after ice crystal stabilization, lower the temperature at a faster rate (e.g., 0.02 °C every 10 s or 0.001 °C every 10 s; *see* **Note 18**).
6. Repeat **step 3**, but produce a larger ice crystal. Repeat **steps 4** and **5** using different temperature ramping rates. Also try taking measurements using a smaller ice crystal.
7. Aside from the TH levels obtained in these assays, invaluable information comes from the morphology of the ice crystal as it forms in the presence of the IBP and after its freezing point is exceeded. With experience it is possible to identify most of the different IBP types (from various organisms or produced after mutagenesis of IBP sequences) simply from the ice crystal habit. *See* Fig. 6 for examples of ice crystal morphologies influenced by extracts from plants or protein preparations.

3.4 Insoluble Recombinant Plant IBPs

Describing the purification of IBPs from plant tissues is beyond the scope of this paper and is dependent upon the characteristics of that particular protein. For example, some IBPs from both monocots and dicots can refold after high temperature treatment and thus boiling crude plant extracts can be a valuable step in a purification protocol. With the increasing availability of partially sequenced genomes, cloning putative open reading frames into expression vectors and allowing the production of recombinant protein in bacteria (normally *Escherichia coli*) may be an alternative strategy to identify IBPs. Although it is possible to efficiently express some plant IBPs in *E. coli*, the lack of appropriate protein folding machinery and post-translational modification can result in low levels of expression and/or the formation of insoluble protein aggregates. This is particularly troublesome with IBPs that are stabilized by disulfide bonds [23]. There are a number of commercially available "specialized" strains that can improve soluble protein expression (*see* **Note 19**), however, when protein solubilisation is not successful, purification using denaturing conditions may be necessary. Outlined is a method using immobilized metal ion affinity chromatography (IMAC) on nickel beads and 8 M urea with recombinant plant IBPs. A detailed protocol on the general purification of poly-histidine-tagged proteins is available in this Methods and Protocol series [24].

1. Add a single *E. coli* colony transformed with the putative IBP coding sequence into 15 mL of LB with antibiotic (e.g., kanamycin 50 μg/mL). Shake in a 50 mL tube at 37 °C, 250 rpm, overnight (~16–18 h). Preheat 1 L of LB in a 37 °C water bath overnight.
2. Inoculate 1 L of preheated LB plus appropriate antibiotic with 5 mL of overnight culture and grow at 37 °C, 250 rpm, until the culture reaches an $OD_{595} = 0.6–0.8$ (~3 h).
3. Cool culture at 24 °C for 1 h while shaking at 250 rpm. Add IPTG (to use at a final concentration of 0.5 mM dissolved in 10 mL of dH_2O). Continue shaking at 24 °C overnight, 100 rpm.
4. Harvest cells by centrifuging at 4 °C, 10,000 × *g* for 25 min and decant supernatant. Store pellet at −20 °C if necessary.
5. If frozen, thaw pellet on ice for ~30 min. Resuspend pellets in 50 mL of lysis buffer (Buffer A), with 1 mini EDTA-free protease inhibitor tablet added just before use.
6. Centrifuge at 15 °C, 52,000 × *g* for 40 min.
7. Resuspend pellet in denaturing buffer (Buffer B). Allow cells to lyse on shaker at room temperature, 100 rpm, for ~4–5 h until culture has a "grainy" appearance.
8. Centrifuge at 15 °C, 52,000 × *g* for 40 min.
9. Add supernatant to 8 mL of Ni-NTA nickel bead resin and shake at room temperature for 1 h, at 100 rpm.

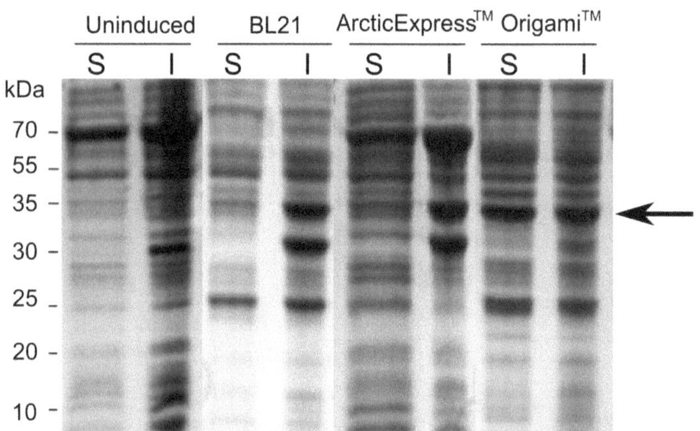

Fig. 7 Denaturing gel electrophoresis (SDS polyacrylamide gel) showing the expression of recombinant *B. distachyon* (purple false brome) IBP in different *E. coli* strains. Proteins were stained to see the expression of the IBP (36 kDa) recovered in soluble (S) and insoluble (I) extracts

10. Pour resin into gravity flow column. Run flow-through over the column a second time. Make sure that the column does not dry out by keeping a head of buffer ~1 mm above the resin top.
11. Wash column twice with 2 mL of wash buffer (Buffer C).
12. Wash twice with 2 mL of Buffer D to elute aggregated proteins and twice with 2 mL of Buffer E to elute non-aggregated proteins.
13. Pour eluted fractions into dialysis tubing. Change buffers every 24 h over 4 days, beginning with Buffer F and ending with Buffer I (*see* **Note 20**). Since urea can crystalize at low temperature, dialysis against 6 M urea (Buffer F) and 4 M urea (Buffer G) must be done at higher temperatures (~15 °C). Dialysis against 2 M urea (Buffer H) and 0 M urea (Buffer I) solutions should be done at 4 °C (*see* **Note 21**).
14. Samples should be taken at each purification step and subjected to denaturing polyacrylamide gel electrophoresis (SDS-PAGE) to verify that the IBP has been successfully solubilized (Fig. 7) (*see* **Note 22**). In order to ensure that proteins have refolded correctly, IRI activity should be monitored.

3.5 Ice Affinity Purification

IBPs can be purified by ice affinity since ice, when grown slowly, excludes solutes and proteins. Two successive "rounds" of ice affinity purification (IAP) can purify IBPs from *E. coli* lysates to homogeneity. This has facilitated the production of milligrams of purified and properly folded proteins for structural analysis including the X-ray structure of the perennial rye grass, *Lolium perenne* IBP [25]. In some organisms, including the sea buckthorn [17] privet [16] and triticum [15], IAP has been a valuable purification step.

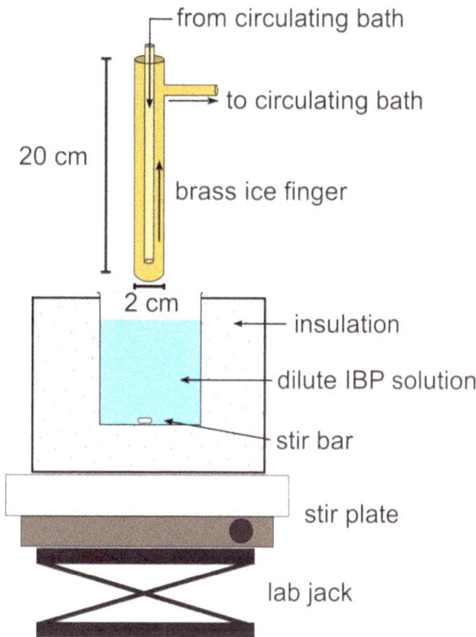

Fig. 8 Apparatus for the purification of IBPs by ice affinity

However, to date, IAP has not proved successful in the efficient purification to homogeneity of several plant IBPs in our labs (e.g., [18]).

1. See the diagram (Fig. 8) for a schematic of the apparatus used for IAP (*see* **Note 23**).
2. Turn on the circulating bath and set the temperature to −0.5 °C. The antifreeze should flow through the brass ice finger allowing it to cool completely (30 min) (*see* **Note 24**).
3. Fill a 100 mL beaker ~3/4 full with deionized cold water. Add granular ice to the water so that the beaker is filled. Nucleate ice growth on the finger with a thin layer of ice. Using the lab jack (under the magnetic stirrer, *see* Fig. 8), raise the beaker in the insulated container so that the brass finger is immersed in the ice water for 10–15 min (*see* **Note 25**). A thin layer of ice should be seen on the finger prior to exchanging the ice water with the sample solution.
4. Prepare the sample solution by diluting it to a volume of ~75–80 mL. The solution should have a salt concentration of <100 mM (*see* **Note 26**). Put the cooled solution into a chilled 100 mL beaker containing a small magnetic stir bar and keep it in an ice bucket.
5. Program the water bath so that the temperature slowly decreases from −0.5 °C over the course of at least 24 h. For example: program a starting temperature of −0.5 °C and an end temperature of −3.0 °C, over a time period and conclude at a convenient time ~24 h–36 h after starting (*see* **Note 27**).

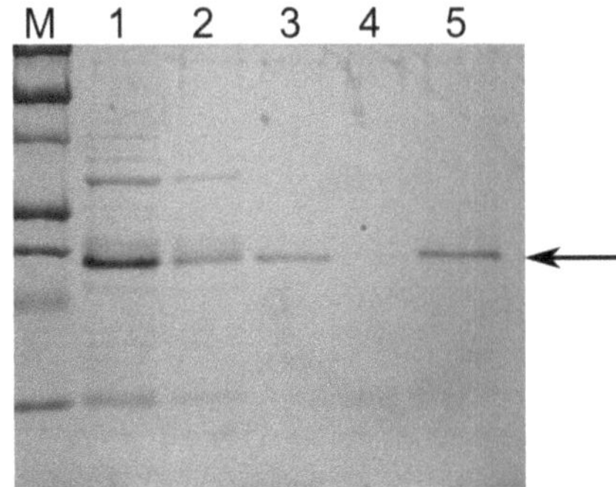

Fig. 9 Denaturing gel showing the expression and purification of recombinant *L. perenne* (perennial rye grass) IBP from *E. coli*. Samples include: lane M, molecular mass marker; *lane 1*, supernatant after lysis of bacterial cells by boiling; *lane 2*, liquid fraction from IAP; *lane 3*, flow through from nickel affinity column using the melted ice fraction; *lane 4*, unbound fractions from nickel affinity column; *lane 5*, melted ice fraction from the nickel column containing purified IBP

6. Remove the ice-water solution from the insulated container after lowering it from the brass finger with the lab jack. Place the diluted protein solution into the insulated container and center onto the stirrer sitting on the lab jack. Move the jack up so the brass rod is inside the beaker approximately 1/4 of the way from the bottom. The magnetic stirrer should be set on a medium stirring speed.
7. After ~24 h (or when about half of the water is frozen), lower the jack so that the ice crystal is no longer immersed in the solution. Allow the ice crystal to melt at room temperature for ~5 min to remove excess solution, or rinse the ice crystal with a small amount of deionized water (at 4 °C). Discard any dripped solution.
8. Remove the ice from the brass finger by increasing the temperature of the circulating bath. Let the ice drop into a clean beaker; this is the ice fraction. Melt the ice rapidly to minimize the transition time to liquid; this can be done on a shaker at room temperature. As the ice melts, gradually add buffer and/or NaCl as required to help prevent the unfolding or precipitation of the IBP.
9. Assess the level of purification (ice fraction vs. liquid fraction) using denaturing gel electrophoresis (Fig. 9). It may be necessary to concentrate the sample for further analysis. This can be done by lyophilizing the sample, but some IBPs are sensitive to this technique. Alternatively use an ultrafiltration concentrating column to decrease sample volume.

3.6 Verification In Planta

In order to demonstrate the function of a protein *in planta*, it is common to express that protein from a transgene and assess the plants for a particular phenotype. When a freeze-susceptible plant, such as *Arabidopsis thaliana*, is transformed with an IBP sequence, the acquisition of a phenotype, including freeze tolerance, can be attributed to the IBP encoded by the transgene. DNA sequences are easily introduced into the germ line of *Arabidopsis* using *Agrobacterium tumefaciens* by the floral dip method. For a detailed protocol, see the contribution in this series [26]. The F0 plants are allowed to proceed to senescence, at which point seed can be harvested and F1 plants screened for the presence of the transgene using antibiotic or herbicide resistance markers and confirmed by PCR analysis or, uniquely for IBPs, by screening crude extracts or guttation fluid for IRI activity and ice shaping. Subsequently, experiments such as the ion leakage assay (see Thalhammer et al. this volume) can be performed.

1. Inoculate 300 mL of LB plus appropriate antibiotic in a 1 L flask with 100 μL of stationary phase *Agrobacterium* liquid culture carrying desired construct (*see* **Note 28**).
2. Grow *Agrobacterium* for approximately 18 h at 28 °C, 100 rpm (*see* **Notes 29** and **30**).
3. Centrifuge *Agrobacterium* for 10 min at 4,000 ×*g*, then remove supernatant.
4. Resuspend pellet by vortexing in 150 mL sterile dH_2O and repeat **step 3**.
5. Resuspend pellet by vortexing in 500 mL inoculation medium (*see* **Note 31**), then add benzylamino purine and Silwet L-77 (*see* **Note 32**).
6. Carefully invert a pot of *Arabidopsis* (*see* **Note 33**) and immerse with gentle agitation in the inoculation medium plus *Agrobacterium* in a shallow container for up to 2 min (*see* **Note 34**). Be sure to immerse the entire aerial portion of the plant. Repeat with each pot.
7. Lay each pot horizontally in a tray and cover with a lid or with plastic wrap to maintain high humidity for 1 day. Return tray to growth chamber.
8. Uncover the tray, return the pots to an upright position and allow plants to grow to maturity and set seed.
9. Once plants have senesced and seed pods are dry, seed can be harvested (*see* **Note 35**). To do this, break open seed pods over a metal sieve on a piece of plain white paper (this can be accomplished by squeezing the aerial portion of the plant) and continue passing plant material through the sieve until the seeds have largely been separated from other plant debris. Transfer the seeds from the paper to a tube. Note that all transgenic plant material is considered biohazardous and should be disposed of appropriately.

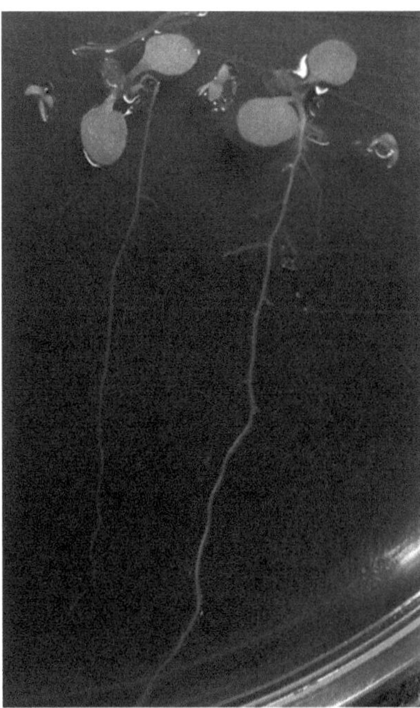

Fig. 10 Photograph of non-transformed versus transformed plants grown on hygromycin (40 μg/mL). The plants were grown vertically in this instance only to more clearly show the differences between transformed (with roots) and three non-transformed plants (shown alternating on the plate)

10. To surface-sterilize seeds using the vapor-phase sterilization method [26], make aliquots of up to approximately 50 μL of seed/tube (*see* **Note 36**). Place open tubes of seed in a sealable glass container in a fume hood, along with the beaker containing 100 mL of bleach. To the beaker, add 3 mL of concentrated HCl (or as much as can be pulled up into a transfer pipet) and immediately seal the container. Incubate seeds for approximately 3–4 h (*see* **Note 37**). After opening the glass jar, allow gas to exhaust for several minutes in a fume hood before removing the tubes.
11. To screen for transformants, sprinkle seed on MS plus appropriate antibiotic plates (up to 50 μL of seed per 100 mm plate) in a laminar flow hood (*see* **Notes 38** and **39**). Seal plates with micropore tape (a porous tape is preferable to parafilm, which prevents gas exchange).
12. Following a 2-day cold treatment in the dark, incubate plates horizontally in a growth chamber or greenhouse for approximately 10–14 days or until there is a clear distinction between transformants and non-transformants (Fig. 10). Putative transformants are characterized by the development of true leaves and a well-established root whereas non-transformants will not

develop true leaves or extensive root systems and may lose color, depending on the antibiotic used (*see* **Note 40**).

13. Very gently, transfer putative transformants to pots of soil using fine forceps and return to the growth chamber or greenhouse. It is sometimes helpful to break up the agar around the seedling to facilitate removal of the root from the plate.
14. A sample of tissue (e.g., a leaf) may be removed after 2 weeks or more to aid in the confirmation of the presence of the gene of interest in these plants. In this case, it may be necessary to adjust chamber settings to 12–15 °C prior to taking a sample (*see* **Note 41**). Crude extracts and/or guttation fluid can then be assayed for ice shaping or IRI (capillary or splat methods). Additionally, genomic DNA can be used for confirmatory PCR amplifications.
15. For extraction of crude total cell lysates, flash freeze a leaf (of known mass) in liquid nitrogen and grind to a fine powder in a mortar and pestle. Add extraction buffer at a ratio of approximately 1 mL per 0.5 g of tissue, transfer to a microcentrifuge tube and centrifuge for 2 min at 17,000 × *g* (*see* **Note 42**). The supernatant can be used for ice shaping or IRI assays.
16. For collection of guttation fluid, place well-watered pots (subirrigated to avoid water drops on the leaves) with 3–4 week old plants in a tray with at least 2.5 cm. of water to maintain a high level of humidity and keep covered overnight in growth chamber (*see* **Note 43**). Collect fluid from leaf edges.
17. In most cases it is not desirable to conduct assays on these plants but instead allow them to proceed for another generation (*see* **Note 44**). After transfer to 12–15 °C (*see* **Note 41**), progeny may then be subjected to a variety of assays to ascertain whether they show any evidence of increased freeze tolerance. This can be done with ion leakage assays, survival assessments etc. (see other appropriate chapters in this volume).

4 Notes

1. The desiccation beads (t.h.e. Desiccant, EMD Millipore) are added to the hexane/isooctane in the bath. This is important under humid conditions.
2. In order to be able to measure TH, protein must be reasonably concentrated. Often with raw cell extracts, TH activity is too low to assess. Although TH assays may be most relevant once IBPs have been purified to homogeneity or recombinant protein is obtained, even with dilute samples, the methods outlined here can often yield TH levels.

3. *E. coli* cultures can be induced at varying concentrations of IPTG (0.1–1 mM). When expressing insoluble proteins it is often favorable to induce with a lower concentration of IPTG, as this may increase the amount of soluble protein, however, this has to be optimized for each IBP.
4. Since urea undergoes an endothermic reaction when dissolved in water, it is easiest to add urea slowly to a 1 L beaker of dH_2O with a stir bar, heated to ~30 °C. Be careful not to heat above 37 °C as this accelerates the formation of isocyanate. This can lead to carbamylation of proteins, covalently modifying lysine residues, and react with the amino-terminus of proteins and thus prevent the use of some downstream analytical procedures. A stock solution of urea can be kept on the shelf if the pH is basic (~pH 8.0).
5. Due to the dissociation of urea the pH of buffers must be adjusted immediately prior to use.
6. Approximately 5–6 of the 1:2 dilutions are usually suitable for an appropriate range of concentrations, but some trial and error may be required, depending on the concentration of the extract.
7. Hexane or isooctane (2,2,4-trimethyl pentane) can be used rather than ethylene glycol inside the chamber. The use of clear ethylene glycol in the circulating bath facilitates photographic imaging.
8. Prepare samples starting with the negative control and then go up to the most concentrated sample (if doing dilutions) in order to minimize the possibility of contamination (*see* Fig. 2).
9. When placing the capillaries on the tape or in the holder, keep them in order (lowest to highest concentration; Fig. 3) so that they are easily identified at the conclusion of the experiment. The marking of a few capillary tubes by filling them with dye or by marking with an indelible pen or paint may also be useful.
10. A temperature of ≤−55 °C can be routinely achieved by placing the beaker of ethanol (100 mL) into a −80 °C chest freezer for ~60 min.
11. Some capillaries may crack when flash frozen. This may not interfere with the assay, but is a good reason to make duplicates of each sample.
12. Individual crystals at time = 0 are not readily apparent; there is an overlying "feathery pattern" to the strained small crystals that is not important. After recrystallization, any large ice crystals are easily seen.
13. It is cumbersome to do more than six samples at a time.
14. The probability of the drop hitting the side of the tube can be minimized by using a screen with a hole in the center, mounted

on the top of the tube. However, if done properly, the drop should fall down the tube and hit the cover slide, freezing instantly.

15. Dipping the pipette tip in mineral oil before dispensing prevents the drop from adhering, however, make sure that excess oil is wiped off since if there is too much, samples will not adhere properly to the cover slip.
16. Sometimes it is hard to initially see ice crystals. A waiting period of 1 h makes visualizing the ice crystals a little easier, however, some caution is advised since after 1 h some samples may start to recrystallize.
17. To ensure consistency when varying the temperature gradient, make sure the ice crystals in each experiment are approximately the same size.
18. The TH gap should increase when the decrease in temperature is ramped more slowly.
19. ArcticExpress™ (DE3) cells (Agilent Technologies) can enhance soluble protein expression by slowing the rate of protein synthesis using psychrophilic chaperonins Cpn10 and Cpn60 that are able to process proteins at low temperatures. Origami™ (DE3) cells (Novagen; EMD Biosciences Inc.) provide an oxidizing environment that allows disulfide bond formation, preventing the degradation and aggregation of IBPs that accompanies the expression of cysteine rich proteins in traditional host cell lines (Fig. 7).
20. If proteins begin to precipitate during dialysis, the denaturant is being removed too quickly.
21. Dialyzing conditions will vary depending on the nature of the IBP, however, it may be important to allow proteins to refold at temperatures <15 °C, since some IBPs have been shown to partially unfold at room temperatures (e.g., [18]).
22. Urea is a mild denaturant, if solubilization is not successful, the use of a 6 M guanidine solution may be necessary. Note that some IBPs are glycoproteins or with primary sequences that may not stain well with Coomassie Brilliant Blue [27]; silver staining can be used as an alternative.
23. Hoses to and from the brass "finger" can be made of pieces of garden hose and should be insulated with foam pipe insulation. The seams of the insulation are easily held together with tape.
24. The circulating bath may be filled with radiator or plumbing antifreeze, which is inexpensive and is not flammable.
25. The insulated container to hold the beaker can be conveniently made from insulated Styrofoam packing boxes or from a small block of polystyrene that has a hole bored in it. An insulated lid

for the top of the beaker that wraps around the ice finger is made separately.

26. If the sample solution has a salt concentration that is higher than 100 mM, a steeper temperature curve must be programmed to allow sufficient ice growth.
27. The aim of the temperature ramp program is to freeze approximately 50 % of the water into the ice crystal. In order to achieve this result with a new sample, periodic checking of the ice crystal will be necessary, allowing for some adjustments to the program (e.g., a steeper/shallower temperature gradient). Although IAP can be done over a shorter period, typically slowly growing ice results in better inclusion of IBPs.
28. Most protocols detailing *Agrobacterium* growth call for a richer medium (e.g., MG, YEB, YEP), which are all perfectly acceptable, however LB is easier to prepare and is sufficient.
29. There is a great deal of flexibility with respect to the density at which the *Agrobacterium* can be allowed to grow (i.e., anywhere between an OD_{600} of ~0.6–2.0).
30. It is not absolutely necessary to grow *Agrobacterium* at 28 °C (e.g., 25 °C appears to work equally well), however it is critical not to allow the temperature to exceed 32 °C, as otherwise the ability of the *Agrobacterium* to transfer its genetic material can be compromised [28].
31. It is often reported that the density of *Agrobacterium* in the inoculation medium should be at $OD_{600} = 0.8$, however there is a great deal of flexibility here and in our experience growing the *Agrobacterium* as described in **step 2** and resuspending in inoculation medium as in **step 5** works very well and determination of the density (e.g., with a spectrophotometer) is not actually required.
32. Benzylamino purine is not absolutely necessary in the inoculation medium but does appear to improve transformation efficiency. Silwet can be toxic to plants at higher than recommended concentrations.
33. The goal is to use *Arabidopsis* plants that have numerous unopened, immature flowers but few siliques (seed pods). It is often helpful to encourage the growth of multiple secondary shoots to a length of 2–10 cm by initially trimming back the primary shoot.
34. Different protocols call for different submersion times (anywhere from a few seconds to a few minutes). The actual time is not as critical as ensuring that as many unopened flowers as possible are coated with the *Agrobacterium*.
35. Since different portions of an *Arabidopsis* plant senesce at different rates, if it is necessary to decrease time between

generations it is possible to harvest dry seed pods from otherwise green plants and sow these seeds immediately. Several weeks may be saved by doing this.

36. Transformation efficiency varies but could be expected at ~1 % (or approximately 1–2 transformants per 50 μL of seed), so number of seeds to be screened depends upon the desired number of transformants. Only sterilize seed that will be plated soon (within a month or so) as the viability of seed that has been surface-sterilized in this manner decreases over time.

37. Longer than 3–4 h might be necessary in some cases, for example when plants have been subjected to fungal contamination. Sterilization may proceed for up to overnight, however rates of seed survival drop dramatically with increased sterilization time. Also be aware that a lengthy sterilization time may result in the bleaching of labels written on tubes with marker. A grease pencil can be used to avoid this problem.

38. Many recipes for growing seedlings on MS medium include sucrose. Omitting sucrose decreases the risk of fungal contamination substantially with little effect on seedling growth.

39. In our experience, 40–50 μg/mL of hygromycin works well with pCambia1305. At this concentration false positives are minimal and transformed seedlings grow very well. Other antibiotics are commonly used as well, depending on the binary vector. Kanamycin is often used but in our experience it can sometimes interfere with normal seedling growth, resulting in false-negatives. Phosphinothricin (or "Basta") is a herbicide commonly used for selection and offers the distinct advantage that it can be topically applied to soil-grown plants, eliminating the need for surface sterilization of seeds and growth on culture medium, however, some care is required to ensure its proper application otherwise the rate of false-positives can be significant.

40. In order to be confident of the phenotype of a non-transformant, it may be helpful to plate some seed from a non-transformed, wild-type plant for comparison.

41. Since some IBPs unfold at elevated temperatures, it may be prudent to move the plant material to lower temperatures before sampling. However, this requirement depends on the properties of the IBP under study.

42. The addition of protease inhibitors may be desirable at this step.

43. Guttation fluid is more easily obtained from younger plants, however if older plants need to be assayed or if there are problems obtaining guttation fluid exuded from hydathodes (leaf pores connected to the vascular system), shoots can be cut near

the base and fluid can often be obtained from these cuts within minutes. Soil must be quite wet to allow water pressure to build up within the plant. High humidity with no air flow is required to prevent the guttation fluid from evaporating from the leaf. Although we have decreased the chamber temperature to 12–15 °C, it was not required to obtain guttation fluid with IRI activity in our hands.

44. One reason for not performing assays on primary transformants is that these plants must be allowed to survive to the point of seed set in order for the line to be perpetuated.

Acknowledgements

We would like to thank Dr. M. Kuiper, along with many undergraduate students who have participated in data collection and "trouble-shooting" these techniques over the years. The research was supported by a CIHR and NSERC (Canada) grants to PLD and VKW, respectively.

References

1. Xu W, Liu M, Shen X, Lu C (2005) Expression of a carrot 36 kD antifreeze protein gene improves cold stress tolerance in transgenic tobacco. Forest Stud China 7:11–16
2. Knight CA, Wen D, Laursen RA (1995) Nonequilibrium antifreeze peptides and the recrystallization of ice. Cryobiology 32:23–34
3. Sidebottom C, Buckley S, Pudney P, Twigg S, Jarman C, Holt C, Telford J, McArthur A, Worrall D, Hubbard R, Lillford P (2000) Heat-stable antifreeze protein from grass. Nature 406:256
4. Urrutia ME, Duman JG, Knight CA (1992) Plant thermal hysteresis proteins. Biochim Biophys Acta 1121:199–206
5. Duman JG (1994) Purification and characterization of a thermal hysteresis protein from a plant, the bittersweet nightshade *Solanum dulcamara*. Biochim Biophys Acta 1206:129–135
6. Hon W-C, Griffith M, Chong P, Yang DSC (1994) Extraction and isolation of antifreeze proteins from winter rye (*Secale cereale* L.) leaves. Plant Physiol 104:971–980
7. Huang T, Duman JG (1995) Purification and characterization of thermal hysteresis protein from cold-acclimated kale, *Brassica oleracea*. Cryobiology 32:577–581
8. Griffith M, Antikainen M, Hon W-C, Pihakaski-Maunsbach K, Yu X-M, Chun YU, Yang SC (1997) Antifreeze proteins in winter rye. Physiol Plant 100:327–332
9. Lu CF, Wang H, Jian LC, Kuang TY (1998) Progress in study of plant antifreeze proteins. Progr Biochem Biophys 25:210–216
10. Hoshino T, Odaira M, Yoshida M, Tsuda S (1992) Physiological and biochemical significance of antifreeze substances in plants. J Plant Res 112:255–261
11. Worrall D, Elias L, Ashford D, Smallwood M, Sidebottom C, Lillford P, Telford J, Holt C, Bowles D (1998) A carrot leucine-rich-repeat protein that inhibits ice recrystallization. Science 282:115–117
12. Aticia Ö, Nalbantoğlu B (2003) Antifreeze proteins in higher plants. Phytochemistry 64:1187–1196
13. Wang W, Wei L, Wang G (2003) Multistep purification of an antifreeze protein from *Ammopiptanthus mongolicus* by chromatographic and electrophoretic methods. J Chromat Sci 41:489–493
14. Moffatt B, Ewart V, Eastman A (2006) Cold comfort: plant antifreeze proteins. Physiol Plant 126:5–16
15. Zhang C, Zhang H, Wang L, Zhang J, Yao H (2007) Purification of antifreeze protein from wheat bran (*Triticum aestivum* L.) based on its hydrophilicity and ice-binding capacity. J Agric Food Chem 55:7654–7658

16. Cai Y, Liu S, Liao X, Ding Y, Sun J, Zhang D (2011) Purification and partial characterization of antifreeze proteins from leaves of *Ligustrum lucidum* Ait. Food Bioprod Process 89:98–102
17. Gupta R, Deswal R (2012) Low temperature stress modulated secretome analysis and purification of antifreeze protein from *Hippophae rhamnoides*, a Himalayan wonder plant. J Proteome Res 11:2684–2696
18. Lauersen KJ, Brown A, Middleton A, Davies PL, Walker VK (2011) Expression and characterization of an antifreeze protein from the perennial rye grass, *Lolium perenne*. Cryobiology 62:194–201
19. Tomczak MM, Marshall CB, Gilbert JA, Davies PL (2003) A facile method for determining ice recrystallization inhibition by antifreeze proteins. Biochem Biophys Res Commun 311:1041–1046
20. Scotter AJ, Marshall CB, Graham LA, Gilbert JA, Garnham CP, Davies PL (2006) The basis for hyperactivity of antifreeze proteins. Cryobiology 53:229–239
21. Takamichi M, Nishimiya Y, Miura A, Tsuda S (2007) Effect of annealing time of an ice crystal on the activity of type III antifreeze protein. FEBS J 274:6469–6476
22. Bar-Dolev M, Celik Y, Wettlaufer JS, Davies PL, Braslavsky I (2012) New insights into ice growth and melting modifications by antifreeze proteins. J R Soc Interface 9:3249–3259
23. Qin W, Tyshenko MG, Doucet D, Walker VK (2006) Characterization of antifreeze protein gene expression in summer spruce budworm larvae. Insect Biochem Mol Biol 36: 210–218
24. Loughran ST, Wells D (2011) Purification of poly-histidine-tagged proteins. In protein chromatography: methods and protocols. Meth Mol Biol 681:311–335
25. Middleton AJ, Marshall CB, Faucher F, Bar-Dolev M, Braslavsky I, Campbell RL, Walker VK, Davies PL (2012) Antifreeze protein from freeze-tolerant grass has a beta-roll fold with an irregularly structured ice-binding site. J Mol Biol 416:713–724
26. Bent A (2006) *Arabidopsis thaliana* flora dip transformation method. In agrobacterium protocols. Meth Mol Biol 243:87–104
27. Møller HJ, Poulsen JH (2009) Staining of glycoproteins/proteoglycans on SDS gels. The protein protocols handbook. Springer Protocols Handbooks, Collana, pp 569–574
28. Jin S, Song YN, Deng WY, Gordon MP, Nester EW (1993) The regulatory VirA protein of *Agrobacterium tumefaciens* does not function at elevated temperatures. J Bacteriol 175: 6830–6835

INDEX

Dirk K. Hincha and Ellen Zuther (eds.), *Plant Cold Acclimation: Methods and Protocols*, Methods in Molecular Biology, vol. 1166, DOI 10.1007/978-1-4939-0844-8, © Springer Science+Business Media New York 2014

O

P

Q

R

S

T

V

W

Z

If you have any concerns about our products,
you can contact us on
ProductSafety@springernature.com

In case Publisher is established outside the EU,
the EU authorized representative is:
Springer Nature Customer Service Center GmbH
Europaplatz 3, 69115 Heidelberg, Germany

Printed by Libri Plureos GmbH
in Hamburg, Germany